12 PHYSICS

PNG UPPER SECONDARY

David Kolkoma
John Boereboom
Athol Binns
Denis Burchill
David Housden
Peter Kinsler

OXFORD

Oxford University Press is a department of the University of Oxford. It furthers the University's objective of excellence in research, scholarship, and education by publishing worldwide. Oxford is a registered trademark of Oxford University Press in the UK and in certain other countries.

Published in Australia by
Oxford University Press
Level 8, 737 Bourke Street, Docklands, Victoria 3008, Australia

First published 2012

Reprinted 2013, 2014, 2017, 2018, 2019, 2020, 2021, 2022, 2023, 2024

This book was originally published by ESA Publications, Auckland, New Zealand. This edition, specially adapted for the Grade 12 syllabus in Papua New Guinea, is published by arrangement with ESA Publications. Authors of the original work were John Boereboom, Athol Binns, Denis Burchill, David Housden, Peter Kinsler, and this edition has been produced by David Kolkoma.

ISBN 978 0 19 557882 9

Illustrations by diacriTech and Birdwing PNG
Typeset by diacriTech
Printed in China by Golden Cup Printing Co. Ltd

Oxford University Press Australia & New Zealand is committed to sourcing paper responsibly.

Contents

Unit 12.5 Radioactivity and Nuclear Energy

Introduction

This book has been published to provide the information required by students in order to successfully complete the Upper Secondary course in Physics at Grade 12 in Papua New Guinea.

The book is written in a manner that best develops an overall understanding of the physics concepts and processes set out in the PNG Grade 12 Syllabus. The order of Units in this book follows the order of Units presented in the Physics syllabus for Grade 12:

Unit 12.1 Fluids

Unit 12.2 Temperature and Heat

Unit 12.3 Waves

Unit 12.4 Electromagnetism

Unit 12.5 Radioactivity and Nuclear Energy

Within each Unit there are a number of Topics which follow the main subheadings and bullet points set out within each Unit in the syllabus. The aim is to provide structure and content in a concise and compact format for students to use as an effective resource to support the classroom experience. It is acknowledged that Physics is more interesting and meaningful when students are exposed to a variety of resources and materials and we encourage students and teachers not to rely on this book as a sole source of information.

If you have any suggestions about how this book might be improved in future editions, please make contact with Oxford University Press:

Fax: 00 61 3 99349 100

Customer Service Email: exportsales.au@oup.com

We wish you every success in your studies.

The authors

Acknowledgments

There are many people to be thanked for their help and assistance in enabling the publication of this series to happen. First, we acknowledge the cooperation and generosity of Mark Sayes at ESA Publications in New Zealand, who responded with interest and support when the proposal to adapt his Study Guide series was put to him.

We would also like to acknowledge many other individuals who have been happy to advise and assist in different ways: Joy Sahumlal, Greg Kapanombo, Anne Sangi, Safak Deliismail.

The author and the publisher wish to thank the following copyright holders for reproduction of their material:

Corbis/Super Stock, p. 263; istockphoto/akiyoko, p. 220; istockphoto/Iain Sarjeant, p. 219; istockphoto/Žiga Četrtič, p. 215; Science Photo Library, p. 253.

Every effort has been made to trace the original source of copyright material contained in this book. The publisher will be pleased to hear from copyright holders to rectify any errors or omissions.

Authorship

Authors of the original editions were John Boereboom, Athol Binns, Denis Burchill, David Housden and Peter Kinsler.

This edition has been produced with the assistance of:

David Kolkoma, lecturer in Physics at Papua New Guinea University of Technology. His qualifications include: Certificate in Medical Physics, International Centre for Theoretical Physics, Trieste, Italy; Postgraduate Certificate in Communication for Science and Technology, Papua New Guinea University of Technology; Bachelor of Science in Electrical Engineering, Papua New Guinea University of Technology. He has also been a tutor for Grade 12 Physics and a member of the Grade 12 Physics National Examination Committee. At the time of publication Mr Kolkoma is completing his Masters in Medical Physics at Queensland University of Technology.

The aim of this book is to provide Grade 12 Physics students in PNG with a book that they can use as a compact summary of content and skills related to the PNG Grade 12 syllabus.

Preliminary Unit
The importance of physics

Physics is important because of its relevance to the present and future technology-driven lifestyle in Papua New Guinea. In all Units, students engage in practical tasks and investigations that will enable them to build the basic physics principles, concepts and skills needed for higher education or self-employment. This Preliminary Unit offers advice and guidelines on:

- Identifying, describing and explaining aspects of physics.
- Solving problems.
- Mathematics used in physics.

What is physics?

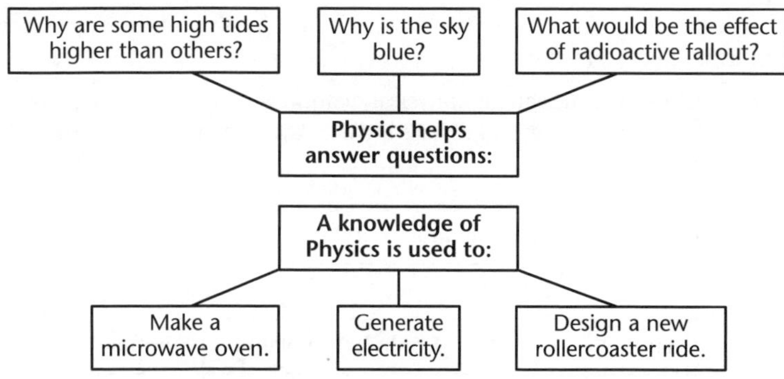

The involvement of physics

Physics helps us to understand our world better and to use this understanding for the benefit of people and the environment.

Example A

A geophysicist studies earthquakes and uses the knowledge gained from observations and experiments to predict where further earthquakes will occur, and what their effect might be on nearby people, wildlife and buildings.

Physics involves a wide range of ideas and areas of investigation – it plays an important part in medicine, industry, transport and communications, as well as in leisure activities and the home.

Physics is involved in many other areas of science. Some recent growth areas involving physics include biotechnology, forensic science, sports medicine, physical chemistry and biophysics.

Example B

An English physicist, Francis Crick, and an American biochemist, James Watson, discovered the structure of the DNA molecule. This was one of the most important advances in biology of the twentieth century.

Physical quantities

Physics involves measuring physical quantities, such as the length of a spring or the change in temperature of a liquid. Every measurement of a physical quantity involves a **unit**, which is a standard amount of that quantity. Most quantities can be measured using several different units.

Example C

The thickness of a sheet of paper can be measured in inches (an older unit) or millimetres (preferred), while **mass** can be measured in such units as ounces, pounds and kilograms.

To avoid using different units when measuring the same physical quantities, a standard system of units has been agreed on.

The system of units used to measure all physical quantities is called the 'Système International d'Unités' (commonly abbreviated to **SI**). Seven quantities and their units form the set of **fundamental units**.

Physical quantities and their units			
Quantity		SI unit	
Name	Symbol(s)	Name	Symbol
Length	l, x, y, d, etc	metre	m
Mass	m, M	kilogram	kg
Time	t, T	second	s
Electric current	I	ampere	A
Temperature	T	kelvin	K
Luminous intensity	I	candela	cd
Amount of substance	N	mole	mol

The symbols for the physical quantities are the symbols that appear in formulae and equations.

Fundamental units

The units for all other physical quantities can be *derived* from the set of units above.

Example D

Speed, *v*, is the rate at which **distance** changes with respect to time. A simple formula for **constant speed** is: $v = \frac{d}{t}$, so the SI unit for distance (metre) divided by the SI unit for time (second) gives the SI unit for speed (metres per second) $\frac{\text{m}}{\text{s}} = \text{m s}^{-1}$.

In just the same way that $\frac{1}{10}$ is written as 10^{-1}, $\frac{\text{m}}{\text{s}}$ can be written as m s^{-1}. The notation m s^{-1} will be used throughout this book.

The SI system is a **metric** system because each unit can be reduced or enlarged using multiples of 10. This is done using a standard set of **prefix multipliers**.

μ is a Greek letter pronounced 'mew'.

There are other SI prefix multipliers, but these are the most common ones.

A non-standard prefix in common usage is 10^{-2} or centi, symbol *c*.

Standard Prefix Multipliers		
Power	**Prefix**	**Symbol**
10^9	giga	G
10^6	mega	M
10^3	kilo	k
10^{-3}	milli	m
10^{-6}	micro	μ
10^{-9}	nano	n
10^{-12}	pico	p

Prefixes

Example E

a. Change 8 km to m.

b. How many micrograms are there in a kilogram?

c. Write 2.8×10^{10} **watts** in gigawatts.

Solution:

a. 8 km $= 8 \times 10^3$ m [since kilo means 10^3]
$= 8\ 000$ m

b. 1 kg $= 1 \times 10^3$ g [since 1 kg = 10^3 g]
$= 1 \times 10^3 \times 10^6$ μg [since 1 g = 10^6 μg]
$= 10^9$ μg

c. 2.8×10^{10} W $= 2.8 \times 10 \times 10^9$ W
$= 28 \times 10^9$ W
$= 28$ GW [since 10^9 W = 1 GW]

The SI unit for mass is the kilogram, kg, not the gram. This unit has the prefix *kilo* and so 1 *kilo*gram = 1 000 gram.

Preliminary Unit Activity A: Physics quantities and units

1. Using figure on the previous page, write down the units for the following combinations of quantities:

a. current × time **b**. $\dfrac{\text{length}}{(\text{time})^2}$ **c**. $\dfrac{\text{mass} \times \text{length}}{(\text{time})^2}$

2. A non SI unit of volume (commonly used in chemistry) is the litre, L. This is the capacity of a cube formed from sides 10 cm long.

Determine:

a. How many cm^3 there are in 1 L.

b. How many litres there are in 1 m^3.

c. How many millilitres (mL) there are in 1 L.

d. How many mL there are in 1 cm^3.

3. Express the following amounts of time in 'everyday' units, ie minutes, hours and days:

a. 10 000 s **b**. 1 kilosecond (1 ks)

c. 1 megasecond (1 Ms) **d**. 1 gigasecond (1 Gs)

4. How many:
   a. Microseconds in 0.1 s?
   b. Microseconds in one millisecond?
   c. **Kilowatts** in one gigawatt?
   d. Amperes in one milliampere?
5. Express:
   a. 10^6 m in kilometres.
   b. 1 km in centimetres.
   c. 1 mA in microamperes.
   d. 10 000 s in megaseconds.
6. A non SI unit for mass is the tonne, t. 1 tonne = 1 000 kg.
   How many tonnes are there in:
   a. 5.2×10^8 kg?
   b. 3.8×10^{10} g?
   c. 500 kg?
   d. 0.5 kg?

Identifying, describing or explaining aspects of physics

Physics is much more than putting data into formulae and 'number-crunching' to find a numerical answer. Your ability to communicate physics understanding and reasoning, through words and diagrams, is just as important as solving problems numerically.

For questions of the 'identify-describe-explain' type, the complexity of the situation will determine the performance standard possible for a question.

- 'Satisfactory achievement' will involve a description of a single aspect related to phenomena, concepts or principles.
- 'High achievement' will involve explanations in terms of phenomena, concepts, principles and/or relationships.
- 'Very high achievement' will involve giving concise explanations that show clear understanding.

Statements, descriptions and explanations can be written, diagrammatic or graphical.

Example F

An 'identify-describe-explain' type question

A used aluminium drink can containing a small amount of water (about 20 mL) is heated on a hotplate.

When the water is boiling, the can is quickly inverted using a pair of kitchen tongs and immediately put (open end down) into a sink of cold water. As it enters the water, the can instantly collapses and fills with water.

a. Describe what is happening inside the can when it is on the hotplate and the water inside the can is boiling.
b. Explain why the can collapses when its open end is put into the water.
c. Explain why the can also fills with water as it collapses.

A couple of seconds afterwards, the can has collapsed.

Solutions:

a. Hot water vapour forms above the boiling water and pushes out the air that was originally in the can.

> This question is at 'Satisfactory achievement' level and requires the insight that the answer has to have something to do with the result of boiling the water.

b. As the can touches the water, the water vapour in the can condenses to a small volume of water, creating a vacuum in the can. The air pressure surrounding the can is no longer balanced by pressure inside the can. The thin and flexible walls of the can rapidly buckle inwards due to the unequal pressures (or forces).

'Satisfactory achievement'	Answer would have a basic statement either about how the air pressure pushing in on the can, or how the vacuum being formed in the can, causes it to collapse.
'High achievement'	Answer would connect the imbalance of pressure (or forces) on the can to the vacuum formed inside the can and the resulting collapse.
'Very high achievement'	Answer would be a concise but comprehensive answer that logically follows the following steps of the explanation: Vapour condenses rapidly → Vacuum forms inside the can → Greater pressure (or force) outside the can than inside → Can collapses since walls are very thin and flexible.

c. Air pressure also acts on the water in the sink and pushes water up into the can at about the same time as the can collapses.

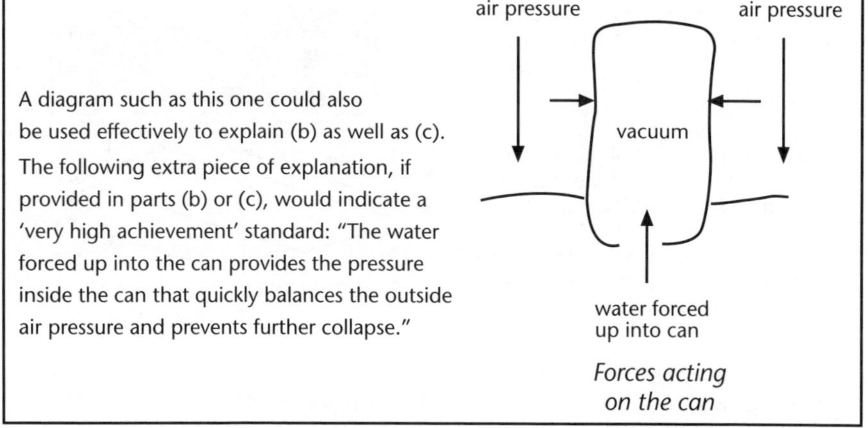

A diagram such as this one could also be used effectively to explain (b) as well as (c). The following extra piece of explanation, if provided in parts (b) or (c), would indicate a 'very high achievement' standard: "The water forced up into the can provides the pressure inside the can that quickly balances the outside air pressure and prevents further collapse."

Forces acting on the can

When answering 'identify-describe-explain' type questions, concentrate on the key ideas and reasons in the answer and write about them in a logical way. Avoid giving information on other aspects that are not central to the answer.

Where possible, practise giving reasons in answers even if only asked to *identify* or *describe* physics situations. This helps you to become proficient in physics explanations.

A standard way to answer an 'explain' question is to give a brief statement followed by a reason. Using this method, part (c) in the example could be answered in the following way: – "The can also fills with water as it collapses because air pressure pushes water from the sink up into the can while a vacuum exists inside the can."

Solving problems

Physics is frequently used in engineering to determine such things as the amount of concrete needed to construct a dam, the size of the **forces** produced by winds on a tall building, the **energy** output of a generator, etc. Mathematics is a tool used in physics to determine final answers.

Solving a 'physics problem' can usually be separated into four parts:

1. Work out what the problem requires. Identify what information and data are given. A sketch of the situation may help organise your thoughts.
2. Organise the way in which the problem can be analysed and solved. Usually this involves writing down a mathematical equation involving the quantities in the problem.
3. Rearrange, simplify and solve the problem.
4. Interpret the result and check that it is reasonable and makes sense.

Example G

Before installing a waterbed in your bedroom, it would be wise (as some, particularly in older houses, have discovered) to know whether the floor will support the weight of the bed. You read in the advertisement for the bed shown alongside that the dimensions of the plastic containing the water are: length 2.0 m, width 1.4 m, depth 0.30 m. A reference book gives the density of water as 1 000 kg m^{-3}, ie 1 cubic metre of water has a mass of 1 000 kg.

Nambawan
Waterbeds
Limited offer
Only K1000
Size
length 2.0 m
width 1.4 m
depth 0.30 m
Shipping Weight - 60 kg

Solution:

Part 1: The problem requires determining the mass of the water in the bed and the mass of the material in the bed's framework. The mass of water can be found from the bed's dimensions and the density of water, both of which are known. The mass of the material in the bed's framework is given as 60 kg.

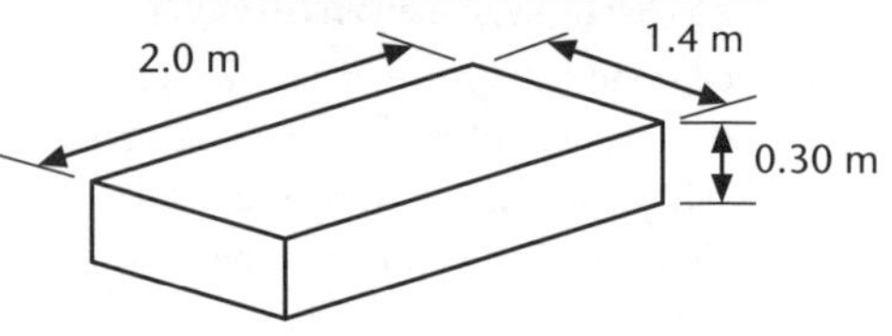

The waterbed's dimensions

Part 2: The volume of the bed is given by $V = \ell \times w \times d$.
Density is given by $\rho = \frac{m}{V}$ (where ρ is a Greek letter pronounced 'roe').

Part 3: The volume is
$$V = \ell \times w \times d = 2.0 \times 1.4 \times 0.30 = 0.84\ \text{m}^3$$

Substituting the values for density and volume in the density formula gives:

$1\,000 = \frac{m}{0.84}$, which can be rearranged to give the mass of water:

$$m = 1\,000 \times 0.84 = 840\ \text{kg}$$

> ***Part 4:*** The mass of the water is 840 kg and the mass of the material in the bed's framework is 60 kg. Therefore, the bed's total mass is 900 kg, which is the same as the mass of a small car. Another way to interpret the result is that it is approximately the mass of 11 adult people! The result suggests that you should talk to a person with building knowledge to find out how well the floor is supported.

For questions of the problem-solving type, a process or processes are used to find a physical quantity. (A process involves recognising the relevant concept or principle; selecting the method (eg formula, **graph**, diagram, logical deduction); selecting relevant information.)

The complexity of the situation and problem-solving process will determine the level of achievement possible for a question.

- A 'satisfactory achievement' level problem at Grade 11 + 12 is one involving a single process. The relevant concept or principle will be transparent, the method will be straightforward (a formula will need no more than a simple rearrangement), and the information will be directly usable.
- For 'high achievement', a problem is typically one in which the relevant concept or principle may not be immediately obvious. The method may involve the use of a complex formula or rearrangement, or the information may not be directly usable or immediately obvious.
- A complex problem at 'very high achievement' level will typically involve more than one process. The recognition of two different concepts must be involved.

Example G would be a 'very high achievement' question since it requires more than one logical or mathematical process to arrive at the solution. Usually, problems like Example G would be broken up into sections and there would be a series of 'satisfactory achievement' and 'high achievement' questions leading to the final 'very high achievement' level question.

Example G could be re-written as:

a. Calculate the volume of the waterbed in m^3.
(A straightforward problem so it is at 'satisfactory achievement' level.)

b. Show that the mass of water needed to fill the waterbed is 840 kg.
(Less straightforward since the formula for density needs to be chosen and rearranged to give mass, so it is set at 'high achievement' level.)

The answer (840 kg) is given in part (b), which means:

- The *process* of your solution and how *clearly* you set out the steps of the solution is important.
- You can see if your working is correct in parts (a) and (b).
- For those who cannot work out the answer to (b) themselves, there is the opportunity to answer any subsequent questions based on the answer to (b).

Mathematics in physics

Mathematics plays an important part in physics, because mathematics is used to express the **relationships** between physical quantities, and to find the size of those quantities.

Many examples and exercise problems in this book require the proper use of a scientific calculator.

The following examples illustrate some of the calculations and rearrangements which you should be able to do confidently.

Example H
Simplifying numerical expressions

a. $$\frac{1}{6}-\frac{1}{9}=\frac{3-2}{18}$$
$$=\frac{1}{8}\ (=0.05\dot{5})$$

b. $$\frac{1}{2}\text{ of }\frac{1}{4}=\frac{1}{2}\times\frac{1}{4}$$
$$=\frac{1}{8}\ (=0.125)$$

c. $$\sqrt{\frac{25}{64}}=\frac{\sqrt{25}}{\sqrt{64}}$$
$$=\frac{5}{8}\ (=0.625)$$

d. $$\frac{325.4\times\sqrt{4\,865}}{(4.962)^2}=\frac{325.4\times 69.75}{24.62}$$
$$=921.8$$

e. $$\frac{10^{12}\times 10^{-6}}{10^5}=\frac{10^6}{10^5}$$
$$=10^1\ (=10)$$

f. $$\frac{3.4\times 10^{10}\times 6.1\times 10^{-4}}{8.3\times 10^{-2}}=\frac{2.074\times 10^7}{8.3\times 10^{-2}}$$
$$=2.5\times 10^8$$

Example I
Solving equations and using formulae

a.
$$2x+5=17$$
$$2x+5-5=17-5 \quad \text{[subtract 5 from both sides]}$$
$$2x=12 \quad \text{[simplify]}$$
$$\frac{2x}{2}=\frac{12}{2} \quad \text{[divide both sides by 2]}$$
$$x=6$$

b.
$$\frac{1}{x}=\frac{1}{2}+\frac{1}{6}$$
$$\frac{1}{x}=\frac{3}{6}+\frac{1}{6} \quad \text{[find the common denominator]}$$
$$=\frac{4}{6} \quad \text{[simplify]}$$
$$\frac{x}{1}=\frac{6}{4} \quad \text{[invert both sides]}$$
$$x=1.5 \quad \text{[simplify]}$$

c. make I the subject of $V = IR$

$$\frac{V}{R} = \frac{IR}{R} \qquad \text{[divide both sides by } R \text{ and cancel]}$$

$$I = \frac{V}{R}$$

d. Make v the subject of the equation $\frac{1}{2}mv^2 = mgh$

$$\frac{1}{2}mv^2 = mgh$$

$$\frac{\frac{1}{2}mv^2}{m} = \frac{mgh}{m} \qquad \text{[divide both sides by } m \text{ and cancel]}$$

$$2 \times \frac{1}{2}v^2 = 2gh \qquad \text{[multiply both sides by 2]}$$

$$v^2 = 2gh \qquad \text{[simplify]}$$

$$v = \sqrt{2gh} \qquad \text{[take the square root of both sides]}$$

e. Change 1 mm² to m².

There are 1 000 mm in 1 m, so there are

$1\,000 \times 1\,000$ mm² in 1 m².

This means that:

$$1\text{ mm}^2 = \frac{1}{1\,000^2}\text{ m}^2$$

$$= \frac{1}{10^6}$$

$$= 1 \times 10^{-6}\text{ m}^2$$

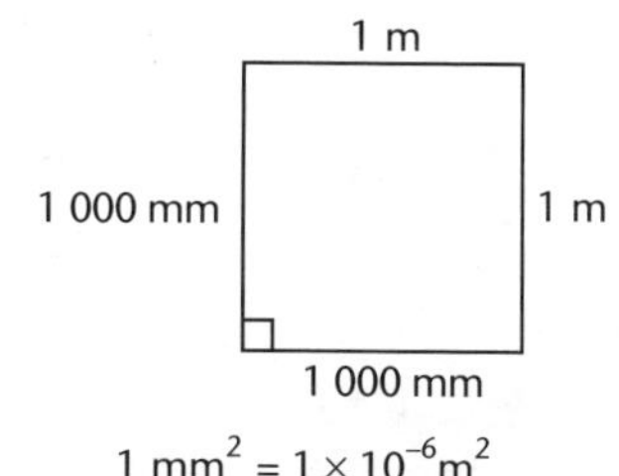

$1\text{ mm}^2 = 1 \times 10^{-6}\text{m}^2$

f. Convert m s⁻¹ to km h⁻¹.

There are 1 000 m in 1 km so 1 m = $\frac{1}{1\,000}$ km, and there are

$60 \times 60 = 3\,600$ seconds in 1 hour (abbreviation h), so

$$1\text{s} = \frac{1}{3\,600}\text{ h}$$

$$\text{So } 1\text{ m s}^{-1} = \frac{\frac{1}{1\,000}}{\frac{1}{3\,600}}\text{ km h}^{-1}$$

$$= \frac{1}{1\,000} \times \frac{3\,600}{1}\frac{\text{km}}{\text{h}} \qquad \text{[multiply by reciprocal]}$$

$$= \frac{3\,600}{1\,000}\frac{\text{km}}{\text{h}}$$

$$= 3.6\text{ km h}^{-1}$$

Example J
Pythagoras and trigonometry

a. Find the angle θ and the side x.

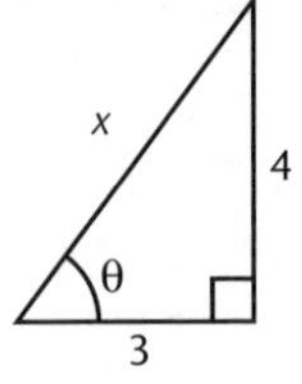

$$\tan\theta = \frac{\text{opposite side to }\theta}{\text{adjacent side to }\theta}$$
$$= \frac{4}{3}$$
$$\theta = \tan^{-1}(1.\dot{3})$$
$$\theta = 53.1°$$

Pythagoras gives $x^2 = 3^2 + 4^2$

$= 25$

$x = 5$ [taking square root]

b. Find the angle θ and the side x.

13
5
θ
x

$$\sin\theta = \frac{\text{opposite side}}{\text{hypotenuse}}$$
$$= \frac{5}{13}$$

$= 0.3846$ [to 4 sig figs]

$\theta = 22.6°$ [taking $\sin^{-1}$]

Pythagoras gives $13^2 = x^2 + 5^2$

$169 = x^2 + 25$

$x^2 = 169 - 25$ [rearranging]

$x^2 = 144$

$x = 12$ [taking square root]

Change in quantities

The change (or difference) in a quantity, Δ ('delta'), is found by subtracting the initial amount of the quantity from the final amount, ie

$$\Delta = \text{final amount} - \text{initial amount}$$

Example K

a. The initial reading of a car's 'speedo' is 3894 km. At the end of a trip, the reading is 4215 km. The distance travelled, Δd, during the trip is:

Δd = final distance – initial distance

= 4215 – 3894

= 321 km

b. The volume of a piece of sponge rubber is 0.54 m³. After compressing the sponge rubber, its volume changes to 0.15 m³. The change in volume, ΔV, is:

$\Delta V = V_f - V_i$ The subscripts, 'f' for final, and 'i' for initial are used with the **variable** V. Thus the final volume is written V_f and the initial volume is V_i.

$= 0.15 - 0.54$

$= -0.39$ m³ The negative sign shows the volume has decreased.

When finding the change in vector quantities, the subtraction becomes a **vector** subtraction and usually involves drawing a **vector diagram**.

The constant of proportionality

An important activity in Physics is finding the mathematical **relationship** between two sets of measurements.

Example L

Two quantities A and B are measured in an experiment and recorded.

A	1	2	3	4	5	6
B	3	6	9	12	15	18

The values of A and B show that there is a relationship between A and B:

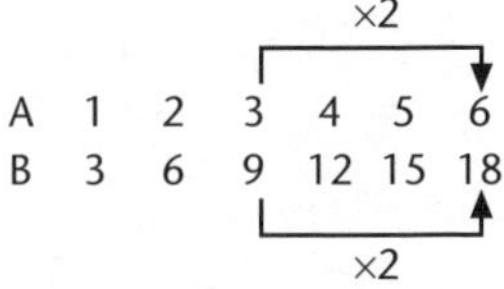

If A is doubled, B doubles.

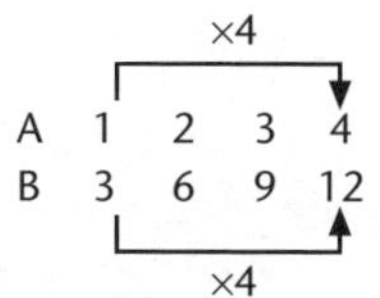

If A increases 4 times, B increases 4 times.

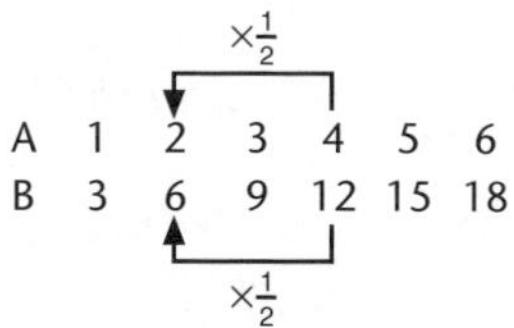

If A is halved, B halves.

The relationship in Example L is described by saying 'B is **proportional** to A'. This is written B ∞ A. There is a constant **ratio** between A and B. The value of this ratio is:

$$\frac{B}{A} = \frac{3}{1} \text{ (or } \frac{6}{2} \text{ or } \frac{9}{3} \text{ etc)}$$

$$= 3$$

$$B = 3A \qquad \text{[rearranging]}$$

The number 3 is a **proportionality constant**.

Often proportionality constants are given symbols such as *c* or *k*.

Example M

The relationship between speed expressed in m s^{-1} and the same speed expressed in km h^{-1} is **constant proportion**. The proportionality constant is 3.6 (as found in Example I, part (f)).

The relationship can be written:

B = 3.6 × A, where B is the speed in km h^{-1} and A is the speed in m s^{-1}.

To change a speed of 15 m s^{-1} to a speed in km h^{-1}, substitute A = 15 m s^{-1} into the equation B = 3.6 × A:

$$B = 3.6 \times 15$$
$$= 54 \text{ km h}^{-1}$$

Conversely, the ratio for converting a speed in km h^{-1} to the corresponding speed in m s^{-1} is the **inverse** (reciprocal) of 3.6:

$$A = \frac{1}{3.6} \times B$$

Preliminary Unit Activity B: Useful mathematics

1. Evaluate each of the following:

   a. $3\times4\frac{1}{2}$ b. $6\frac{2}{3}\div4$ c. $\sqrt{\frac{147}{363}}$ d. $\frac{8.03\times10.57}{0.732\times1.42}$

   e. $\sqrt{154.1}\times\left(\frac{0.316}{0.232}\right)^2$

2. Evaluate each of the following:

   a. $\frac{41.6\times\sin 43.2^\circ}{\cos 119.6^\circ}$ b. $\frac{9.96\times10^9\times6.25\times10^{-6}}{\left(2.73\times10^{-3}\right)^2}$

3. Solve each of the following for x:

   a. $\frac{2}{3}x=8$ b. $3x-7=17$ c. $\frac{1}{x}=\frac{1}{5}+\frac{1}{10}$ d. $\frac{x}{3}=\frac{x}{6}+\frac{2}{3}$

4. Simplify the following ratios:

   a. 25 : 15 b. 48 : 18 c. 5.1 : 3.4 d. $8\frac{1}{3}:3\frac{3}{4}$

5. Rearrange each of the following equations to make:

   a. f the subject in $v = f\lambda$. b. I the subject in $P = I^2R$.

   c. m the subject in $E_k = \frac{1}{2}mv^2$. d. P_1 the subject in $P_1V_1 = P_2V_2$.

   e. V_2 the subject in $P_1V_1 = P_2V_2$. f. t the subject in $d = \left(\frac{v_i + v_f}{2}\right)t$.

   g. T_1 the subject in $\frac{V_1}{T_1}=\frac{V_2}{T_2}$.

6. Find the length of the side (x) and the size of the angle (θ) in each of the following right-angled triangles:

a.

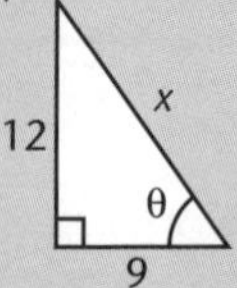

b.

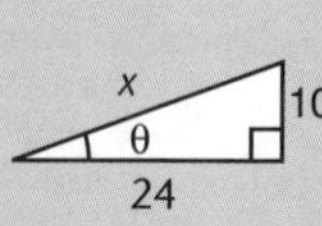

c.

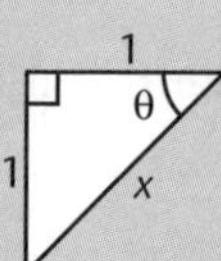

d.

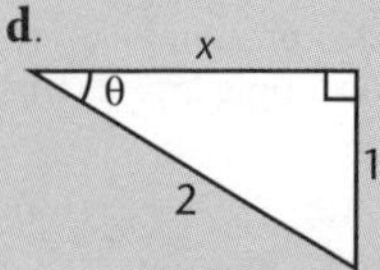

e.

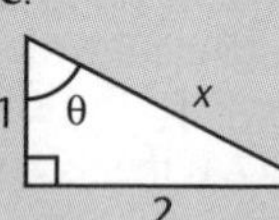

f.

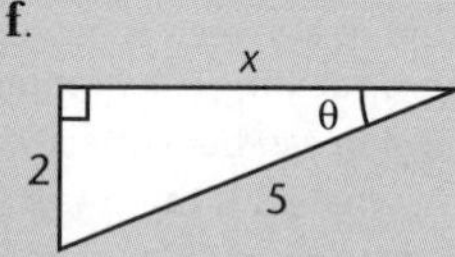

7. Calculate the change in the following quantities:

   a. A child grows from a height of 73.2 cm to a height of 84.1 cm.

   b. The temperature changes from 12.5 °C to –4.8 °C.

   c. Time changes from 8.45 am to 3.25 pm.

Unit 12.1 Fluids

Topic 1: Fluid statics

In this Topic we look at fluid statics (Syllabus p. 24), in particular:

- Density in solids and liquids and specific gravity.
- Pressure in solids and liquids.
- Pascal's Law.
- Buoyancy and Archimedes' Principle.
- Surface tension and capillarity.

Introduction

Matter comes in three states that are distinguished by the strength of the bonds holding the molecules of the matter together. The three states of matter are:

- **Solids:** these have strong bonds between their molecules making them very difficult to deform.
- **Liquids:** these have relatively weak bonds between molecules which allow them to be deformed without effort. Liquids have a fixed volume, but their shape is determined by the shape of the container holding them.
- **Gases:** these have virtually no bonds existing between their molecules, so can spread into any available space. The volume of a gas is determined by the size of the container holding it.

This Unit on fluids covers Topic 1, fluids at rest or fluid statics, and Topic 2, fluid in motion or fluid dynamics.

Density in solids and liquids and specific gravity

The **density** (ρ) of an object of total mass M and volume V is given by $\rho = \frac{M}{V}$.

The units are always in kg/m^3. The **specific gravity** of a substance is the ratio of the density of the substance to the density of pure fresh water ($\rho_{water} = 1 \times 10^3$ kg/m^3) at 4 °C. It is only a ratio and has no units. An object with specific gravity less than 1 will float and an object with a specific gravity greater than 1 will sink. Similarly, an object will float in water if its density is less than the density of water and sink if its density is greater than that of water. The table below gives specific gravity values of some few common substances.

Substance	Specific gravity value
Gold	19.3
Mercury	13.6
Alcohol	0.7893
Benzene	0.8786

Pressure in solids and liquids

When an object is immersed in a fluid or gas it experiences a uniform inward force over its entire surface, mainly due to the **weight** of the fluid sitting on top of it or surrounding it. This applies to solids as well. The **pressure** on the object is defined to be the force per unit area.

$$P = \frac{F}{A}$$

The unit of pressure is **Pascals** (Pa) and 1 Pa = 1 N/m^2. Example: At the earth's surface, the pressure felt by an object due to the weight of the atmosphere above it is one atmosphere which is equivalent to 1.01×10^5 Pa. This is also equivalent to 760 mm of mercury. Atmospheric pressure can also be measured in millibars using a barometer.

Pressure is force perpendicular to a surface/area of the surface. In a static fluid (liquid or gas), the pressure increases with depth due to the weight of the fluid. It is easy to show that the pressure at depth h is given by

$$P = P_0 + \rho g h,$$

where ρ = density (mass/volume), g = acceleration due to gravity and P_0 = pressure at the surface. The same equation would work for pressure in a stack of rectangular plywood of the same width and length. The difference is that the pieces of plywood do not push sideways but straight down vertically on top of one another.

If a tank of liquid is open to the atmosphere, P_0 is atmospheric pressure. You should also be aware that if you travel downward a distance Δh, the pressure increase is $\Delta P = \rho g \Delta h$. This is also true for gases.

Unit 12.1 Activity 1A

1. Rank the following objects in order from highest to lowest average density.
   a. mass 4.00 kg, volume 1.60×10^{-3} m^3.
   b. mass 8.00 kg, volume 1.60×10^{-3} m^3.
   c. mass 8.00 kg, volume 3.20×10^{-3} m^3.
   d. mass 2560 kg, volume 0.640 m^3.
   e. mass 2560 kg, volume 1.28 m^3.
2. Find the mass and weight of the air in a living room at 20 °C with a 4.0 m × 5.0 m floor and a ceiling 3.0 m high. The density of the air is 1.2 kgm^{-3}.
3. What is the total downward force on the surface of the floor due to air pressure of 1.00 atmosphere in (b) above?

Principles of hydraulic lift using Pascal's Law

Pascal's Law states that any pressure applied to an enclosed fluid is transmitted undiminished to every point of the fluid. Thus, in the figure below:

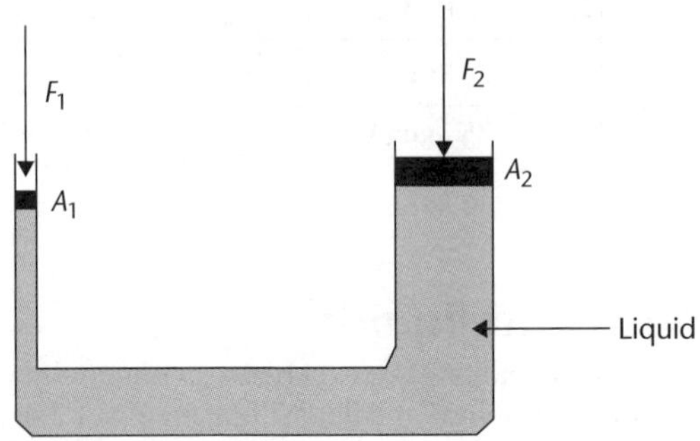

A pressure of $P_1 = \frac{F_1}{A_1}$ applied downward to the surface on the left of the container gets transmitted as an equal pressure upward of $P_2 = P_1$ on the surface on the other side of the container. The force on the other side is given as:

$$F_2 = P_2A_2 = F_1\left(\frac{A_2}{A_1}\right)$$

If A_1 is less than A_2 , the transmitted force, F_2, is greater than the applied force, F_1. This is the principle behind the hydraulic press. For example, the transmitted force F_2 is used to balance the weight of a car in the hydraulic lift.

The figure below shows an example of Pascal's Law in application. A car mechanic is jerking a car using a hydraulic lift by stepping on the smaller surface with his foot.

Buoyancy and Archimedes' Principle

The behaviour of an object submerged in a fluid is governed by **Archimedes' Principle**. Archimedes (a Greek mathematician who lived during the period of 212 BC–287 BC) determined that a body that is completely or partially submerged in a fluid experiences an upward force called the **Buoyant Force**, *B*, which is equal in magnitude to the weight of the fluid displaced by the object. This principle can be used to explain why ships loaded with millions of kilograms of cargo are able to float in the ocean and large rivers.

Unit 12.1 Activity 1B: Research activity

1. The following instruments operate on the principles of Archimedes' Principle: (i) open-tube manometer and (ii) mercury barometer. Look around inside your physics laboratory for these two instruments. Research the physics behind the operation of these instruments.
2. The pressure at the top of a cylindrical water tank of height h and cross-sectional area A open to atmosphere is given by $P = P_O + \rho gh$. Research how this formula was obtained.

Surface tension and capillarity

Have you been to a fish pond? You would observe that an insect called a dragonfly is able to sit on top of the water. How is it possible? Similarly, a razor blade made of aluminium metal can float on the surface of water. This seems to violate Archimedes' principle because the density of aluminium is twenty-seven times greater than the density of water. In fact, the reason why the razor blade does not sink is not the buoyant force, but the water's **surface tension**. If the razor blade penetrates that surface, it will sink.

The molecules in a liquid are attracted to one another due to an elastic force which exists between the molecules. This attraction is not as strong as in a solid, and decreases as the temperature increases. This attraction plays itself as a tension on the surface of a fluid, similar to the **tension** in a pulled rope or on the membrane of a drum. The surface tension, γ, is defined as a tension force divided by the length, L, where the force acts:

$$\gamma = \frac{F}{L}$$

The unit of surface tension is in N/m. Since a force times a length is a given energy, the surface tension can also be defined as an energy per unit area, J/m^2. For a cylindrical tube with radius r we get $L = (2\pi r)$.

The capillary action

If an object on the liquid made contact with liquid surface at an angle (θ) less than 90 degrees, the surface tension force has a vertical **resultant** component pulling the surface upwards. In a very thin tube, the weight of the column of liquid is so small that the surface tension pushes the column of liquid upwards. This rising is called **capillary action**. The capillary action is responsible for transporting nutrients and water from the soil through the roots up to the trunk and eventually to branches and leaves of plants and trees.

The word *capillary* means air in Latin. The rise would be very large if the tube were air thin. The figure below shows an illustration of capillary action. The diagram is enlarged for illustration of capillary rise.

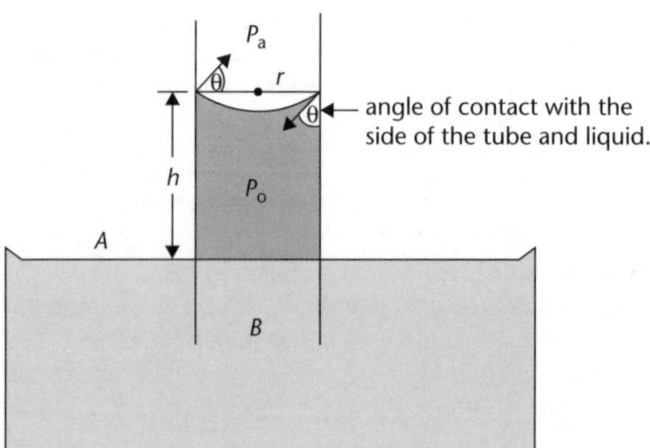

Consider a vertical capillary tube of a circular cross-section of radius r inserted into an open vessel of water as shown in the figure above. The contact angle between the tube and water is less than 90°. The surface of the water in the capillary is **concave**, thus creating a pressure difference between the two sides of the top surface. This pressure difference is given by:

$$P_{\text{INSIDE}} - P_{\text{OUTSIDE}} = (P_a + P_o) - P_a = \frac{2S}{r \sec} = \frac{2S}{r} = \cos\theta = pgh$$

where ρ = density, h = capillary rise, S = surface area and g = gravitational pull, taking into account the fact that A and B must be at the same pressure.

Consider a vertical capillary tube of a circular cross-section of radius r inserted into an open vessel of water as shown in the figure above. The contact angle between the tube and water is less than 90°. The surface of water in the capillary is concave, thus creating a pressure difference between the two sides of the top surface. This pressure difference is given by:

$$P_{\text{INSIDE}} - P_{\text{OUTSIDE}} = (P_a + P_o) - P_a = \frac{2\gamma}{r \sec} = \frac{2\gamma}{r}\cos\theta = pgh$$

where ρ = density, h = capillary rise, γ = force per unit length of the tube and g = gravitational pull, taking into account the fact that A and B must be at the same pressure.

Unit 12.1 Activity 1C

1. Three tubes of different diameters with equal heights are placed into a liquid solution at the same time, as shown in the figure below. In which of the three tubes will the liquid solution rise the highest? Explain your answer.

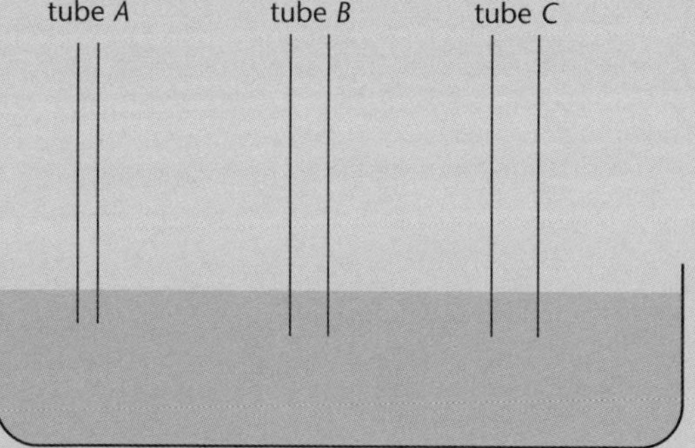

2. A tap root of a plant with a diameter of 60 μm standing near a water pond is pulling water molecules from the pond. The density of the water in the pond is 1040 kg/m^3. The plant has a height of 20 cm including depth to water level, γ = 0.0728 N/m, and g = 9.81 m/s^2. Calculate the angle of contact the tap root made with the surface of the water for capillary action.

Unit 12.1 Fluids

Topic 2: Fluid dynamics

In Topic 1 we looked at fluids at rest or fluid statics and in Topic 2 we will look at fluid in motion or fluid dynamics (Syllabus p. 24). This Topic covers:

- Flow rate and equation of continuity.
- Bernoulli's equation.

Introduction

Since the motion of fluids can be very complicated due to turbulence, frictional forces between molecules and layers of the fluid and other influences, we need to make some assumptions to make our analysis easier. These assumptions are:

i. No internal **friction** between layers of fluid (*zero viscosity*).

ii. The density of the fluid remains the same throughout the fluid (*non-compressibility*).

iii. Fluid **velocity** and pressure at each point in the fluid do not change with time (*steady state*).

iv. The fluid particles follow smooth, predictable flow lines (*streamline or laminar flow*).

Flow rate and equation of continuity

The **flow rate** is defined as the volume of fluid passing through the pipe at that point per unit time in a non-uniform cross-section such as a pipe or hose. It is mathematically defined as:

$$\textit{Flow rate} = \frac{\Delta V}{\Delta t} = A\frac{\Delta x}{\Delta t} = Av.$$

where A is the cross-sectional area of the pipe at that point and $v = \dfrac{\Delta x}{\Delta t}$ is the fluid velocity. The unit of flow rate is m^3/s.

The **continuity equation** states that if the fluid is incompressible, the flow rate must be the same everywhere along the cross-section of the pipe, according to the conservation of fluid particles which states that 'what goes in must come out'. Mathematically, this is defined as:

$$Av = \text{constant.}$$

Bernoulli's Equation

From conservation of energy the **work** done in moving a volume of fluid through a portion of pipe must go into **kinetic energy** and potential energy of the fluid. In this situation, assume that there are no frictional forces. Consider a fluid moving through a pipe as in the figure below.

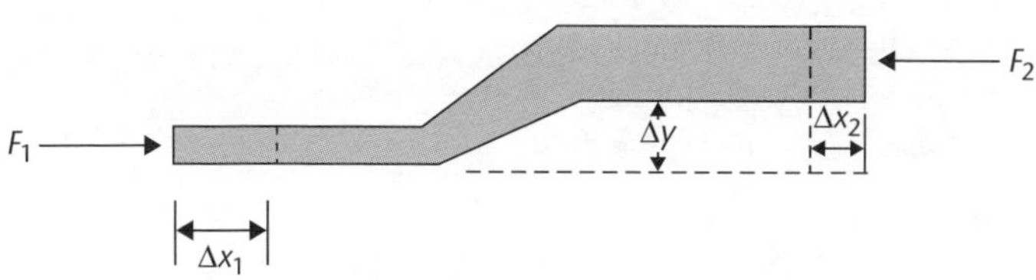

Fluid motion

The conservation of energy in this situation is kinetic energy, and potential energy, which is the **mechanical energy** of the fluid, must be conserved. We begin by analysing the mechanical energy of the fluid as it moves through the cross-section of the pipe. The work done by the pressure P_1 on the left in moving the fluid a distance Δx_1 is:

$$W_1 = F_1 x_1 = P_1 A_1 (\Delta x_1).$$

On the right, the work done by the pressure: $W_2 = -F_2(\Delta x_2) = -P_2 A_2 (\Delta x_2)$

The minus sign occurs because the force is in the opposite direction to the **motion** in this case. By the continuity equation, the amount of fluid entering the pipe on the left must equal the amount leaving on the right. This is to say.

$$A_1(\Delta x_1) = A_2(\Delta x_2).$$

This volume of fluid, which we call V, can undergo both a change in kinetic energy and a change in potential energy as it is pushed through the pipe, given by:

$$\Delta KE = \frac{1}{2}\rho V v_2^2 - \frac{1}{2}\rho V v_1^2$$
$$\Delta PE = \rho V g y_2 - \rho V g y_1.$$

Now we equate the work done ($W = W_1 + W_2$) to the change in mechanical energy ($\Delta KE + \Delta PE$) leads us to what we call a **Bernouilli's Equation** (note that the volume $V = A_1 \Delta x_1 = A_2 \Delta x_2$ cancels):

$$P_1 + \frac{1}{2}\rho v_1^2 + \rho g y_1 = P_2 + \frac{1}{2}\rho v_2^2 + \rho g y_2$$

In the equations above, P_1, v_1 and y_1 are the pressure, fluid velocity and height of the fluid on the left side of the equation and P_2 , v_2 and y_2 are the same parameters on the other side. If there is no height difference, Bernouilli's equation relates the pressure at two points along the flow to the fluid velocities and reduces to:

$$P_1 - P_2 = \frac{1}{2}\rho (v_2^2 - v_1^2).$$

Therefore pressure in the fluid decreases whenever the fluid velocity increases. This results can be observed in every day application such as the lift produced by an airplane wing, the curving of a spinning baseball and the fact that the shower curtain gets sucked in when you first turn on the shower.

Unit 12.1 Activity 2A

1. Which is the most accurate statement of Bernoulli's principle?

A. fast-moving air causes lower pressure

B. lower pressure causes fast-moving air

C. both (a) and (b) are equally accurate.

2. A maintenance crew is working on a section of a three-lane highway, leaving only one lane open to traffic. The result is much slower traffic flow (a traffic jam). Do cars on a highway behave like

A. the molecules of an incompressible fluid

B. the molecules of a compressible fluid?

3. As part of a lubricating system for heavy machinery, oil of density 850 kg/m^3 is pumped through a cylindrical pipe of diameter 8.0 cm at a rate of 9.5 litres per second. What is
   a. the speed of the oil?
   b. the mass flow rate?

Unit 12.1 Activity 2B

1. A boy tries to use a garden hose to supply air for a swim at the bottom of a 50 m-deep pool. What is most likely to go wrong?
2. Mercury is less dense at high temperatures than at low temperatures. Suppose you move a mercury barometer from the cold interior of a tightly sealed refrigerator to outdoors on a hot summer day. You find that the column of mercury remains at the same height in the tube. Compared to the air pressure inside the refrigerator, is the air pressure outdoors (i) higher, (ii) lower, or (iii) the same? (Ignore the very small change in the dimensions of the glass tube due to the temperature change.).
3. A storage tank 12.0 m deep is filled with water. The top of the tank is open to the air. What is the absolute pressure at the bottom of the tank and the gauge pressure?
4. You place a statue made of wood and note the reading on the scale. You now suspend the statue in the water. How does the scale reading change?
   A. it increases
   B. it decreases
   C. it remains the same;
   D. none of these.
5. A car weighing 1.2×10^4 N rests on four tyres. If the gauge pressure in each tyre is 200 kPa, calculate the area of each tyre in contact with the road.
6. A rubber hose is attached to a funnel, and the free end is bent around to point upward. When water is poured into the funnel, it rises in the hose to the same level as in the funnel, even though the funnel has a lot more water in it than the hose does. Why? What supports the extra weight of the water in the funnel?
7. The same car as in problem 5 above sits on a hydraulic press as shown below. The area of the cylinder holding the car up is four times greater than the area of the cylinder on the other side of the press. Calculate the required force that must be applied to the other side of the hydraulic press.

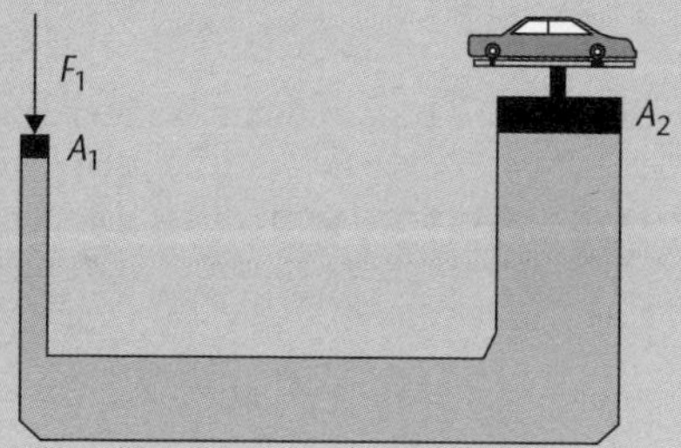

8. An aluminum object has a mass of 27.0 kg and a density of 2.70×10^3 kg/m^3. The object is attached to a string and immersed in a tank of water. Determine:
   a. the volume of the object
   b. the tension in the string when it is completely immersed.
9. An object is weighted with a spring balance in the air and then totally immersed in water. The readings on the balance are 0.98 N and 0.88 N respectively. Calculate the density of the object.
10. A fisherman in Sissano Lagoon is floating in a canoe in the middle of the lagoon. He catches a 10 kg monster red emperor and loads it onto his canoe. Does the water level in the lagoon rise or fall? Explain.
11. A surfer from Vanimo surfing club is surfing on a surfboard made of wood under the surface of an ocean wave in Lido beach. After the surfboard is completely submerged, he keeps pushing and riding on the wave surface. As he does this, what will happen to the buoyant force on it? Will the force keep increasing, stay the same, or decrease? Why?
12. A garden hose of diameter 2 cm is used to fill a 20-litre bucket. It takes one minute to fill the bucket.
   a. Calculate the speed at which the water enters the hose.
   b. The open end of the hose is then squeezed to a diameter of 5 mm. What is the speed at which water comes out of the hose?
13. A wind blows at the speed of 30 m/s across a 175 m^2 flat roof of a house.
   a. What is the pressure difference between the inside of the house and the outside of the house just above the roof? (Assume that the air pressure inside the house is atmospheric pressure.)
   b. What is the force on the roof due to the pressure difference?
14. Water flows through the pipe shown below at a flow rate of 0.10 m^3/s . The diameter at point 1 is 0.4 m. At point 2, which is 3.0 m higher than point 1, the diameter is 0.20 m. If the end at point 2 is open to the air, determine the gauge pressure at point 2.

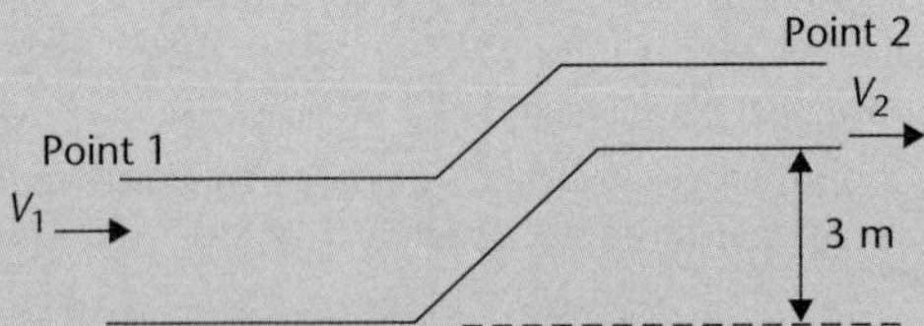

15. Water enters a house from a ground pipe with an inside diameter of 2.0 cm at an absolute pressure of 4.0×10^5 Pa. A pipe that is 1.0 cm in diameter leads to the second-floor bathroom 5.0 m above the ground floor. The flow speed at the ground pipe is 1.5 m/s. Find the flow speed, pressure, and volume flow rate in the bathroom.

16. **a**. Explain why the blood pressure in humans is greater at the feet than at the brain.

b. Explain why water with detergent dissolved in it has a small angle of contact.

17. Give two reasons why the upper arm of a human is often used to measure the blood pressure of a human.

18. During a blood transfusion the needle is inserted in a vein where the gauge pressure is 1500 Pa. At what height must the blood container be placed so that blood may just enter the vein? Density of whole blood is 1.06×10^3 kg/m^3.

19. Explain why grease and oil stains can only be removed from dirty clothes made of cotton by adding detergents such as soap to water.

20. Why are drops and bubbles spherical?

21. Research the topic 'What is blood pressure?'.

Unit 12.2 Temperature and Heat
Topic 1: Temperature, thermal expansion and heat transfer

Unit 12.2 deals with temperature and heat. Topic 1 looks at temperature, thermal expansion and heat transfer (see Syllabus pp. 26–27), specifically:

- Definition of temperature and its units.
- Mechanical, electrical and radiation types of temperature-measuring instruments.
- Conversion of one temperature scale to another.
- Linear, superficial and volume thermal expansions.
- Specific heat capacity.
- Thermal conductors and insulators.

Heat and **temperature** are different quantities. The figure below shows a 1 L container of boiling water and a 500 mL container of boiling water. The two containers are at the same temperature, but the 1 L container contains more heat.

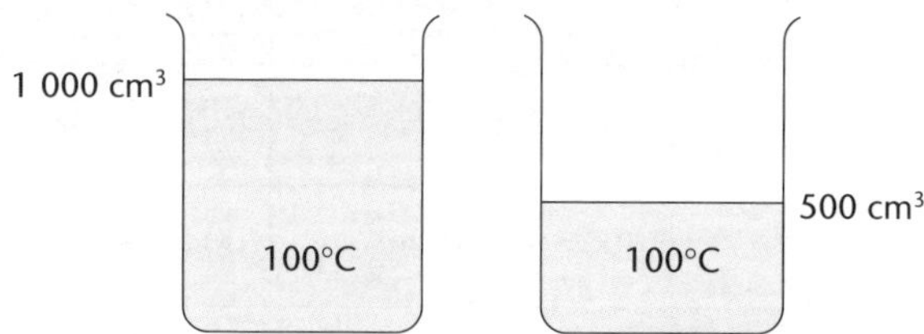

The difference between heat and temperature.

Heat is a measure of the total internal energy of the particles that make up the object.

The temperature of an object is a measure of the average kinetic energy of the particles that make up the object. In a hot object, the particles move faster than in a cold object. When two objects that have different temperatures are touching, heat energy will flow from the hotter object to the cooler object. Eventually the two objects will reach the same temperature and no heat will flow between them. Temperature can be measured using a **thermometer**. Two temperature scales are important:

- Celsius (Centigrade) temperature scale.
- The **Kelvin scale**.

The Celsius (Centigrade) temperature scale

The lowest fixed point on this scale is the temperature of pure melting ice – assigned a value of 0 °C.

The highest fixed point on the scale is the temperature of boiling water – assigned a value of 100 °C.

The temperature difference between these fixed points is divided up into 100 equal intervals, called *degrees centigrade*.

The Kelvin scale

The lowest value on the Kelvin scale is 0K (zero Kelvin). This is the temperature at which all particle motion ceases. It is called *absolute zero*. Absolute zero is equivalent to –273 °C. One degree on the Kelvin scale is the same as one degree on the Celsius scale.

Example A

Convert 50 °C to K.

Answer:

50 °C = 50 + 273 = 323K

The Fahrenheit scale

Apart from the Celsius (Centigrade) and Kelvin scale, the Fahrenheit scale is also recognised and used throughout the world. The table below shows how one can convert Celsius and Kelvin temperatures to and from Fahrenheit temperatures.

	From Fahrenheit	To Fahrenheit
Celsius (°C)	$[(°F) - 32] \times \frac{5}{9}$	$(°C) \times \frac{9}{5} + 32$
Kelvin (K)	$[(°F) + 459.67] \times \frac{5}{9}$	$(K) \times \frac{9}{5} - 459.67$

Thermometers

The most common thermometer is made of a length of *capillary tube*, sealed at one end and expanded to a small bulb at the other. The bulb is usually filled with alcohol or mercury. When the bulb is heated, the mercury or alcohol inside expands and rises up the tube. The height it rises depends on the temperature of the bulb.

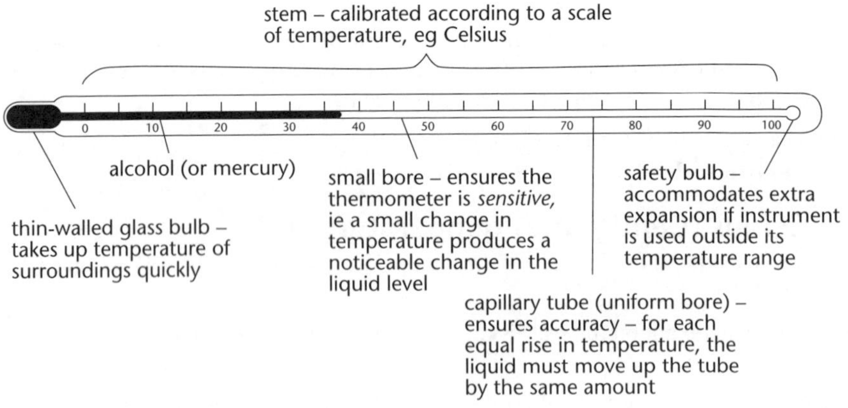

Celsius thermometer.

A *clinical thermometer* is designed to measure human body temperature. It measures temperature over a limited range, 35–43 °C. The capillary tube has a constriction to prevent the mercury from falling after a temperature reading is taken, so the temperature inside the body is still shown when the clinical thermometer is withdrawn from the body. This thermometer needs to be shaken to reset it. The 'normal' temperature of the human body is 37 °C.

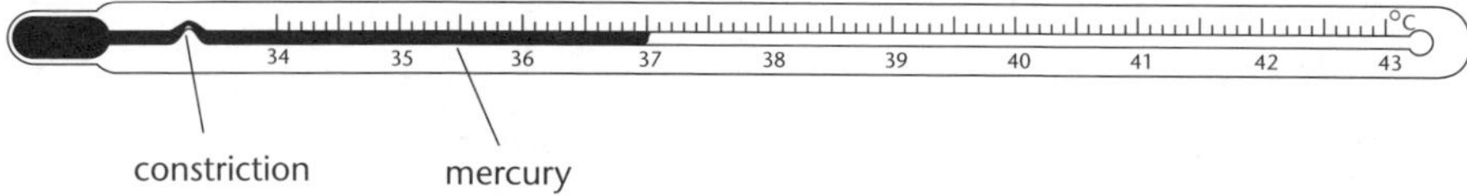

Clinical thermometer.

Types of temperature measuring instruments

The temperature of a substance can be measured using various thermometric systems and properties. The main ones are given in the table below.

System	Property	Examples
Fixed gas volume	Pressure	Mercury, alcohol
Helium vapour	Pressure	Helium thermometer
Platinum wire	Resistance	Platinum sensor
Thermocouple	Electromagnetic force	Electric thermometer
Paramagnetic salt	Magnetic susceptibility	Magnetic thermometer
Black body	Radiant emission	Pyrometers (different types)

Unit 12.2 Activity 1A

1. Rank the following temperatures from lowest to the highest:
- **a**. 0.00 °C
- **b**. 0.00 °F
- **c**. 0.00 K
- **d**. 20.00 K
- **e**. 20.0 °F
- **f**. 20.0 °C

2. Which of the following thermometer(s) is most suitable for measuring body temperature in PNG hospitals?
- **A**. helium thermometer
- **B**. mercury in glass
- **C**. alcohol in glass
- **D**. bio-metallic strip
- **E**. electric thermometers.

Thermal expansion

When a substance is heated it generally expands. An object has an initial length, L_O at some temperature. When it undergoes a change in temperature ΔT, its length changes by an amount ΔL, which is directly proportional to its initial length and temperature. The increase in length is given as:

$$\Delta L = \Delta T \alpha L_O$$

where α = average coefficient of **linear** expansion; it is different for different materials.

Linear expansion

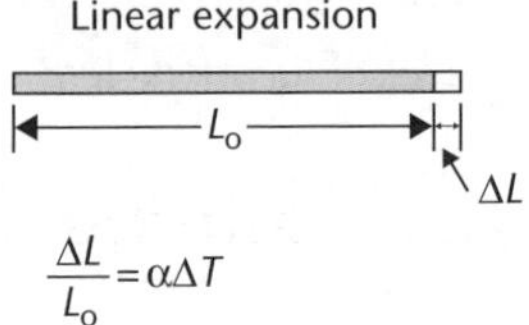

$$\frac{\Delta L}{L_o} = \alpha \Delta T$$

Likewise the volume also expands when subjected to a temperature change. The increase in volume is given as:

$$\Delta V = \Delta T \beta V_O$$

where β = coefficient of volume expansion = 3α.

Volume expansion

$$\frac{\Delta V}{V_o} = 3\alpha \Delta T$$

Like the volume and length, the area also expands when subjected to a temperature change. The increase in area is given as:

$$\Delta A = A - A_O = \gamma \Delta T A_O$$

where γ = average coefficient of area expansion = 2α.

Area expansion

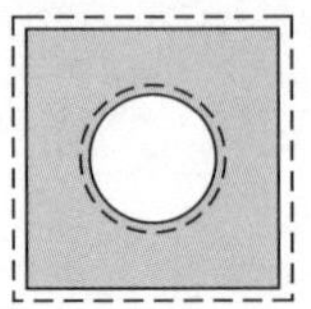

$$\frac{\Delta A}{A_o} = 2\alpha \Delta T$$

Unit 12.2 Activity 1B

1. A surveyor in Lae City uses a steel measuring tape that is exactly 50.00 m long at a temperature of 20 °C. What is its length on a hot sunny day when the temperature is 35 °C? The average coefficient of linear expansion α for steel is $1.2 \times 10^{-5}\ K^{-1}$.
2. An aluminum ring has a diameter of 400.0 mm at 20 °C. Calculate the increase in volume at a temperature of 240 °C. The average coefficient of linear expansion of aluminum in this temperature range is $2.4 \times 10^{-5}\ K^{-1}$.
3. The rod below is made of copper and has a cross-sectional area of 1.20 cm^2 and a length of 20 cm. It is insulated as shown and T_H = 100 °C and T_L = 0 °C. The thermal conductivity of copper k = 385 w/m.°C.

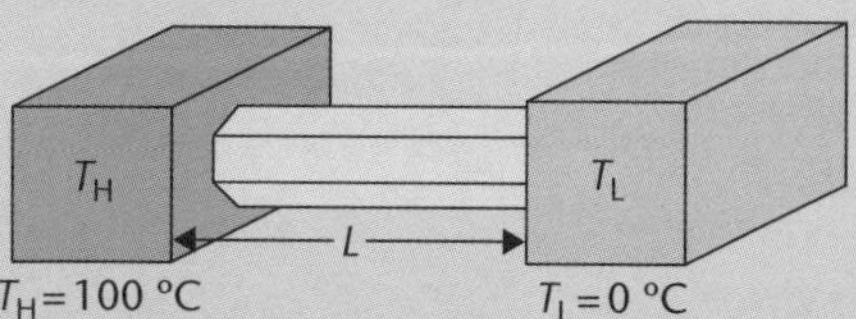

   a. What is the final steady state temperature gradient along the rod?
   b. What is the heat current in the rod in the final steady state?
   c. What is the final steady state temperature at a point 6.00 cm from its left end?
4. A glass flask with volume 100 cm^3 is completely filled with mercury at 20 °C. How much mercury will overflow when the temperature of the glass and mercury is increased to 100 °C? The coefficient of linear expansion of the glass is $8.5 \times 10^{-6}\ K^{-1}$. The volume expansion coefficient β for mercury = $181 \times 10^{-6}\ K^{-1}$.
5. Research thermal expansion in water and its effects on plants and animals in an ecosystem such as a pond or an isolated lagoon situated on a mountain top.

Specific heat capacity

Heat energy is measured in **joules** (J). When an object gains heat, its temperature rises. The increase in temperature depends on the **specific heat capacity** of the substance. The specific heat capacity of a substance (c) is the number of joules of heat required to raise the temperature of 1 kg of the pure substance by 1K.

Heat capacity is the amount of energy required to raise the temperature of a pure substance by 1°C. Specific heat capacity involves mass; mass is not considered in the calculation of heat capacity.

The figure to the right shows the specific heat capacities of some common substances.

Substance	c (J kg–1 K)
Aluminium	900
Brass	380
Copper	400
Glass	470
Ice	2 100
Iron	460
Lead	130
Mercury	140
Water	4 200

Specific heat capacities.

The following formula is useful to solve problems involving heat and temperature:

$$Q = mc\Delta T$$

Where:

Q is the amount of heat absorbed or released (J)
m is the mass of the object (kg)
c is the specific heat capacity (J kg^{-1} K)
ΔT is the temperature change (K)

Example B

How much heat energy is required to raise the temperature of 3 kg of water from 15 °C to 100 °C?

Answer:

$Q = mc\Delta T$
$Q = 3 \times 4\,200 \times (100 - 15)$
$Q = 1\,071\,000$ J (1 071 kJ)

Heating and cooling curves

If a cube of ice, H_2O, at a temperature well below 0 °C is heated slowly and the temperature is plotted at regular time intervals, the following graph would be obtained:

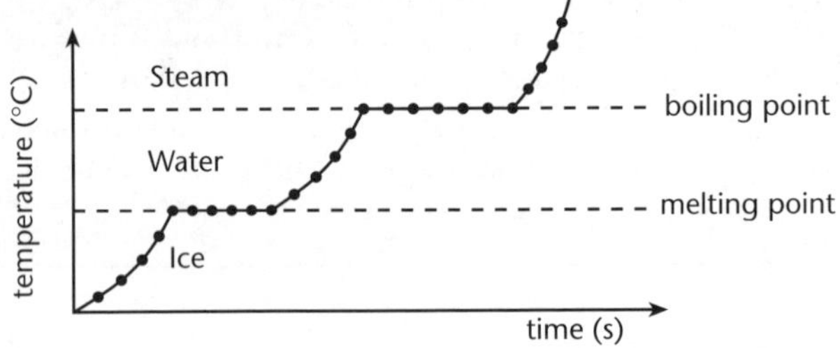

The heating curve for water.

At both the *melting point* and *boiling point* of the H_2O the temperature does not rise (ie stays constant), even though the H_2O is being heated.

At the melting point and boiling point temperatures, heat energy is used to change the *state* (or **phase**) of a substance (ie solid, liquid or gas) without increasing the temperature of the substance. The heat required to change the state is called **latent heat**. Latent heat is sometimes called *hidden heat* because the heat is absorbed without its presence being shown by a rise in temperature.

The specific **latent heat of fusion** is the heat required to change 1 kg of a substance from solid to liquid at its melting point. It is measured in J kg^{-1}. The latent heat of fusion of ice is 336 000 J kg^{-1}.

The specific **latent heat of vaporisation** is the heat required to change 1 kg of a substance from liquid to *vapour* (gas) at the boiling point. It is measured in J kg^{-1}.

The amount of heat absorbed or released when a substance *melts* or *solidifies* can be calculated using the formula:

$$Q = mL$$

Where:

Q is the amount of heat energy absorbed or released (J)
m is the mass of the object (kg)
L is the latent heat of fusion or vaporisation (J kg^{-1})

Example C

How much heat energy is required to melt 3 kg of ice?

Answer:

$Q = mL$
$Q = 3 \times 336\,000$
$Q = 100\,800$ J (100.8 kJ)

Some substances, such as carbon dioxide and iodine, change state directly from a solid to a gas when heated without going through the liquid phase – this is called **sublimation**.

The following graph shows what happens when water (as steam) is cooled until it solidifies. (The water passes through all three phases.)

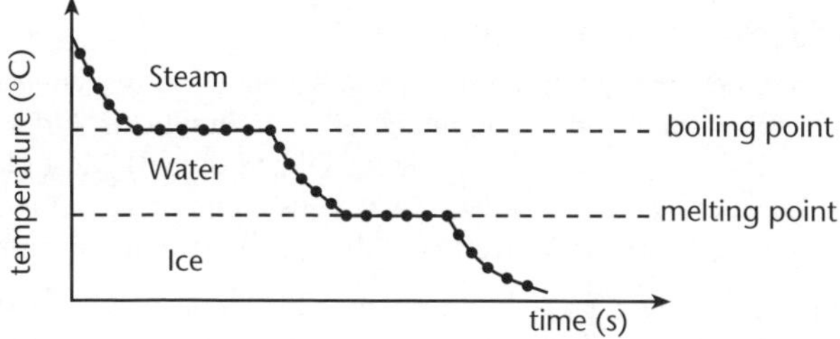

The cooling curve for water.

Heat transfer

The transfer of heat is normally from an object at high temperature to an object at lower temperature. Heat transfer changes the internal energy of both systems involved according to the Zeroth Law of Thermodynamics.

An example of the Zeroth Law of Thermodynamics in a PNG context is when we cook large quantities of food using the mumu method. This involves using very hot stones to slowly cook the food that is wrapped and placed close to the stones or distributed over the red hot stones with some covering in between to prevent overcooking.

The heat is transferred from the red hot stones on one side of the wrapping to the food inside the wrapping and then to the stones on the other side until all are at the same temperature. At this point we say they are in thermal **equilibrium**. Under the equilibrium condition there is no more heat transfer as the food, the wrapping and the stones are all at the same temperature. In the process of heat transfer from red hot stones to food and wrapping, the food is cooked.

Unit 12.2 Activity 1C

1. Which of the following types of thermometers have to be in thermal equilibrium with the mumu in order to give accurate readings of the mumu?
   A. a bimetallic strip
   B. a resistance thermometer
   C. a temporal artery thermometer
   D. both A and B
   E. all of A, B and C.
2. You put a thermometer in a pot of hot water and record the reading. What temperature have you recorded?
   A. the temperature of the water
   B. the temperature of the thermometer
   C. an equal average of the temperatures of the water and thermometer
   D. a weighted average of the temperatures of the water and thermometer, with more emphasis on the temperature of the water
   E. a weighted average of the water and thermometer, with more emphasis on the temperature of the thermometer.

Thermal conductors and insulators

A material that allows heat to pass through it easily is a good thermal conductor. A material that does not is a poor thermal conductor and is called an **insulator**. Most metals are good thermal conductors as they allow heat to conduct well and quickly. Most **elastic** materials and ceramics are insulators, and are used as thermal insulation in covering handles of cooking and eating utensils, covering electrical cables, and many more applications.

In physics, thermal conductivity, k, is the property of a material's ability to conduct heat. A material that conducts heat well and quickly has a relatively high thermal conductivity. Thermal conductivity of materials is temperature dependent. We can generally say that materials become more conductive to heat as the average temperature increases. The reciprocal of thermal conductivity is thermal resistivity.

Unit 12.2 Activity 1D

1. Research how a refrigerator works. Explain why the freezing part of the fridge is placed above the storage compartment in terms of convection currents.
2. Research fire dancing in East New Britain Province. Explain the physics part of the phenomena of why the dancer's feet do not get blisters and burns.

Unit 12.2 Temperature and Heat

Topic 2: Heat and energy

Following on from Topic 1, Topic 2 (see Syllabus p. 27) looks at:

- Heat transfer by means of conduction, convection and radiation.

The transmission of heat

Heat can be transmitted in three ways – **conduction**, **convection** and **radiation**.

The **transmission** of heat by conduction and convection can only occur through a **medium**. Heat can be transmitted through a *vacuum* by radiation.

Conduction

Heat energy can be passed from particle to particle by conduction. Metals are generally good **conductors** and non-metals are generally poor conductors of heat. Materials that are poor conductors are called *insulators* and can be used to prevent heat from escaping. Fibreglass 'Batts' are used in walls, ceilings and around hot water cylinders to reduce heat loss. Electric jugs and cooking pots have handles made from an insulating material to prevent burns.

Example D

The conductivity of different metals can be shown using different metal rods with a drop of wax holding a drawing pin at the end of each rod. The end of the rod away from the wax is heated. When the wax melts the drawing pin drops. The first drawing pin to drop is attached to the metal that is the best heat conductor.

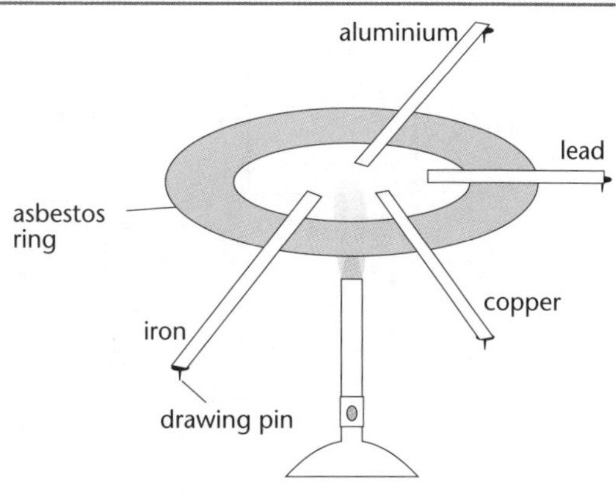

Comparing the rate of heat conduction in different metals.

Convection

When a liquid is heated, the particles in the liquid speed up and move apart. As a result, the density of the warm liquid decreases and the warm liquid increases in volume. Colder, more dense liquid sinks to take the place of the warm liquid. This sets up a *convection current* that transmits heat from the bottom of the container to the top of the container.

Example E

The pattern of purple dye that results when some potassium permanganate is dropped into a beaker of water that is being gently heated shows the convection currents in the water.

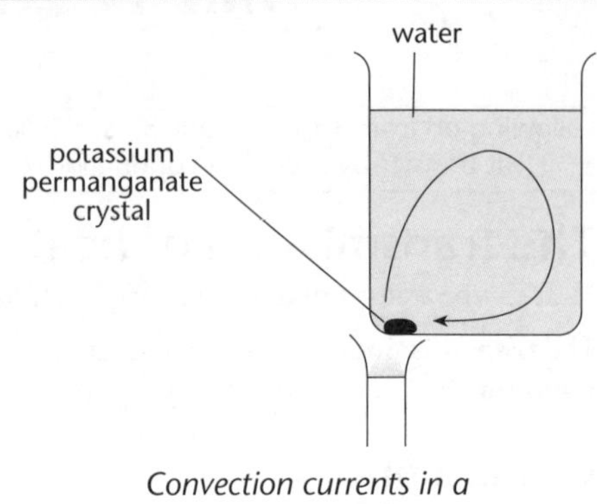

Convection currents in a beaker of water.

In a typical domestic hot water system, water is heated by an electrical element at the bottom of the water cylinder. The hot water rises and cold water sinks to take its place. Hot water is drawn from the top of the cylinder to supply the hot water taps around the house.

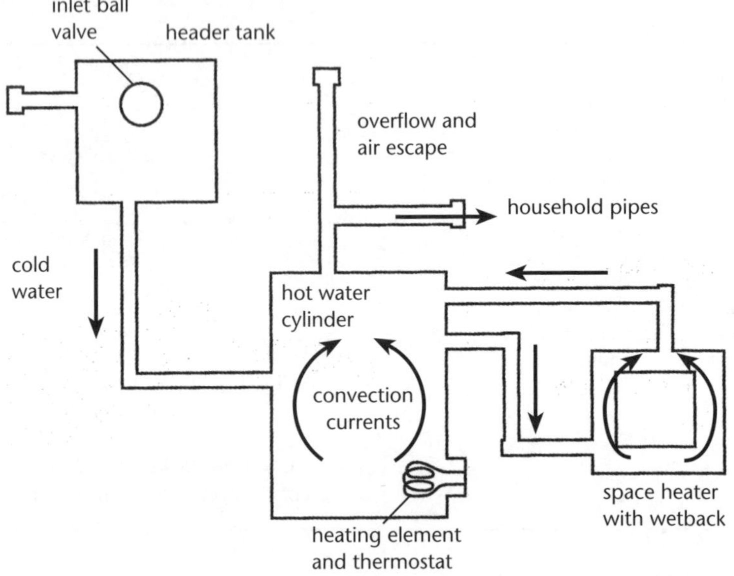

A domestic electric hot water system.

Convection currents exist in the magma in the mantle of the Earth and may be responsible for the Earth's magnetism.

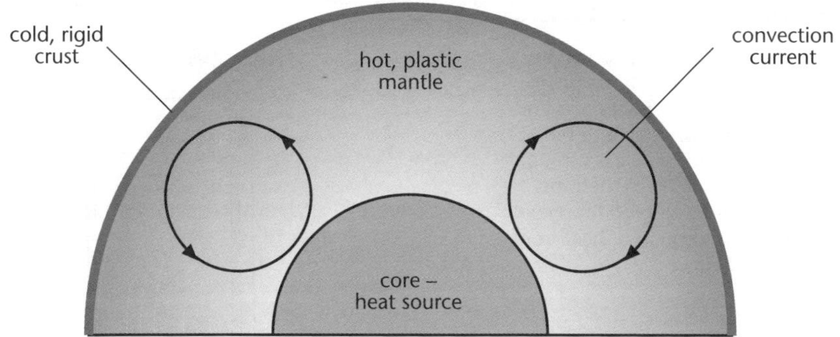

Convection currents in the mantle of the Earth.

During convection, the particles of the medium travel from one place to another, taking the heat energy with them; whereas, during conduction, the particles only vibrate back and forth and pass the heat energy from particle to particle.

Radiation

Radiant heat consists of *infrared radiation* and is part of the **electromagnetic spectrum**. It travels at the **speed of light** (3×10^8 m s^{-1}) and can travel through a vacuum. When radiant heat falls on a surface, some of the energy is absorbed and some of the energy is **reflected**. The absorbed energy heats the surface and causes the surface temperature to rise.

The best absorber is a matt black, rough surface. Surfaces that are good absorbers are poor reflectors and good radiators, or emitters, of heat.

The worst absorber is a smooth, shiny silver surface. Surfaces that are good reflectors are poor absorbers and poor radiators, or emitters, of heat.

Example F

If a white tin and similar black tin are filled with the same volume of cold water and left exposed to the sun, the temperature of the water in the black tin will increase faster because the black surface is a better absorber of radiant energy than the white surface.

If a white tin and similar black tin are filled with hot water and left to cool in a dark place, the temperature of the water in the black tin will drop faster because the black surface is a better emitter of radiant energy than the white surface.

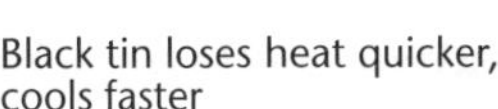

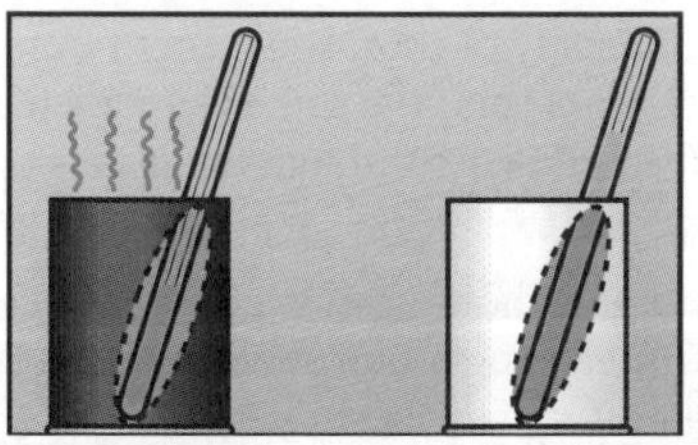

White tin loses heat slowly, cools slowly

Comparing the cooling rates of black and white tins.

In hot climates, people often wear white clothes (good reflectors) to keep cool. Shiny chrome teapots keep tea warm for longer because chrome is a poor radiator of heat.

Example G

A vacuum or *thermos flask* can be used to keep food or drinks warm because the silver surfaces prevent heat loss from radiation and the vacuum prevents heat loss through conduction. Likewise, a thermos flask can be used to keep food cool by slowing absorption of heat from the surroundings.

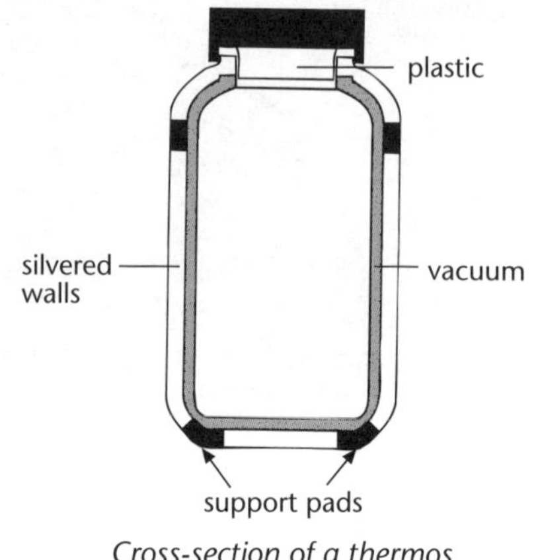

Cross-section of a thermos flask.

The illustration below shows how heat is lost in the home and how heat loss can be reduced.

Heat loss in the home and how to reduce it.

In a greenhouse, the radiant energy of short wavelength from the Sun passes through the glass and is absorbed by the plants and the soil. Heat radiated from the plants and soil is of a longer wavelength and cannot pass through the glass – therefore the greenhouse traps heat.

The increase in the so-called greenhouse gases (principally carbon dioxide, CO_2, and methane, CH_4) is presumably causing a similar effect in the Earth's atmosphere – the Earth's atmosphere is predicted to heat up significantly, which will have serious effects upon the Earth's climate and sea levels.

Unit 12.2 Activity 2A: Heat and temperature

1. Four identical copper cubes are painted with different types of paint. All the cubes are heated to the same temperature and left to cool in identical situations. Which of the following cubes would initially lose heat most rapidly?

Matt (low gloss) white. Shiny white.

Matt (low gloss) black. Shiny black.

The following information relates to Questions 2–5.

A balloon has been designed to fly around the world. It is filled with a very light gas called helium. The balloon carries two men in a metal cabin and flies at a height of many thousands of metres.

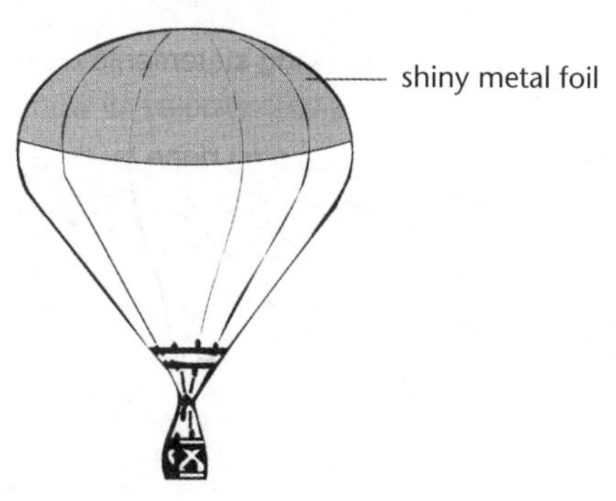

2. One problem with the balloon is that the Sun shines on it during the day and heats up the helium. Describe what would happen to the helium and the effect this will have on the balloon.

3. To help solve the problem indicated in Question 2, the top of the balloon is covered with shiny metal foil glued tightly to the fabric. Why does this stop thermal energy from reaching the fabric?

4. By what process does thermal energy from the Sun reach the atmosphere?

5. To help keep the two men warm, the metal cabin is covered with a thick layer of insulation. The insulation contains a lot of small air bubbles. What is the main reason the insulation stops energy escaping from the cabin?

6. Water has five times the specific heat capacity of sand. One kilogram of hot sand is poured into one kilogram of water at 20 °C. The temperature of the mixture is 40 °C. What was the original temperature of the hot sand, in °C?

7. Inside some indicator bulbs in cars is a bimetallic switch (see diagram). The switch automatically turns the light on and off so that the bulb flashes. Explain how this automatic switching works.

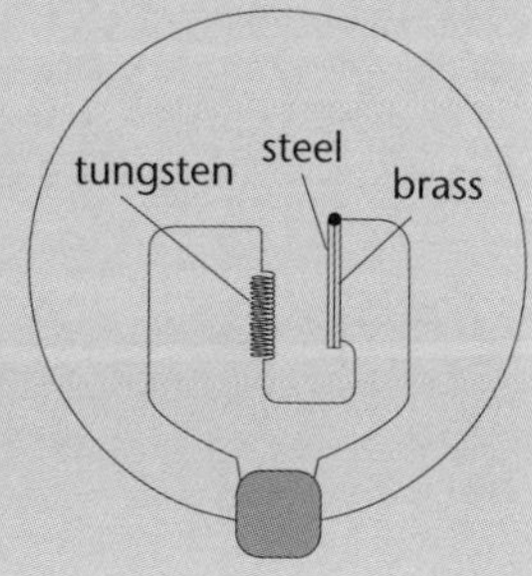

8. Tina is finding it difficult to open a new glass jar of jam. She holds the metal lid under hot running tap water for a while and then finds it easier to remove the lid. What is the most likely explanation for this?
9. Emergency blankets consist of a plastic sheet with a shiny surface coating. What is the reason these blankets keep people warm?
10. Temperature (°C)

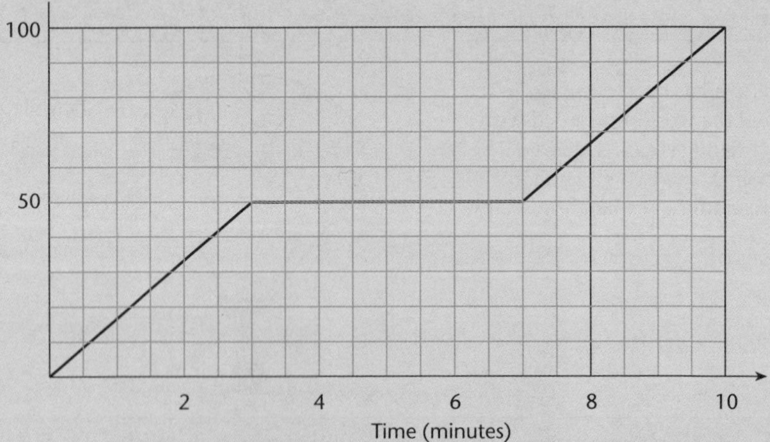

The graph above shows how the temperature of wax changes when heated steadily. Jane makes the following statements about the graph:

I. The melting point of the wax is 50 °C.
II. Both solid and liquid wax are present in the time between 3 and 7 minutes.
III. The specific heat capacity is larger for solid wax than for liquid wax.

Which of the statements are true?

11. The beaker shown contains water at 4 °C. After its temperature has been checked, it is placed on a thick metal pad which has its temperature maintained at 0 °C. Which of the following correctly describes what will happen in the next few minutes?

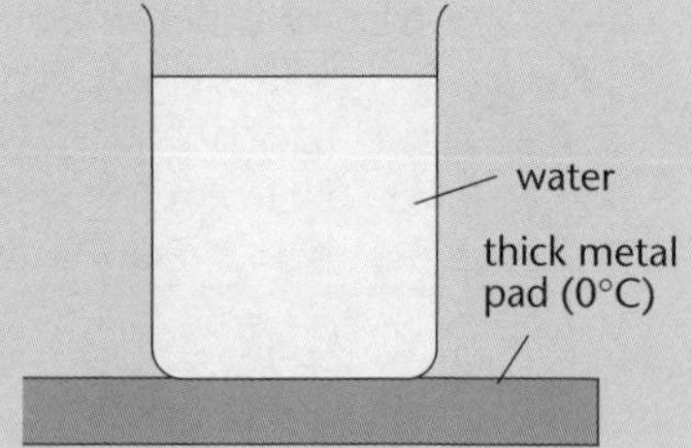

12. Joe has installed a solar panel on his roof to help cut down the cost of heating water in his home, as shown below.

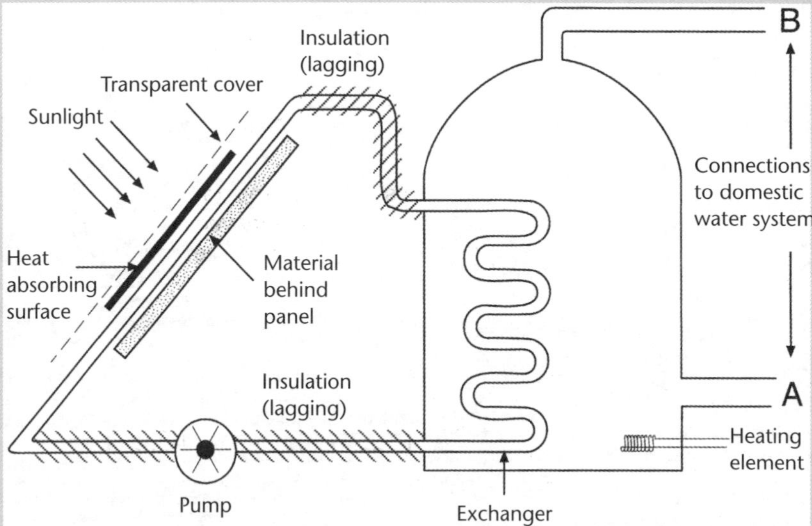

a. Give two reasons why the surface of the solar panel is covered with a transparent cover.

b. What colour should Joe make the heat-absorbing surface?

c. Does the water flow in the solar heating system in a clockwise or anticlockwise direction?

d. Which pipe would feed the hot water taps in the house, A or B? Explain your answer.

e. **i**. Give one reason why copper is used to make the spiral in the exchanger.

ii. Give one reason why a spiral is used instead of a straight tube.

f. The pipes between the solar panel and the hot water tank have been insulated (lagged) by material wrapped around them.

i. Name a suitable material for the insulation.

ii. Give one reason why insulation is important.

g. If there was no pump available, how could the water be made to circulate naturally?

h. What sort of material is behind the panel?

i. Explain why the solar panel is tilted at an angle and not laid in a horizontal position.

j. The hot water tank contains 200 kilograms of water. How much energy is needed to heat all the water from 15 °C to 55 °C? (The specific heat capacity of water is 4 200 J kg^{-1} $°C^{-1}$.)

13. Explain why:

a. In summer it is better to wear light- rather than dark-coloured clothing.

b. Silver teapots keep tea hotter than dark tea pots.

c. Newspaper is good at keeping fish and chips hot yet is also good at keeping ice cream cold.
d. In winter, the manufacturers of cooling fans advertise that the fans will save energy in winter.
e. When standing on a stone floor, it feels colder than carpet.

14. The diagram shows an electric jug being used to heat water.

a. Why is the heating element placed close to the bottom of the jug?
b. How does the shiny outside coating on the jug reduce heat loss?
c. Why does the jug have a plastic handle rather than a metal one?

15. Three methods of heat transfer are:

I conduction II convection III radiation

Which of these forms of heat transfer need a medium (solid, liquid or gas) in which to travel?

16. Why does sweating cool the body?

17. A new kind of coffee mug is being sold which is made of ordinary glass. It is sold together with a cork 'coaster'. If boiling hot water is poured into the glass without the coaster underneath, the glass is likely to crack, but if the coaster is underneath the glass, the glass will not crack. Why do you think this is? (Hint – ordinary glass is a poor conductor and expands when it is heated, unlike Pyrex glass, which does not expand when heated.)

18. Explain why:

a. People are not burned when the white-hot sparks (over 1 000 °C) from a sparkler hit their hands on Guy Fawkes Night, yet can be burned by a teaspoon of boiling water (only about 100 °C).
b. On a cold morning, the steel head of a hammer feels very cold, while the rubber handle does not.
c. Hot water poured into a glass may make the glass crack.
d. A saucepan is made from aluminium with a copper base and an insulating handle.
e. On a hot sunny day, metal roofs often creak.
f. Concrete roads are laid in sections with joints between the sections.

19. Why does a motor cycle engine have fins on its side?

20. What is the process whereby a substance changes its phase from gas to a liquid?

21. Explain how a hot air balloon works and how the operator can adjust the height of the balloon.

22. Different metals expand to different extents for the same temperature change.

a. How is this fact put to use in an oven thermostat?
b. Sketch and label a diagram to show how expansion makes a thermostat work.

23. Denise has gone to the gym for a workout. Before she starts to use the equipment, she is sitting on the bench. The room is cooler than Denise and is well ventilated.

a. The following table shows how heat is lost from her body to the surroundings. Complete this table by filling in the FOUR blank spaces.

Heat Transfer		Percentage of total heat loss
By	**To**	
	surroundings	65%
conduction		5%
evaporation		20%
	air currents	10%

b. Which type of heat loss would increase if:

i. Denise did some weightlifting so she perspired more?

ii. an electric fan was switched on to cool her body?

iii. Denise lay on the cold floor after excercising?

Denise takes a shower after her workout. The water is heated by an electric immersion heater as shown below:

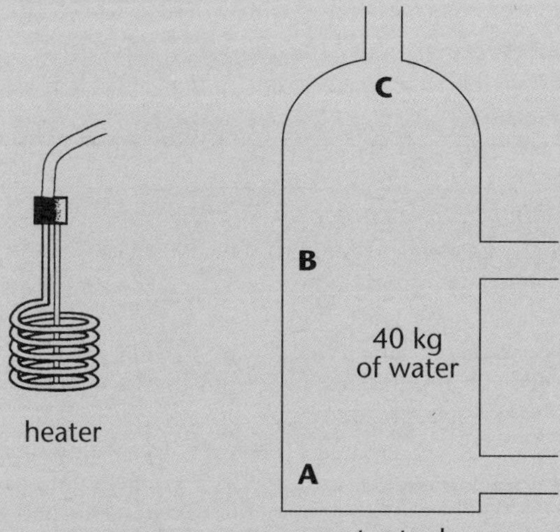

c. i. Where should the heater be placed in the tank? Circle the correct answer below.

A B C

ii. Explain your answer to i. above.

The graph below shows temperature versus time for the water in the tank after the heater has been switched on. The heater is switched off at point A on the graph.

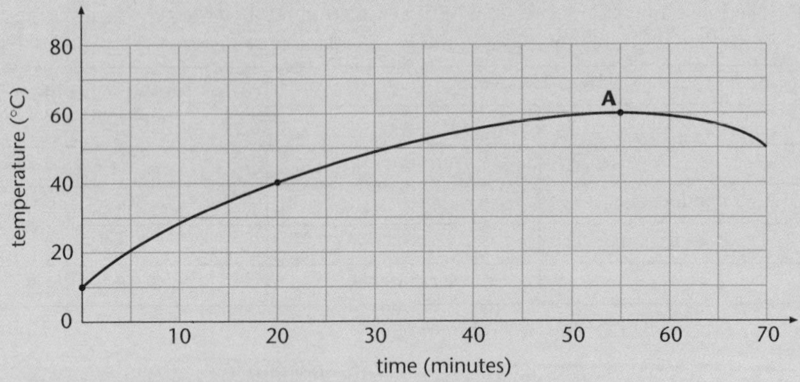

d. What is the rate at which the temperature of the water in the tank increases over the first 20 minutes?

e. The mass of water in the tank is 40 kg. What is the amount of heat supplied by the heating coil in the first 20 minutes?

f. What is the **power** of the heater?

g. **i**. How could the heat loss from the cylinder be reduced?

ii. Why is it desirable to reduce heat loss from the cylinder?

(Source: NCEA Examination paper)

Unit 12.2 Activity 2B: Multiple choice questions

1. Which of the following statements best gives the difference between latent heat of fusion, L_f and latent heat of vaporisation?

A. L_f is the amount of energy required to change the state of one kilogram of a pure substance from liquid to solid state at the same temperature. L_v is the amount of energy required to change the state of one kilogram of a pure substance from solid to gas state at the same temperature.

B. L_f is the amount of energy required to change the state of one kilogram of a pure substance from solid to liquid state at the same temperature. L_v is the amount of energy required to change the state of one kilogram of a pure substance from liquid to gas state at the same temperature.

C. L_f is the amount of energy required to change the state of one kilogram of a pure substance from solid to liquid state as the temperature changes. L_v is the amount of energy required to change the state of one kilogram of a pure substance from liquid to gas state as the temperature changes.

D. L_f is the amount of energy required to change the state of one kilogram of a pure substance from liquid to solid state as the temperature changes. L_v is the amount of energy required to change the state of one kilogram of a pure substance from gas to liquid state at the same temperature.

2. How much heat energy is required to raise the temperature of 2.0 kg of pure water from 0 °C to 30 °C? Specific heat capacity of water = 4200 J/kg °C.

A. 2520 J
B. 25,200 J
C. 252,000 J
D. 2,520,000 J

3. How many stages are involved in calculating the amount of heat required to raise the temperature of ice from –10 °C to 100 °C?

A. One
B. Two
C. Three
D. Four

4. An electric jug with power of 11.2 kw is used to boil water for making tea. The jug is filled with 2 kg of water at 20 °C which is then heated to 100 °C. How much energy is added to the water to get it boiling? Specific heat capacity of water is 4200 J/kg °C.

A. 672 J
B. 6720 J
C. 67,200 J
D. 672,000 J

5. How much time is required to get the water to boil at 100 °C in question 4 above?

A. 60 seconds
B. 80 seconds
C. 100 seconds
D. 120 seconds

6. Which of the following statements best gives the difference between heat and temperature?

A. Heat is a form of energy and temperature is how hot or cold a body is.
B. Heat is a form of kinetic energy which gives rise to temperature and temperature tries to cool the body down.
C. Heat is a measure of how cold or hot a body is by using a thermometer and temperature is to maintain equilibrium.
D. A, B, and C are all correct.

7. The process by which a material changes state from solid to gas is called.

A. Vaporisation
B. Sublimation
C. Condensation
D. Radiation

8. Which of the following methods is not a method of heat transfer?

A. Conduction
B. Radiation
C. Ionisation
D. Convection

9. The average distance from Sun to Earth is 150 million km. Through which method is the Sun able to give out energy to the Earth?
   A. Conduction
   B. Radiation
   C. Ionsation
   D. Convection
10. A room has one wall made of concrete, one wall made of copper, one wall made of steel and one wall made of wood. All of the walls are the same size and at the same temperature of 20 °C. Which wall feels coldest to the touch?
   A. the wood wall
   B. the concrete wall
   C. the copper wall
   D. the steel wall.

Unit 12.3 Waves

Topic 1: Types of waves

Topic 1 deals with the properties and types of waves (see Syllabus pp. 28–29). It covers:

- Longitudinal waves.
- Transverse waves.
- Electromagnetic waves.
- Sound waves.
- Earthquake waves.

When a soccer player kicks a ball into the goal net, kinetic energy carried by the ball is transferred to the net. This is an example of energy transfer by a particle.

A **pulse** or a **wave** may also transfer energy.

A pulse is a disturbance that travels between two points in a **medium**. The energy carried by a pulse is transferred from one point to another, but the particles of the medium do not travel between these points – they vibrate about their equilibrium (at rest) position and pass on the energy to neighbouring particles. A wave consists of a series of pulses.

There are two types of **wave motion** – **longitudinal waves** and **transverse waves**.

- Longitudinal waves are waves where the particles of the medium vibrate **parallel** to the direction of energy transfer, eg sound waves.

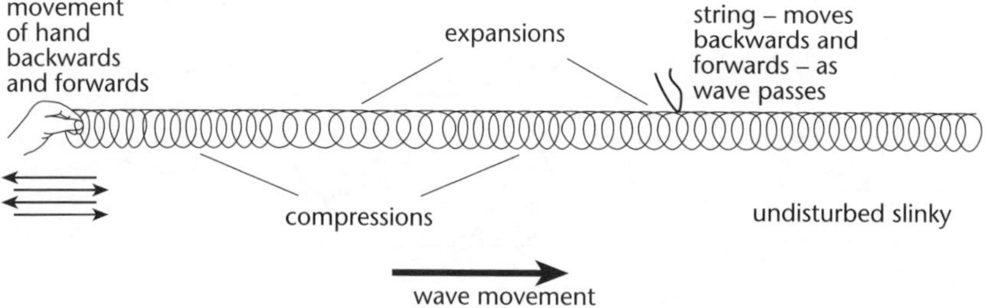

Longitudinal wave moving along a slinky.

- Transverse waves are waves where the particles of the medium vibrate at right angles to the direction of energy transfer, eg water waves and waves along strings.

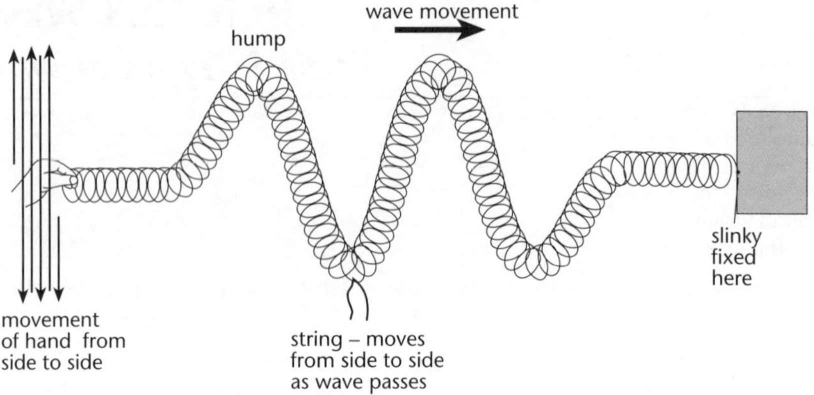

Transverse wave moving along a slinky.

A wave can be described by the following properties:

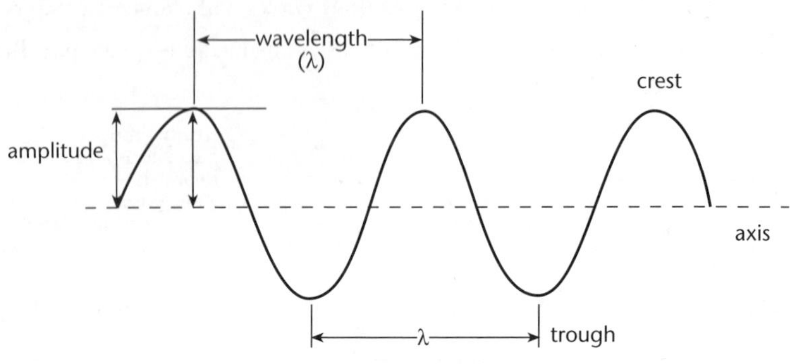

Wave properties.

- The **amplitude** of a wave, A, is the maximum **displacement** of a particle from its equilibrium position.
- The **wavelength**, λ, is the distance between two corresponding points on the wave – measured in m.
- The **frequency** of a wave, f, is the number of waves that pass a point every second – measured in Hz.
- The **period** of a wave, T, is the time it takes for one complete wave to pass a point (s).

$$T = \frac{1}{f}$$

The speed of a wave, v, can be calculated using the **wave equation**:

$$v = f\lambda$$

Example A

A vibrating tuning fork produces a sound of 540 Hz. The speed of sound in air is 330 m s^{-1}. Calculate the wavelength of the sound.

Answer:

$$\lambda = \frac{v}{f} = \frac{330}{540} = 0.61 \text{ m}$$

In nature, there are many examples of energy transfers involving waves, such as electromagnetic waves, sound waves and **earthquake** waves.

Electromagnetic waves

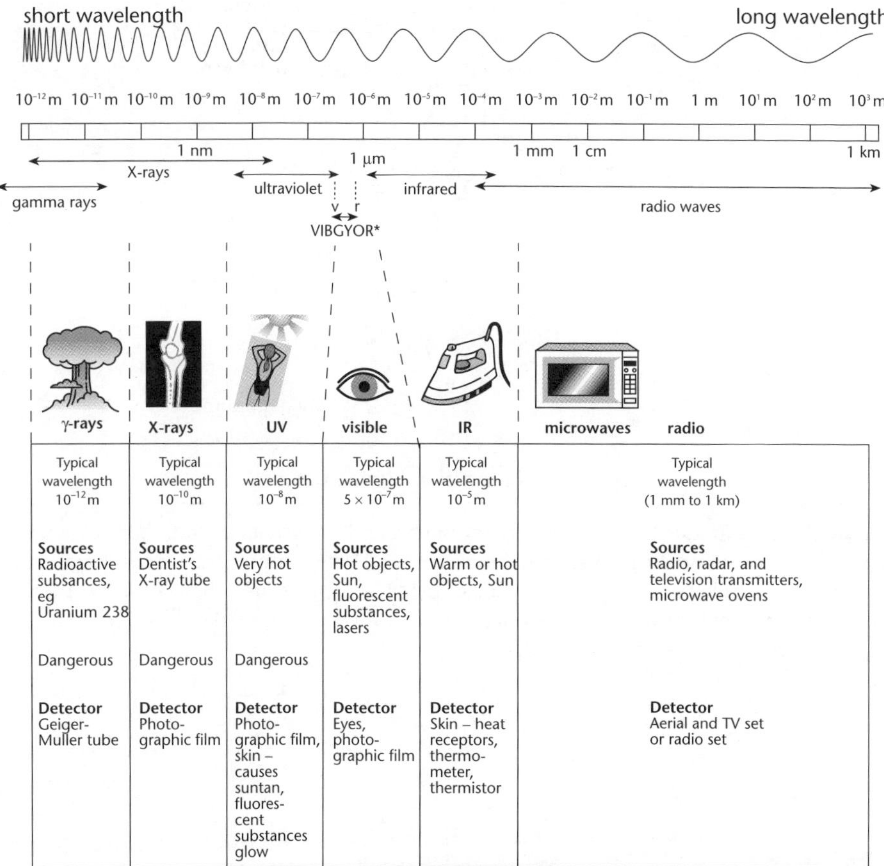

γ-rays	X-rays	UV	visible	IR	microwaves / radio
Typical wavelength 10^{-12} m	Typical wavelength 10^{-10} m	Typical wavelength 10^{-8} m	Typical wavelength 5×10^{-7} m	Typical wavelength 10^{-5} m	Typical wavelength (1 mm to 1 km)
Sources Radioactive subsances, eg Uranium 238	**Sources** Dentist's X-ray tube	**Sources** Very hot objects	**Sources** Hot objects, Sun, fluorescent substances, lasers	**Sources** Warm or hot objects, Sun	**Sources** Radio, radar, and television transmitters, microwave ovens
Dangerous	Dangerous	Dangerous			
Detector Geiger-Muller tube	**Detector** Photo-graphic film	**Detector** Photo-graphic film, skin – causes suntan, fluores-cent substances glow	**Detector** Eyes, photo-graphic film	**Detector** Skin – heat receptors, thermo-meter, thermistor	**Detector** Aerial and TV set or radio set

*VIBGYOR stands for the colours Violet, Indigo, Blue, Green, Yellow, Orange and Red.

The electromagnetic spectrum.

The **electromagnetic spectrum** is a family of waves. Electromagnetic waves can travel through a vacuum. The speed of all electromagnetic waves in a vacuum is 3×10^8 m s^{-1}.

Sound waves

Sound needs a medium to travel through and cannot travel through a vacuum. Sound travels as a longitudinal wave. The speed of sound in air is 300 m s^{-1}. Sound travels quicker in solids than in liquids or gases.

The source of a sound is a vibrating object, such as a string on a musical instrument, a tuning fork or a loudspeaker cone. A sound in air is transmitted by the vibration of air particles – sound therefore cannot travel through a vacuum.

Example B

If all the air is evacuated from a jar containing an electric bell, the sound volume of the ringing bell reduces to zero.

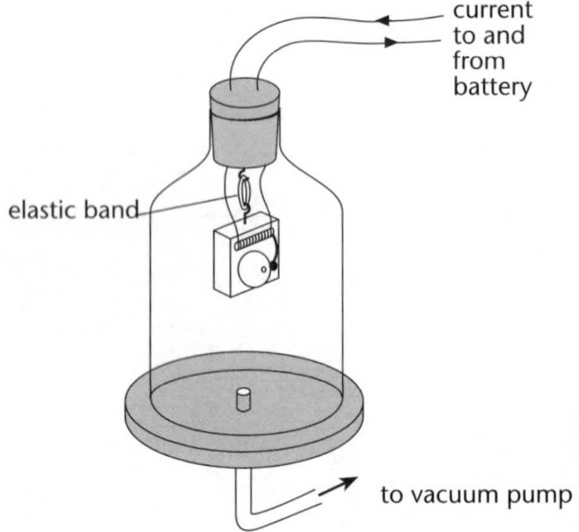

Sound cannot travel through a vacuum.

The **pitch** of a sound is the frequency of the sound and represents the number of sound waves passing a point per second. The human ear can detect sounds ranging in frequency from 20 Hz to 20 kHz. Sounds of frequencies below this range are called subsonic; those above this range are called ultrasonic. Dogs and dolphins can hear sounds too high in frequency for the human ear to detect.

Sound waves can be observed by connecting a microphone to an **oscilloscope**.

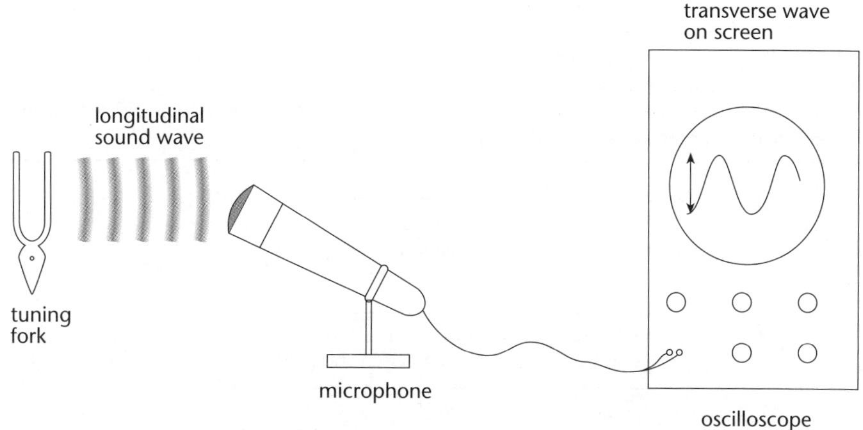

Using an oscilloscope to display a sound wave.

The volume of a sound depends on the amplitude of the sound wave.

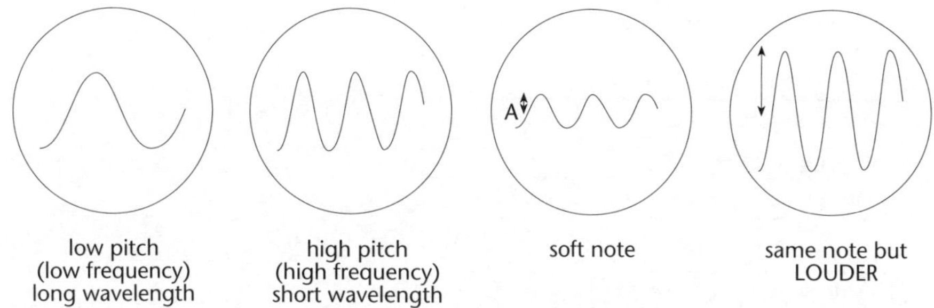

Different types of sound waves – as displayed on an oscilloscope.

Earthquake waves

Earthquakes originate as a result of movement along a **fault line** deep below the surface of the Earth. The point of origin of an earthquake is called the **focus**. The **epicentre** is that point on the Earth's crust directly above the focus. Earthquake waves are recorded on a **seismograph**.

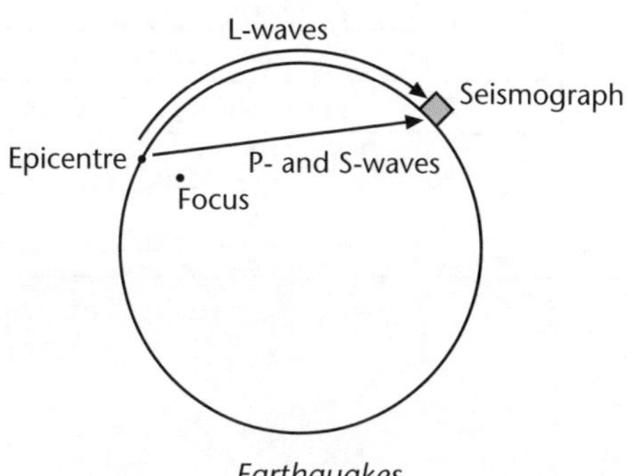

Earthquakes.

S-waves or *secondary waves* (also called *shake waves*) are transverse waves. S-waves are slower than P-waves and travel at approximately 4 500 m s^{-1}. S-waves cannot travel through the liquid outer core of the Earth.

P-waves or *primary waves* (also called push waves) are longitudinal waves. P-waves travel at approximately 8 000 m s^{-1} and are the first waves to be recorded by a seismograph.

Types of earthquake waves

L-waves (*love waves*) are the slowest waves and travel along the surface of the Earth.

Unit 12.3 Activity 1A: Waves

1. Copy and complete the table by stating whether each of the following is a transverse wave or a longitudinal wave:

	Transverse/longitudinal
Sound	a
Water waves	b
Wave on a string	c
P-earthquake wave	d
S-earthquake wave	e

2. Use the wave equation to complete the following table:

Velocity	Wavelength	Frequency
1 500 m s^{-1}	a	2 000 Hz
b	40 000 mm	25 kHz
330 m s^{-1}	18 m	c
3 × 108 m s^{-1}	6 × 10^{-7} m	d
400 cm s^{-1}	e	80 kHz
f	3.5 m	15 Hz

3. James has recorded the following sounds on his oscilloscope:

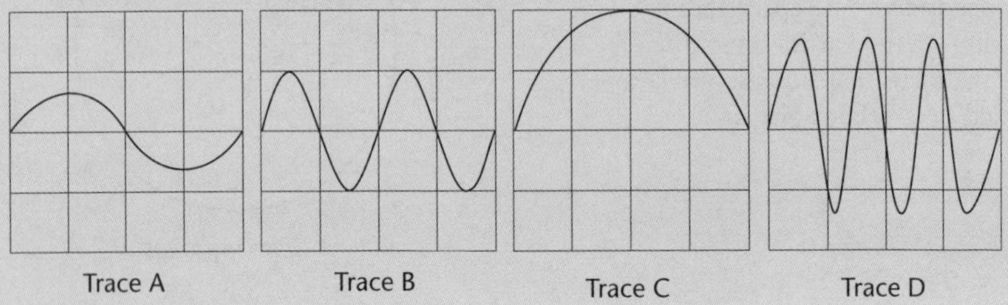

Arrange the waves in order of:

a. Increasing frequency (from low to high).

b. Decreasing volume (from loud to soft).

4. Sort the following types of **electromagnetic radiation** in order of decreasing frequency: Radio, infrared, visible light, UV, X-rays.

5. The frequency of microwaves in a microwave oven is 2 500 MHz. The speed of microwaves is 3×10^8 m s^{-1}. Calculate the wavelength of the microwaves produced by the microwave oven.

6. Radio waves travel at 300 000 km s^{-1}. Radio station 'A' broadcasts at a frequency of 650 kHz. Radio station 'B' broadcasts waves of wavelength 315 m. Calculate:

   a. The wavelength of radio station 'A'.

   b. The frequency of radio station 'B'.

7. Light travels at 300 000 km s^{-1}. The distance from the Sun to the Earth is 1.5×10^8 km. How long does it take for light to travel from the Sun to the Earth?

8. State one property that light and sound have in common and one property they don't.

9. The speed of sound is 330 m s^{-1}. If thunder is heard 15 seconds after a lightning flash, how far away was the lightning flash produced?

10. A radiographer uses an **ultrasound** scanner of frequency 1 MHz to view a baby inside the womb. The velocity of ultrasound in tissue is 1 570 m s^{-1}.

    a. Calculate the wavelength of the ultrasound waves.

    b. Why is ultrasound (uses MHz) used to view a baby inside the womb rather than X-rays?

11. You are a distance of 600 km away from the epicentre of an earthquake. What time delay is there between receiving P-waves and S-waves?

12. A seismograph detects P-waves 5 minutes after an earthquake has occurred. What is the distance between the seismograph and the epicentre of the earthquake?

13. The diagram shows a graph of a sound wave from an oscilloscope.
    Draw similar diagrams to illustrate:

    a. A sound wave of higher frequency but of the same loudness.

    b. A louder sound of the same frequency.

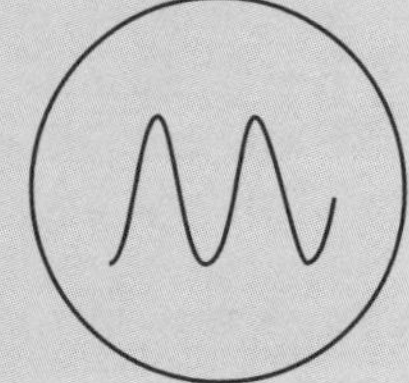

14. The table below contains information about the electromagnetic spectrum.

Radiation	Short wavelength						Long wavelength
	Gamma rays	X-rays	UV rays	Visible light	i	Microwaves	ii
Detector	Photo film	iii	Fluorescent material	iv	Photo transistor	Aerial	Aerial

   a. What speed do all the radiations travel at?

   b. What are the two missing radiations – (i) and (ii) – and the two missing detectors – **iii** and **iv**?

15. A hunter fires a gun and hears the echo from a nearby mountain 4 seconds later. The speed of sound is 330 m s^{-1}. How far away is the mountain?

16. A leaf repeatedly touching the surface of a pond creates one water wave every 0.5 s. If the wavelength of the waves is 3 cm, calculate the wave speed (in cm s^{-1}).

17. The diagram shows an accurate scale diagram of a transverse wave travelling to the right.

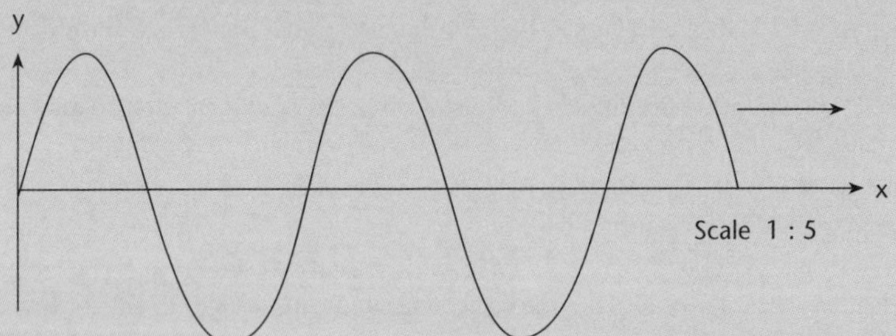

Calculate the wavelength of the wave.

18. A seismograph recorded the following trace of an earthquake:

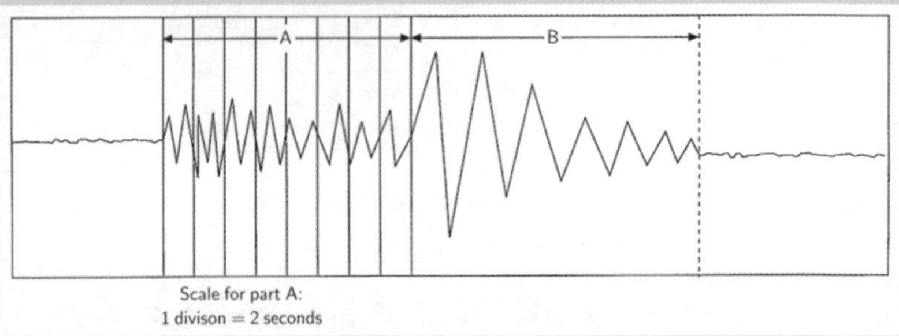

a. For each section of the trace, state whether:

i. It represents a P- or an S-wave.

ii. It represents a longitudinal or a transverse wave.

iii. The wave travels slow or fast.

The P-waves were recorded on the seismograph approximately 4 minutes after the earthquake occurred. The average speed of P-waves is 8 000 m s^{-1}.

b. What is the distance between the seismograph and the centre of the earthquake? Express your answer in kilometres.

c. Using the scale on the seismograph trace, calculate the time difference between the arrival of waves A and B.

d. Assuming waves A and B have travelled the same distance, calculate the average speed of wave B.

e. How can the strength of an earthquake be determined from a seismograph trace?

19. Silas stands 100 m from a low building which has a taller building 40 m behind it. When he shouts, he hears two echoes 0.25 seconds apart.

Calculate the speed of sound in air.

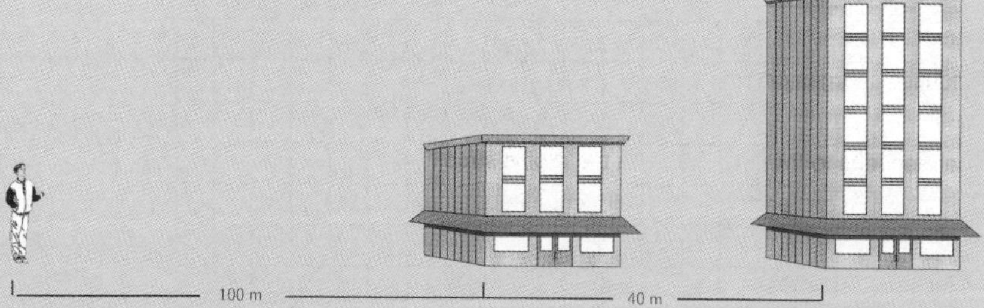

20. The human ear can detect sounds in the frequency range 20 Hz to 20 kHz. Calculate the range of the wavelengths which can be heard. (The speed of sound in air is 330 m s^{-1}.)

21. Liam has just set up his new 'surround sound' system in his home. He has drawn a diagram to explain how one of the loudspeakers works.

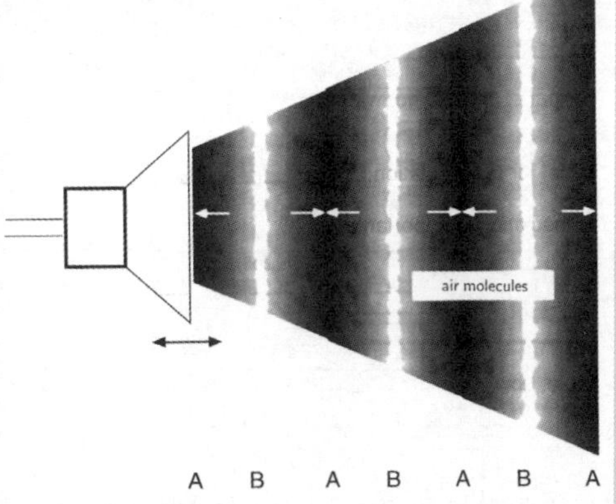

a. Before sound is produced, what must the loudspeaker do?

b. What type of wave is a sound wave?

c. State how the spacing of the air molecules at regions A and B differ.

d. **i.** What do the regions labelled A represent?

ii. What do the regions labelled B represent?

e. In a loudspeaker, energy is transformed from ______ **i** ______ into ______ **ii** ______ energy.

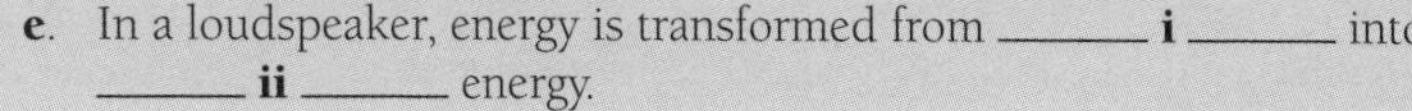

22. The sound system that Liam has set up can be used for karaoke. Imagine that his friend Julie sings into a microphone connected to the system, and that it also has an oscilloscope attached to it. When Julie sings a single pure note, the oscilloscope displays the following waveform.

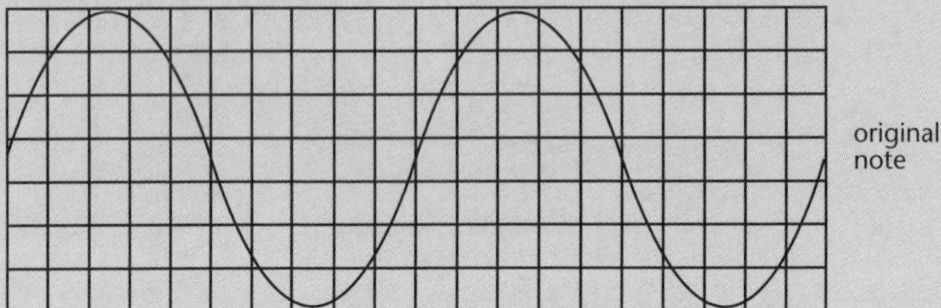

a. Copy the diagram above, and draw on it:

i. What the oscilloscope would display if Julie sang the same note more quietly.

ii. What the oscilloscope would display if Julie sang a different note of the same volume but lower in pitch than the original note.

iii. What the oscilloscoope would display if Julie spoke into the microphone.

b. In a microphone, energy is transformed from ______ **i** ______ energy into ______ **ii** ______ energy.

23. A sound system has a built-in radio receiver. The diagrams below show two alternative radio tuning dial displays. The positions of the pointer to receive radio stations A, B, C, and D are shown for both systems.

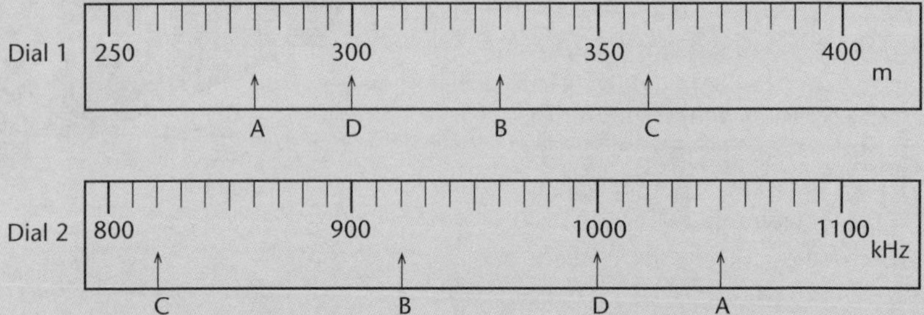

a. Copy and complete the table below using the information displayed on Dials 1 and 2.

Radio station	Dial 1 (Units (?)	Dial 2 (Units (Hz)	Speed Units (m s^{-1})
Station A	i	ii	iii
Station B	330	iv	v
Station C	vi	vii	viii
Station D	ix	$1\,000 \times 10^3$	3×10^8

b. State the relationship between the numbers on Dial 1 and the numbers on Dial 2.

c. Radio station Z broadcasts on a frequency of 1 100 kHz. What wavelength would this have?

The following information relates to Questions 24 and 25.

The diagram, which is *not* drawn to scale, shows a wave travelling along a spring at 6 m s^{-1}.

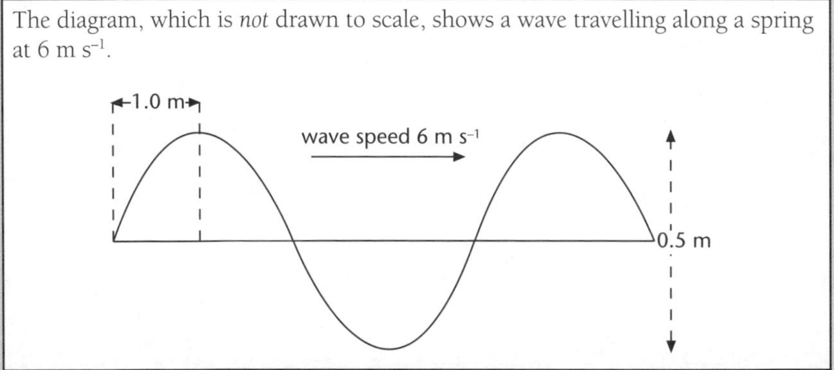

24. What is the wavelength of the wave?

25. What is the frequency of the wave?

26. A sound wave moves from air into water. The speed of sound is 330 m s^{-1} in air, and 1 200 m s^{-1} in water. Consider the following statements, and state whether they are true or false.

I. The wavelength of the sound is longer in water than it is in air.

II. The loudness of the sound is greater in water than it is in air.

III. The frequency of the sound is larger in water than it is in air.

27. During a storm, Helen sees a flash of lightning and a few seconds later hears the corresponding thunder. Explain why she hears the thunder later.

The following information relates to Questions 28, 29 and 30.

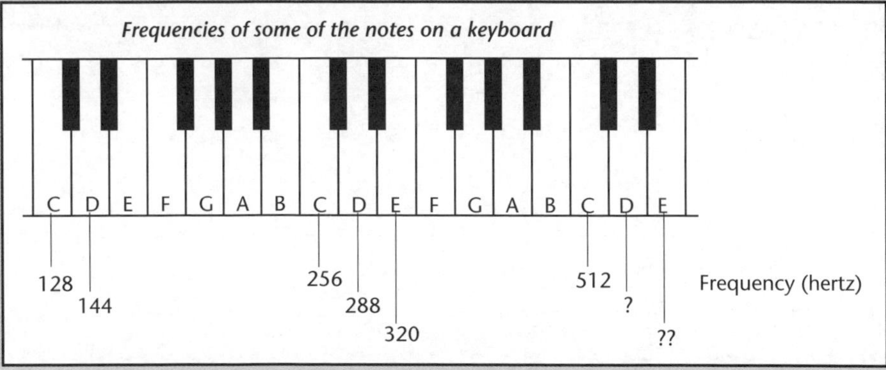

28. The frequencies on a musical scale are always in the same ratios. What are the frequencies of the 'D' and 'E' notes at the right of the keyboard?

29. The lowest frequency which humans can hear is 20 hertz. What is the frequency of the lowest 'C' note which can be heard by humans?

30. The speed of sound in air is 330 m s^{-1}. Which of the notes on the keyboard will have a wavelength closest to one metre?

31. Mere listens to her favourite radio station. The radio is tuned to 96 MHz. Convert this frequency to Hz.

32. Modern fishing boats often use SONAR to detect shoals of fish. A brief pulse of SONAR waves is emitted from a fishing boat and the echo from the waves is detected 0.1 s later. The sonar waves travel through the water at 1 600 m s^{-1}.

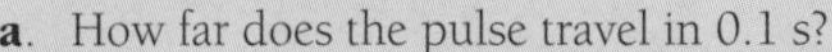

a. How far does the pulse travel in 0.1 s?

b. How far below the boat is the shoal of fish?

c. Is it possible to tell the thickness of a shoal of fish from the reflected pulse? State a reason for your answer.

33. State whether each of the following comparisons of the properties of sound waves and light waves is correct or false.

Light waves:

A. travel faster and are longitudinal waves.

B. travel slower and are transverse waves.

C. travel faster and can travel through a vacuum.

D. travel slower and can travel through a vacuum.

34. A violinist plays a note of 1 650 Hz. The speed of sound in air is 330 m s^{-1}. Calculate the wavelength.

35. A sound wave of frequency 60 Hz travels in air with a speed of 300 m s^{-1}. What is its wavelength?

36. A ray of light passes from air to water. What happens to its wavelength and frequency?

37. Which one of the following waves has the lowest frequency and the longest wavelength?

Gamma. Radio. Infrared. Ultraviolet.

38. The diagram represents a wave travelling along a rope.

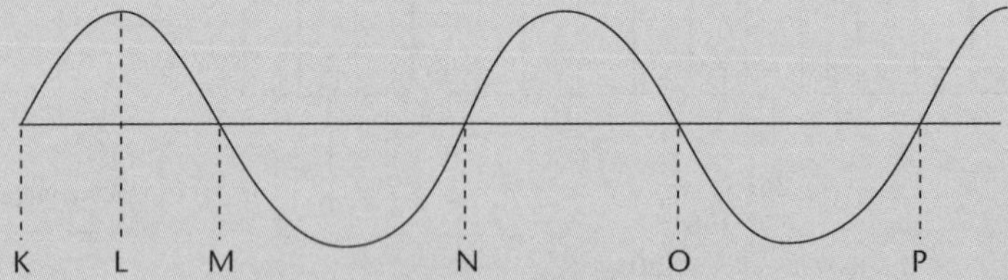

What is its wavelength?

39. Dolphins are able to navigate by emitting ultrasonic waves and listening to the echo.

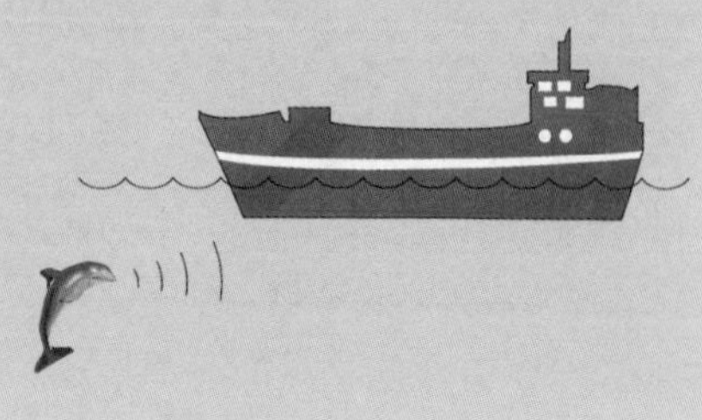

a. Is an ultrasonic wave transverse or longitudinal?

b. A dolphin emits ultrasonic waves that have a frequency of 66 000 Hz. If the speed of sound in water is 1 500 m s^{-1}, calculate the wavelength of the waves.

40. Water waves are generated in a ripple tank. The waves move to the right as shown in the diagram, and meet a barrier.

a. Copy and complete the diagram to show four wavefronts and their direction(s) after the waves are reflected off the barrier.

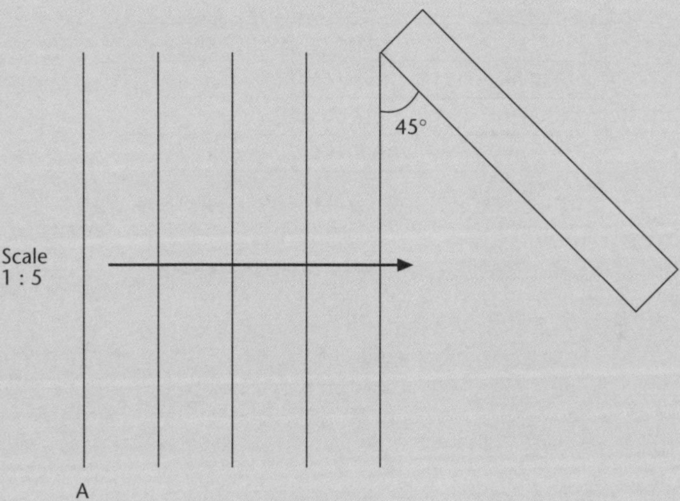

b. What is the wavelength of the incident waves?

c. Five wavefronts passed point A in 3 seconds. What is the frequency of the waves?

41. Two types of shock waves produced by earthquakes are primary and secondary waves. The table shows features of these waves. Complete the table.

	Primary waves	Secondary waves
Speed	Fast	Slow
Earth movement	To-and-fro in the direction of the wave	Up and down at right angles to the wave
Wave type		

42. For the sketch of the wave, which distance indicates the wavelength of the wave?

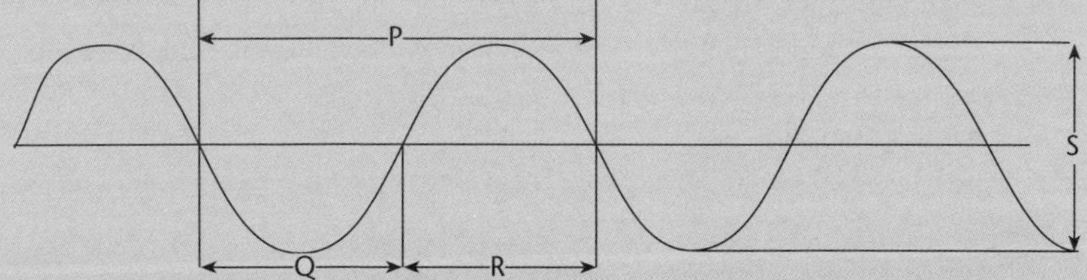

43. Waves are generated in a school swimming pool. Three complete waves are made in two seconds. Calculate the frequency of these waves (the units are waves per second).

44. Copy and complete the following table by calculating the missing details:

Speed, v	Frequency, f	Wavelength, λ
a	1 000 Hz	0.33 m
b	165 Hz	2 m
c	8 kHz	4 cm
330 m s^{-1}	15 Hz	d
1 500 m s^{-1}	10 kHz	e
3.8 m s^{-1}	20 000 Hz	f

45. The speed of a transverse wave is 6 m s^{-1}. The wave frequency is 2 Hz. Calculate the wavelength of the wave.

46. A lighthouse keeper sees an emergency flare. Five seconds later she hears the noise from the flare being fired. If the speed of sound is 330 m s^{-1}, calculate how far from the emergency flare the lighthouse is.

47. An echo sounder in a ship produces a sound pulse which is transmitted through the water. It is reflected off the sea floor. The echo is detected by the ship 0.4 s after the sound pulse was sent. What is the depth of the sea bed? (Sound travels in water at 1 500 m s^{-1}.)

48. The diagram shows a tuning fork 'sounding' in air and a cross-section of the air waves produced by the tuning fork reaching an ear.

a. Use the following words and letters to answer the questions below. (Note that more words and letters are in the list than required to answer the questions.)

List – audible, frequency, longitudinal, pitch, transverse, ultrasonic, vibrations, wavelength, P to Q, Q to R.

The sound was produced by the _____ **i** _____ of the fork.

The part of the wave labelled G was produced while the fork was moving from position _____ **ii** _____ .

The wave produced in air is an example of a _____ **iii** _____ series of pulses.

The fork had 220 Hz stamped on it. This value is the _____ **iv** _____ of the tuning fork.

The distance from D to F on the diagram measures the _____ **v** _____ of the sound waves.

The note produced was heard by the ear; thus it is an _____ **vi** _____ frequency.

b. A student reads that the velocity of sound in air is about 330 m s^{-1}. For the tuning fork, calculate the wavelength of the sound transmitted.

49. A boy placed his ear against a long iron water-pipe. A girl 100 m away hit the pipe once with a hammer. The boy heard two sounds, one slightly after the other.
   a. How did the two sounds travel to the boy's ear?
   b. How could the boy and girl increase the time interval between the two sounds?
   c. Sound waves in the metal pipe have a speed of 2 500 m s^{-1}. Their wavelength is 2 m. Calculate the frequency of the waves in the pipe. Give the unit.
   d. Why does a small rock dropped onto the Moon's surface make no sound when it hits?

50. To calculate the speed of sound in air, a person claps their hands together loudly at a frequency of 3 Hz while standing 55 m in front of a vertical flat cliff. The echoes of each clap are not heard, since each echo returns at the exact moment the next clap is made. What was the speed of sound in air on the day?

Unit 12.3 Activity 1B: Multiple choice questions

1. A fisherman from Ali island is fishing in his canoe in the ocean. He noticed the crest of an ocean wave passes every 1.5 s. If the distance between the crest and the nearest trough is 2.0 metres, what is the speed of a surface wave?
   A. 3.0 ms^{-1}
   B. 2.85 ms^{-1}
   C. 2.67 ms^{-1}
   D. 2.76 ms^{-1}
2. The 1998 Papua New Guinea earthquake was a magnitude 7.0 earthquake that took place in the early evening of Friday, 17 July 1998. The area worst hit was a 30 km (19 mi) coastal strip running north-west from Aitape to the village of Sissano. The total death was put to 2200 lives. What type of wave(s) do you think causes the most destruction?
   A. Secondary wave (S wave)
   B. Primary wave (P wave)
   C. Love waves (L wave)
   D. All of the above are correct.
3. The Primary wave (P-Wave) of the tsunami in question 2 above travels with the velocity of approximately 8000 ms^{-1}. If a crest and the trough of the wave covers 5 km of the coastline, what was the frequency of the tsunami wave?
   A. 16 Hz
   B. 160 Hz
   C. 1600 Hz
   D. 1.60 Hz
4. The earthquake waves are classified according to their speed. Which of the following correctly puts the waves in increasing order according to their speed?
   A. P-waves, L-waves and S-waves
   B. L-waves, P-waves and S waves
   C. L-waves, S-waves and P-waves
   D. S-wave, P-waves and L-waves.

5. A wave in the ocean can be classified as:
   - **A**. Transverse
   - **B**. Longitudinal
   - **C**. Electromagnetic
   - **D**. Both longitudinal and transverse.
6. A wave on a string of a guitar when the length of the string is varied to produce various harmonics can be classified as:
   - **A**. Transverse
   - **B**. Longitudinal
   - **C**. Electromagnetic
   - **D**. Both longitudinal and transverse.
7. In most remote schools in PNG, the starting and stopping times for classes, work parade, assemblies, lunch break and most school activities are controlled by the ringing of a bell. The bell is usually a cylindrical hollow metal which produces sound waves when struck with a solid metal rod. What type of wave is propagated through the air when the bell is struck by the rod?
   - **A**. Transverse
   - **B**. Longitudinal
   - **C**. Electromagnetic
   - **D**. Transverse and longitudinal.
8. In Lumi Provincial High School in Sandaun Province, the furthest dormitory is a male dormitory which is about 400 m from the mess hall. The bell is located within the vicinity of the mess. How long does it take the sound to reach the ears of a student at the furthest dormitory when the bell is struck for dinner? Apply speed of sound for a normal day's conditions of 340 ms^{-1}.
   - **A**. 12 seconds
   - **B**. 1.2 seconds
   - **C**. 0.12 seconds
   - **D**. 120 seconds
9. In question 8 above, a student sitting outside the dormitory on a hot sunny day heard two sounds of the bell in an interval of two seconds when the bell was ringing for a lunch break. What would be the best explanation for hearing two sounds?
   - **A**. The bell produced two waves in the air with one travelling faster than the other.
   - **B**. During hot sunny days sound waves in the air split into many waves.
   - **C**. The sound wave travels through air and solid earth with one travelling faster than the other.
   - **D**. The sound has to travel a long distance so it splits into many wavelengths.

10. Margaret stands at Sipalol Cliff and shouts to her husband Tulex in the garden at Henus, 400 m from the cliff. She heard her own voice 0.5 seconds later. What is the best explanation for this?

A. Her voice split into many sound waves
B. Her voice bounced off the cliff and returned
C. It was her husband's voice she heard
D. The spirits living at the base of the cliff imitated her.

11. Which of the following is able to detect an x-ray wave?

A. TV set
B. Eyes
C. Geiger-Muller tube
D. Photographic film

12. Which of the following is able to detect a gamma ray?

A. Photographic film
B. Geigher-Muller tube
C. Radio antenna
D. Microwave antenna

13. Which one of the following wavelengths of the electromagnetic wave spectrum is for the visible light?

A. 1×10^{-12} m
B. 1×10^{-5} m
C. 1×10^{-3} m
D. 1×10^{-7} m

14. Which of the following wavelengths of the electromagnetic wave spectrum is for the Ultraviolent?

A. 1×10^{-12} m
B. 1×10^{-5} m
C. 1×10^{-8} m
D. 1×10^{-7} m

15. Which of the following is a source of an infrared electromagnetic wave?

A. Radioactive substance
B. Television transmitter
C. X-ray tube
D. Sun

16. Which of the following correctly lists the rays from the most dangerous rays to the least dangerous ones in the electromagnetic wave spectrum?

A. UV, X-ray, γ ray, IR
B. IR, X-ray, γ ray, UV
C. γ ray, X-ray, UV, IR
D. X-ray, γ ray, IR, UV

17. Which of the following waves can travel through a vacuum?

A. Sound wave

B. Infrared (IR)

C. Earthquake waves

D. Water waves

18. Digital is a new mobile company in PNG. What frequency range in the electromagnetic spectrum is the digital mobile likely to be operating on?

A. 3×10^{11} Hz

B. 3×10^{13} Hz

C. 3×10^{15} Hz

D. 3×10^{17} Hz

19. The human ear can detect sound waves in the frequency range of:

A. 1 to 20 Hz

B. 20 to 20 kHz

C. 20 to 60 kHz

D. 60 to 100 kHz

20. The loudness of a sound wave depends on which of the following wave parameters?

A. Frequency

B. Wavelength

C. Period

D. Amplitude

Unit 12.3 Waves

Topic 2: Reflection of light

Topic 2 deals with the reflection of light. It covers:

- Reflection.
- The rules of reflection.
- Lateral inversion.

Light is a form of **electromagnetic radiation**. Sometimes, light behaves as if it were a stream of particles; at other times, it behaves like a wave. The speed of light is 300 000 km s^{-1}. Light always travels in straight lines, called **rays**. An arrow on a ray indicates the direction in which the light travels.

The figure below demonstrates that light travels in straight lines.

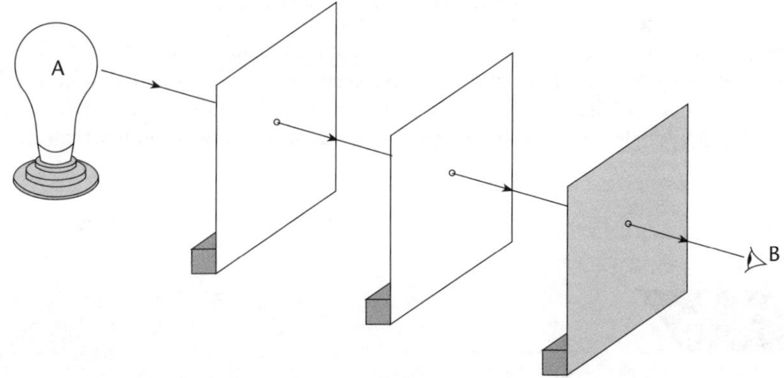

The holes in the pieces of card are lined up with a piece of string. The fact that a spot of light appears on the screen is evidence that light has travelled in a straight line through both holes.

Demonstration that light travels in a straight line.

Shadows occur because *light travels in straight lines* and *part of the light is blocked off by an object in its path*.

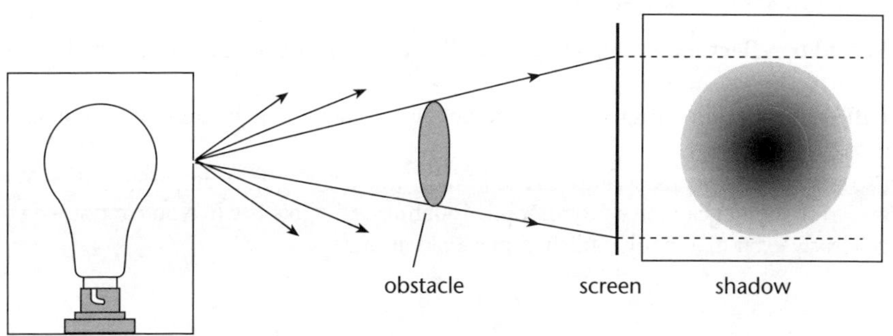

Shadow formation.

The fact that light travels in straight lines also explains how images are formed in a **pinhole camera**. The **image** is always inverted.

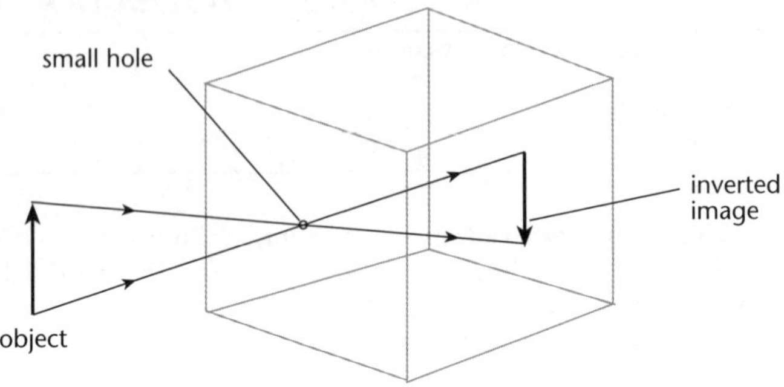

Image formation by a pinhole camera.

When light falls on a surface, it can be either reflected, transmitted or absorbed. An **opaque** object does not transmit light. A **transparent** object transmits light and can be clearly seen through. A **filter** is transparent but only lets a particular colour of light pass through. The figure below shows some examples of what happens when light falls on different surfaces.

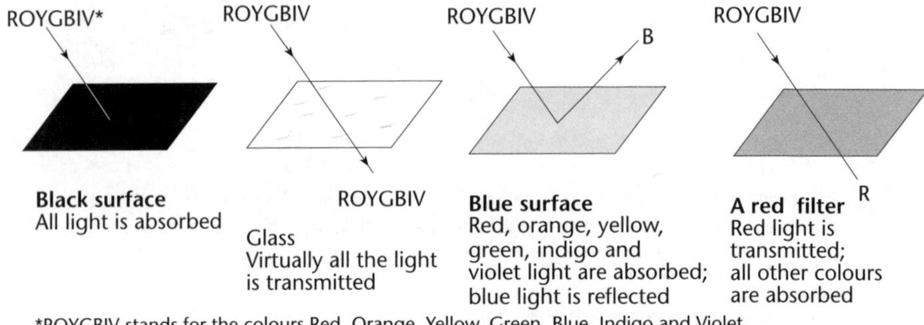

Different effects from light falling on different surfaces.

There are two kinds of **reflection**:

- **Specular reflection** – occurs when parallel rays of incident light are reflected to give parallel reflected rays.
- **Diffuse reflection** – occurs when parallel rays of incident light are reflected in all directions.

Diffuse reflection happens because the reflected surface is rough. Even surfaces that feel smooth to the touch can have microscopic 'roughnesses' – this is why you cannot see your image clearly even in a highly polished piece of flat brass.

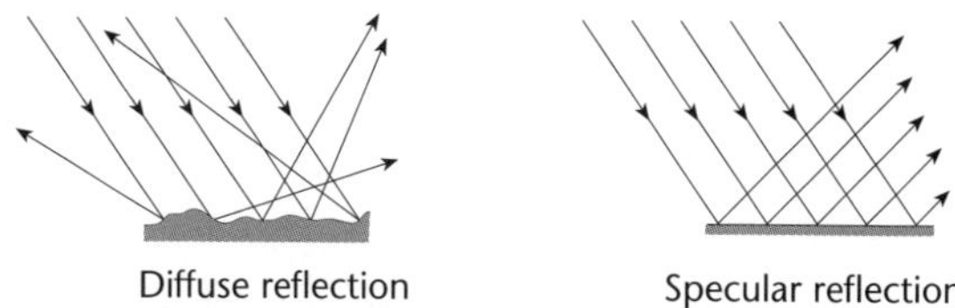

Specular and diffuse reflection.

The rules of reflection

The type of reflection most people are familiar with is reflection in a plane mirror.

Rules of reflection in a plane mirror:

1. The image (height Hi) and object (height Ho) are the same; ie Hi = Ho (see figure on the next page).
2. The image appears the same distance behind the mirror (di) as the object is in front of the mirror(do); ie di = do (see figure on the next page).
3. The image is **virtual** – it appears to be behind the mirror even though it is not actually physically located there; it cannot be projected onto a screen (see figure on the next page).
4. The incident ray, reflected ray and **normal** (line drawn at right angles to the mirror) are all in the same plane (see figure below).
5. The angle of incidence ($\hat{\imath}$) is equal to the **angle of reflection** ($\hat{r}$) (see figure below).

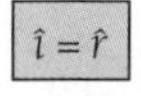

$\hat{\imath} = \hat{r}$

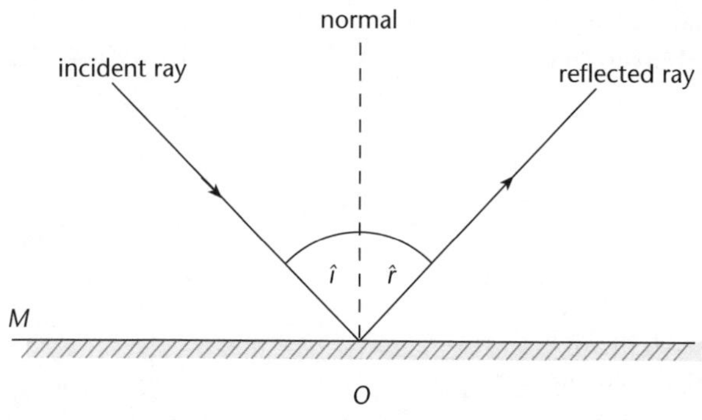

Reflection in a plane mirror.

A **ray diagram** can be used to locate the image of an object reflected in a plane mirror. Hatched lines are used to show the back of the mirror. The reflected rays are extended behind the mirror to locate the position of the image. The lines behind the mirror are not light rays, but construction lines, used to indicate where the reflected light rays appear to originate from. Construction lines are always dotted in ray diagrams.

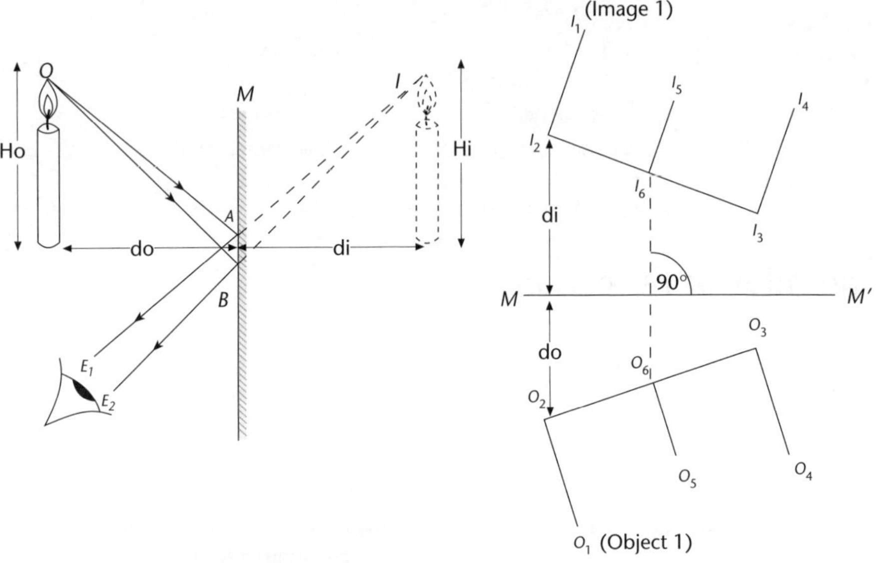

Locating the image in a plane mirror.

Lateral inversion

When you look into a mirror and pull your right ear, the image looking back at you 'seems to be pulling its left ear' – this is an example of **lateral inversion**.

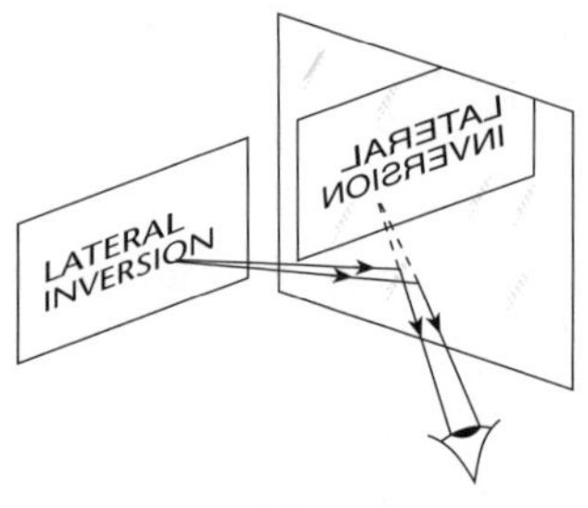

Lateral inversion.

The reflections of each of the letters in the figure to the right are laterally inverted. Laterally inverted letters can be used in mirror writing. The word 'ambulance' is often written in mirror script on the bonnet of an ambulance so that the word 'ambulance' can be read normally by a driver of a car in front of the ambulance looking in their rear vision mirror.

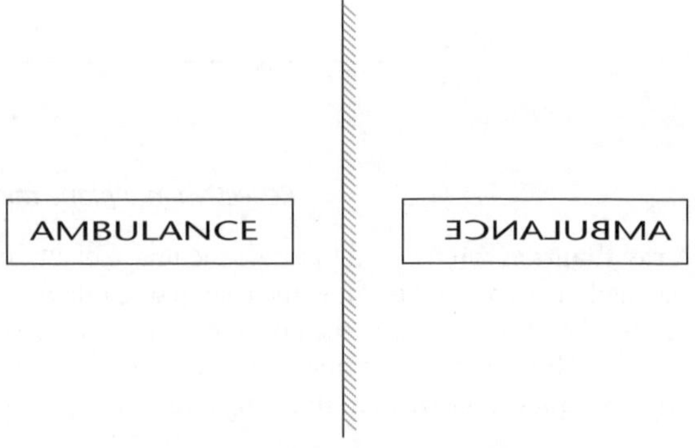

'Reverse' writing of the word 'ambulance'.

Unit 12.3 Activity 2A: Reflection

1. a. What is the speed of light in a vacuum?
   b. A light-year is the distance travelled by light in one year. Calculate this distance in kilometres.
   c. The second closest star to Earth is Alpha Centauri, 4.5 light-years away. Express this distance in kilometres.
2. Draw a diagram of apparatus that can be used to show that light travels in straight lines. Explain how the apparatus is used.
3. Explain why shadows are evidence that light travels in straight lines.
4. A builder has to construct a level foundation for a building. Explain how they could use the properties of light to ensure that the foundation will be level.
5. The diagram shows a ray of light being reflected off a plane mirror. State what each of the following lines and angles represents:
   XO = Angle XON =
   ON = Angle YON =
   OY =

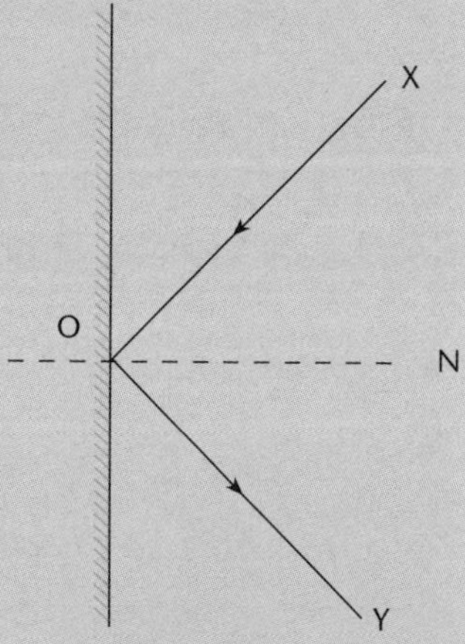

6. Draw a ray diagram to show that the image of an object reflected in a plane mirror is the same distance behind the mirror as the object is in front.
7. A ray of light hits a plane vertical mirror and is reflected. Use a ray diagram to show that when the mirror is rotated through an angle of 30°, the reflected ray is rotated through 60°.
8. Draw a ray diagram to show that a vertical mirror need not be 170 cm high for a 170 cm tall boy to see a full-length image of himself.
9. Describe an experiment that could verify the laws of reflection of light in a plane mirror.
10. Use a ray diagram to show how the eye sees the image of the letter N when it is reflected in a plane mirror.

11. a. The diagram shows a simple periscope. Copy the diagram and draw a ray of light which passes from the periscope to the eye.

b. Why was the periscope a lifesaving device for World War One soldiers in the trenches?

12. Rahul was rehearsing his part in the school play on the stage. The stage light in front of him casts his shadow on the floor behind him.

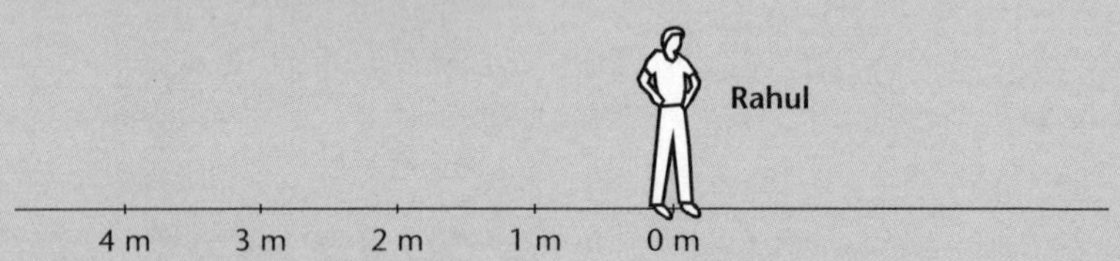

a. On the diagram above, draw in the light rays to find the length of his shadow on the floor.

b. The formation of shadows can be explained using a certain property of light. State this property of light.

Rahul now stands in front of a mirror to check out his costume.

His distance from the mirror is 1.0 metre as shown in the diagram.

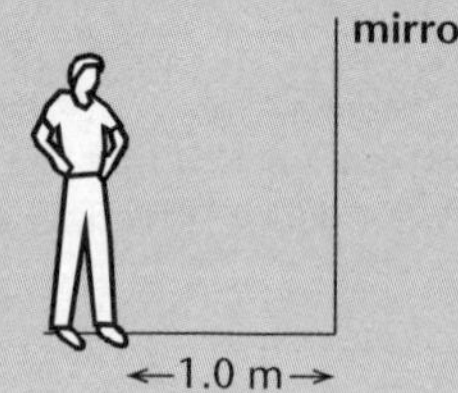

c. What is the distance between Rahul and his image?

d. Rahul's image in the mirror is laterally inverted. Describe what 'laterally inverted' means.

A beam of light from the spotlight falls at the point P on the mirror, at an angle of 35° to the mirror as shown in the diagram below.

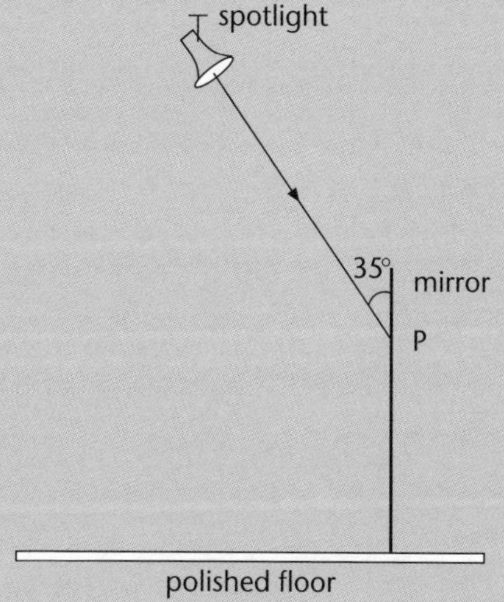

e. Calculate the angle of incidence at the point P.

f. After reflection at the point P, the beam of light travels towards the polished floor and reflects off it as if the floor is a mirror.

- **i.** Complete the diagram to show the path of the beam of light.
- **ii.** Use your diagram to calculate the angle of reflection at the point where the beam of light reflects off the floor.

(Source: NCEA Examination paper)

Unit 12.3 Waves

Topic 3: Refraction of light

Topic 3 deals with the refraction of light. It covers:

- Optical density.
- Refraction.
- Dispersion.

The speed of light in a vacuum, c, is 3×10^8 m s^{-1}; this speed is approximately the same as the speed of light in air. When light passes from a vacuum or air into any other substance, it slows down. The speed of light in a substance depends on the **optical density** of the substance. The higher the optical density of a substance, the slower the light travels in the substance. The following substances are listed in order of increasing optical density:

Vacuum > Water > Oil > Glass > Diamond

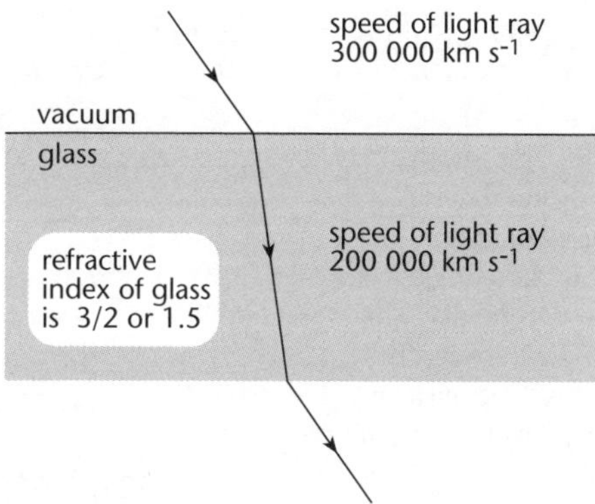

The speed of light reduces in an optically more dense medium.

The **absolute refractive index**, n, of a medium *(the medium is the substance that light travels through)* is a number obtained by calculating the ratio of the speed of light in a vacuum to that of the speed of light in the medium.

$$n_{\text{medium}} = \frac{\text{Speed of light in a vacuum } (c)}{\text{Speed of light in a medium } (v)}$$

The absolute refractive indices of a different media are given in the below table.

Substance	Refractive index (n)
Air (vacuum)	1.00
Water	1.33
Paraffix	1.44
Perspex	1.49
Glass	1.50
Diamond	2.40

Absolute refractive indices of different media.

Example A

The refractive index of water is 1.33. What is the speed of light in water?

Answer:

$n = 1.33$, $c = 300\ 000\ \text{km s}^{-1}$, v_{water} is unknown.

$$n = \frac{c}{v}$$

$$1.33 = \frac{300\ 000\ \text{km s}^{-1}}{v_{water}}$$

Thus, $v_{water} = \dfrac{300\ 000\ \text{km s}^{-1}}{1.33} = 250\ 000\ \text{km s}^{-1}$.

Refraction is the change of direction when a ray (narrow beam) of light travels from one medium into another. A ray diagram can be used to describe what happens when a ray of light is refracted. In the diagram to the right, ray AB is the incident ray ($\hat{\imath}$) which passes from medium 1 into medium 2. Ray BC is the refracted ray ($\hat{r}$). The *normal* is a construction line drawn at right angles to the interface between the two media. The *angle of incidence* is the angle between the normal and the incident ray. The *angle of refraction* is the angle between the normal and the refracted ray.

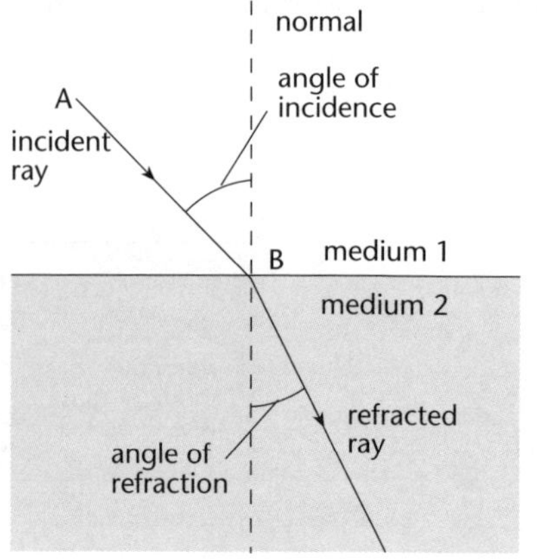

Ray diagram for showing refraction.

The direction in which a refracted ray bends depends on the media involved. When a ray of light passes from an optically less dense substance into an optically more dense substance (ie one of higher refractive index), the light ray slows down and bends towards the normal.

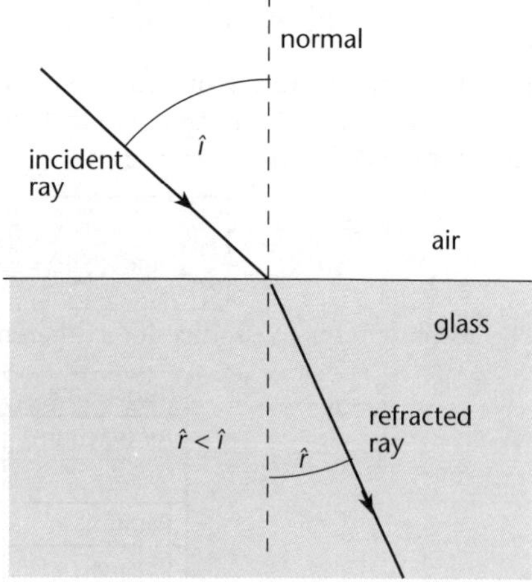

Refraction towards the normal.

When a ray of light passes from an optically more dense substance into an optically less dense substance (ie one of lower refractive index), the light ray speeds up *and* bends away from the normal.

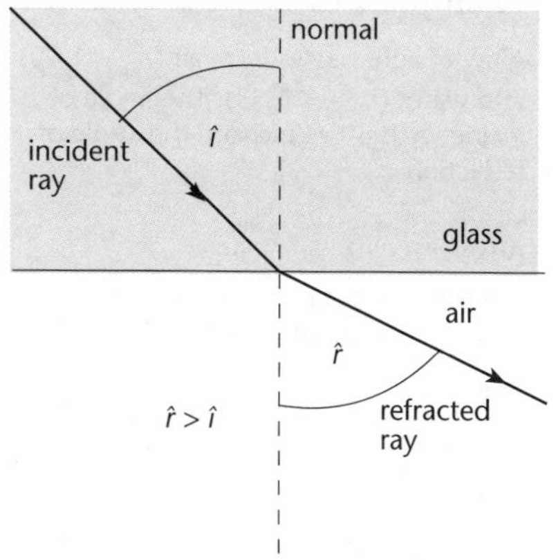

Refraction away from the normal.

In 1621, Willibrod Snell discovered a relationship between the angle of incidence and the angle of refraction. **Snell's law** states that when light passes from one medium (medium 1 of refractive index *n1*) into another medium (medium 2 of refractive index n_2):

$$n_1 \sin \theta_1 = n_2 \sin \theta_2$$

In assessment in Year 11 Physics, one of the media is always air.

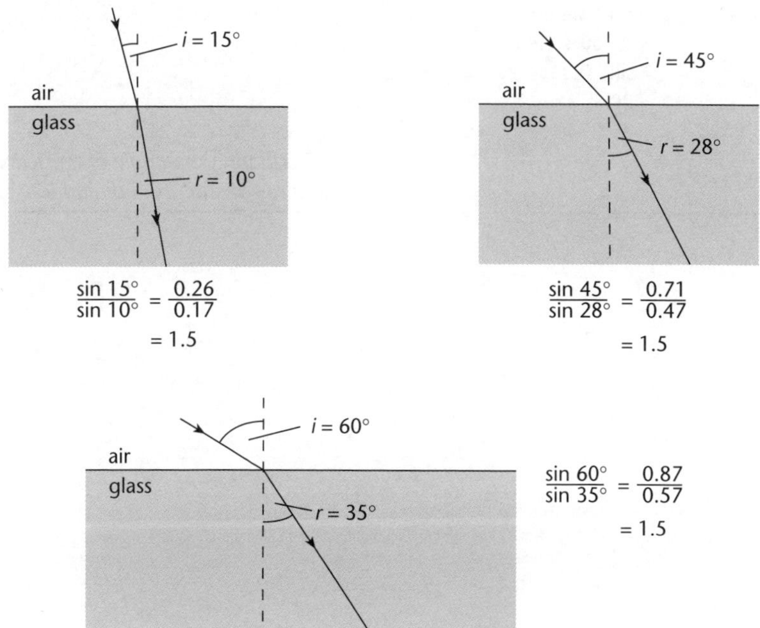

The ratio 'sin i divided by sin r', stays constant.

Example B

A ray of light passes from air (n_{air} = 1.00) into water (n_{water} = 1.33). The angle of incidence is 30°. Calculate the angle of refraction.

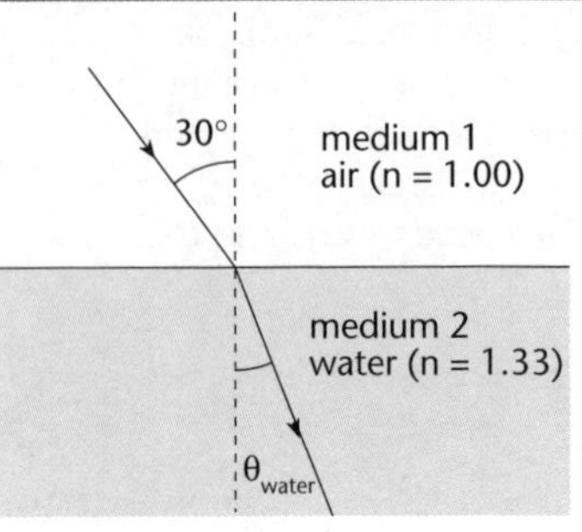

Answer:

$n_1 \sin \theta_1 = n_2 \sin \theta_2$

$1 \sin \theta_{water} = \sin 30 \div 1.33$

$\theta_{water} = \sin^{-1} 0.376$

$\theta_{water} = 22°$

Calculating the angle of refraction for a ray of light travelling from air into water.

Example C

A ray of light passes from diamond ($n_{diamond}$ = 2.44) into air (n_{air} = 1.00). The angle of refraction is 60°. Calculate the angle of incidence.

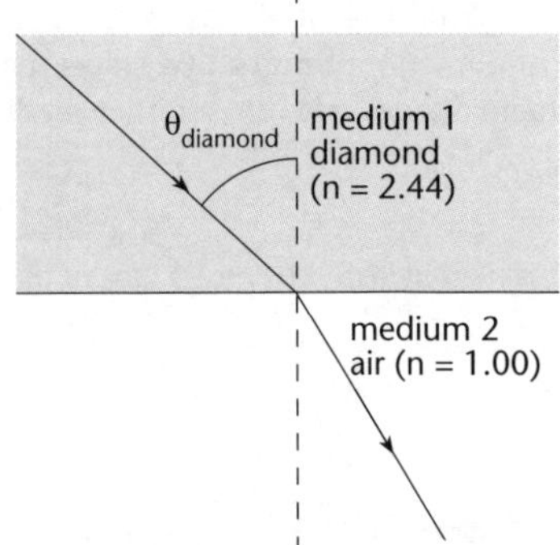

Answer:

$n_1 \sin \theta_1 = n_2 \sin \theta_2$

$2.44 \sin \theta_{diamond} = \sin 1 \sin 60$

$\theta_{diamond} = \sin 60 \div 2.44$

$\theta_{diamond} = \sin^{-1} 0.355$

$= 20.8°$

Calculating the angle of refraction for a ray of light from diamond → air.

There are many everyday occurrences of refraction.

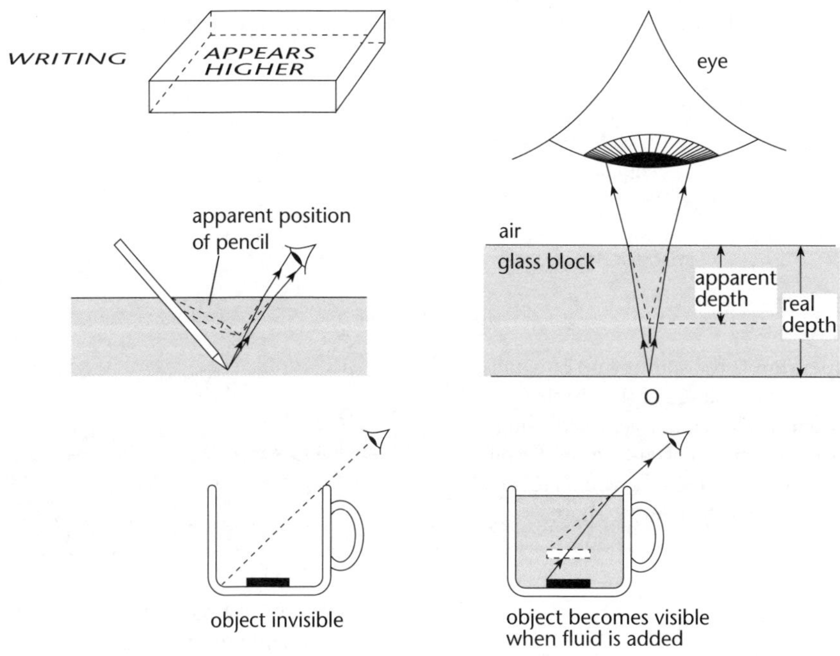

Examples of refraction.

Refraction can make water seem shallower than it really is.

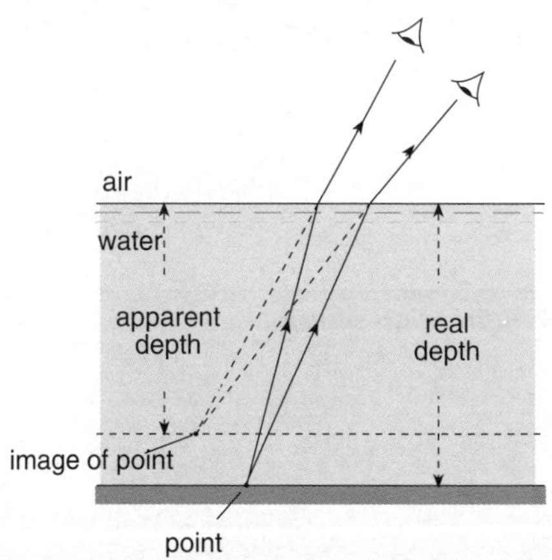

Effect of refraction on apparent depth of water.

Looking along a tarsealed road on a hot summer's day, a **heat shimmer** can often be seen above the road surface. This is because just above the road surface the air is hot and its density is low. Light from low in the sky passes through cooler, dense air towards the hot, less-dense air above the road and is refracted away from the normal. To an observer, the light appears to come from a point below the surface of the road. The observer sees an image of the sky, which looks like the reflection produced by a pool of water.

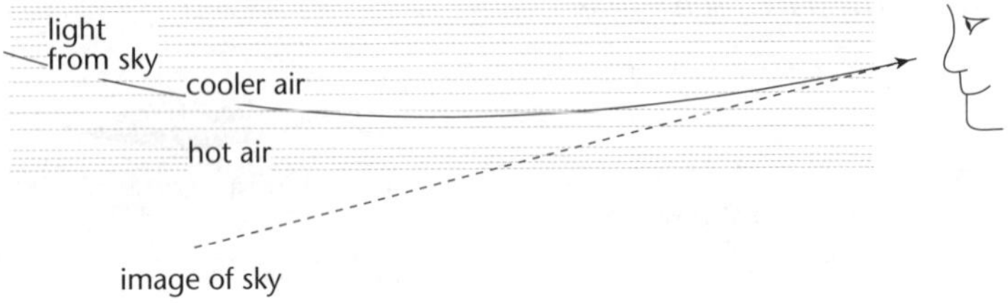

Heat shimmer.

Dispersion is the splitting up of white light into the colours of the visible **spectrum**. This can be achieved using a **prism**. Dispersion occurs because each of the colours of the spectrum is refracted through a slightly different angle.

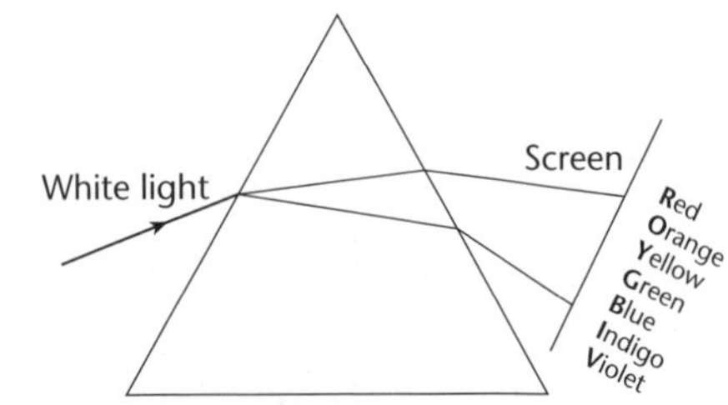

The dispersion of light passing through a prism.

Newton used two prisms to show how white light can be dispersed into the colours of the spectrum then be recombined to form white light.

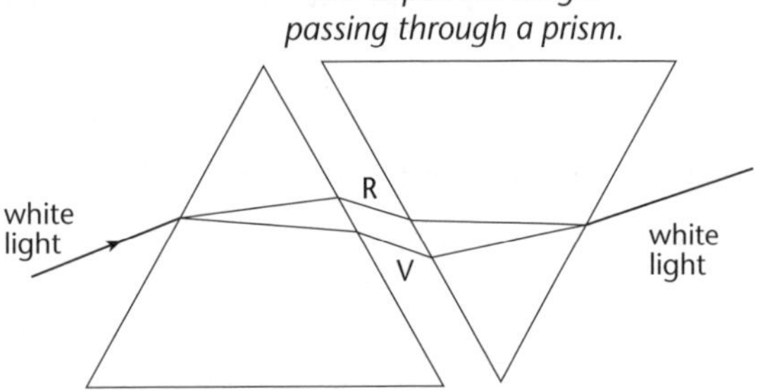

Combining the colours of the spectrum into white light using two prisms.

Unit 12.3 Activity 3A: Refraction of light

1. The diagram shows a ray of light travelling from water to air.
   a. Copy the diagram and clearly:
      i. Indicate which substance is air and which is water.
      ii. Label the incident and refracted rays and normal.
   b. Why does the light ray bend as shown in the diagram?

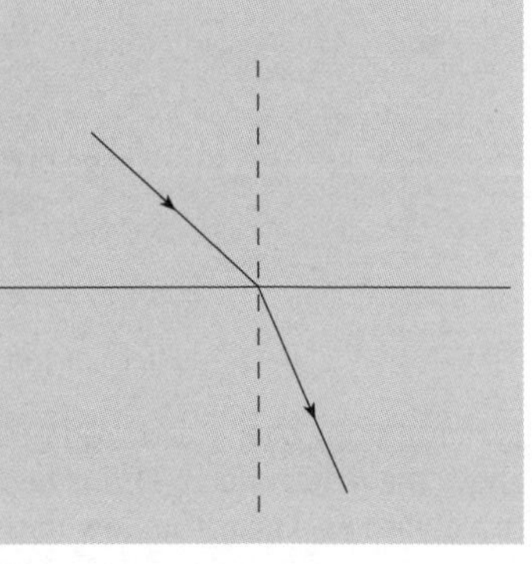

2. Copy and complete each of the following diagrams by clearly showing the path of the light through the glass and how it emerges. Draw in normals wherever the light ray changes direction.

a.

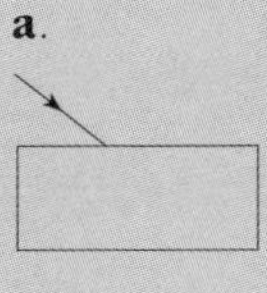

b.

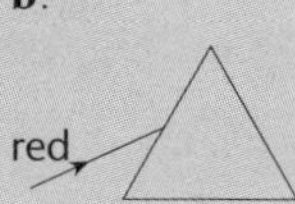

c. **d.**

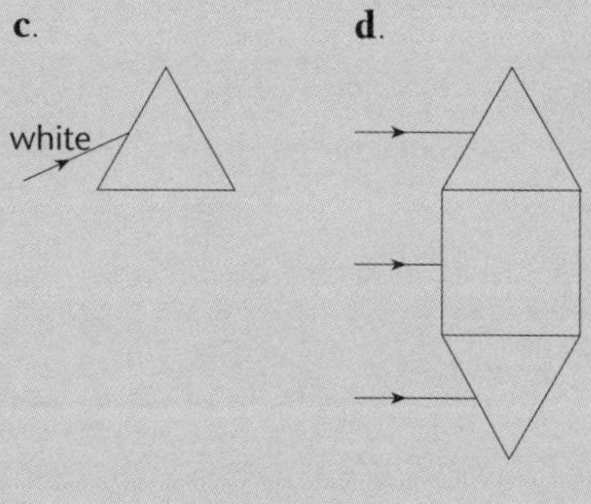

3. The diagram shows a ray of light incident on a plate of glass that is covered by layers of water and oil. Copy the diagram and complete the path of the light ray.

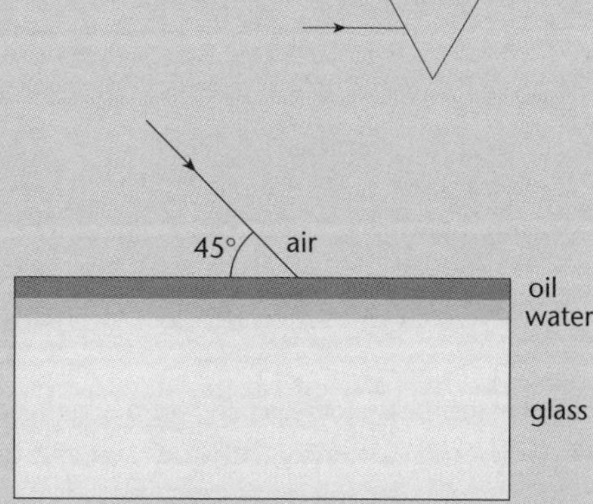

4. The diagram shows Rangi, who has spotted an eel in the river. He wants to spear the eel.

a. Copy the diagram and draw two rays from the eel to Rangi's eye.

b. Draw the apparent position of the eel on your diagram.

c. What advice would you give Rangi so he can spear the eel?

5. Rowena reads the date on a newspaper. She next looks at the same date through a rectangular glass block. Explain how the two views are different. Use diagrams to support your explanation.

6. Red and blue light rays travel from air into a glass prism, as shown. Copy and complete the diagram to show the rays passing through and leaving the prism. Clearly show the difference between the red and blue rays.

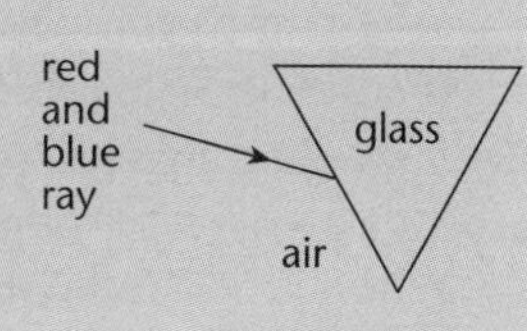

7. James observes a narrow beam of light passing through a glass prism. With his eye in the position shown, James sees yellow light. What is the next colour he observes if his eye is moved slightly up?

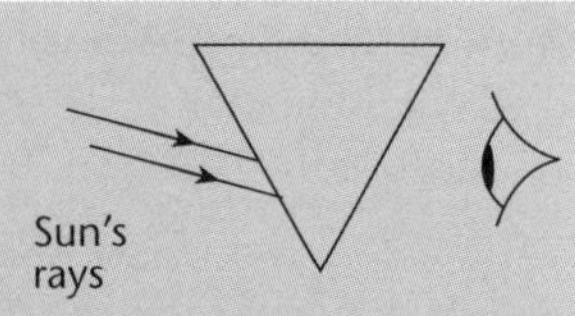

8. Complete the diagram to correctly show the path of a ray of light through a glass block.

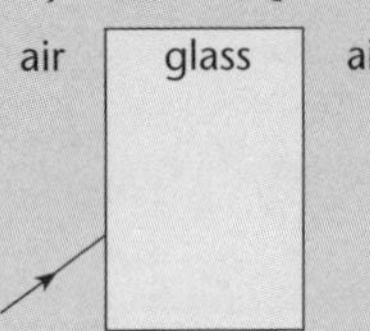

9. Copy and complete the diagram to show refraction and dispersion of white light by a prism.

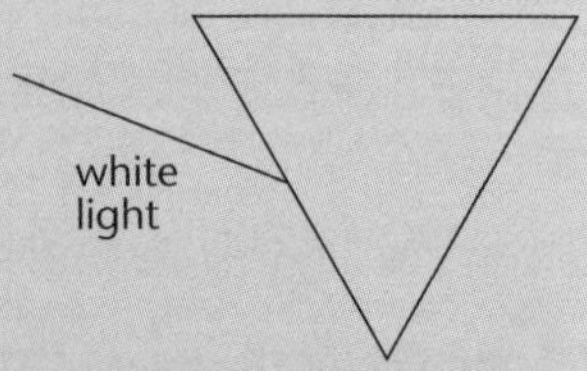

10. Explain the statement 'The refractive index of glass is 1.5'.

11. James and Rowena carried out an experiment with a rectangular block of Perspex. They took the following readings:

Angle of incidence ($\hat{r}$)	22°	28°	33°	38°	41°
sin $\hat{\imath}$	0.3746	i	ii	iii	iv
sin $\hat{r}$	0.2588	v	vi	vii	viii
Angle of refraction ($\hat{r}$)	15°	18°	21°	27°	26°

a. Complete the table for **i–viii** and plot a graph of sin $\hat{\imath}$ versus sin $\hat{r}$

b. Use your graph to determine:

i. The refractive index of Perspex.

ii. The angle of refraction when the angle of incidence is 45°.

12. In a vacuum, light travels at 300 000 km s^{-1}. Calculate the speed of light in:

a. Glass (n = 1.5).

b. Paraffin (n = 1.44).

c. Diamond (n = 2.40).

13. Draw a ray diagram to show how a ray of light is refracted when it passes into:

a. A transparent substance of higher refractive index.

b. A transparent substance of lower refractive index.

14. A ray of light passes from air into water (n = 1.33). For each of the following angles of incidence, calculate the angle of refraction.

a. 24° b. 35° c. 53°.

15. A ray of light passes from air into glass ($n = 1.50$). For each of the following angles of refraction, calculate the angle of incidence.

d. 24° **e**. 35°

16. Red light passes from one medium to another, as shown in the diagram. Describe what happens to the velocity, wavelength and frequency when red light passes from medium 1 to medium 2.

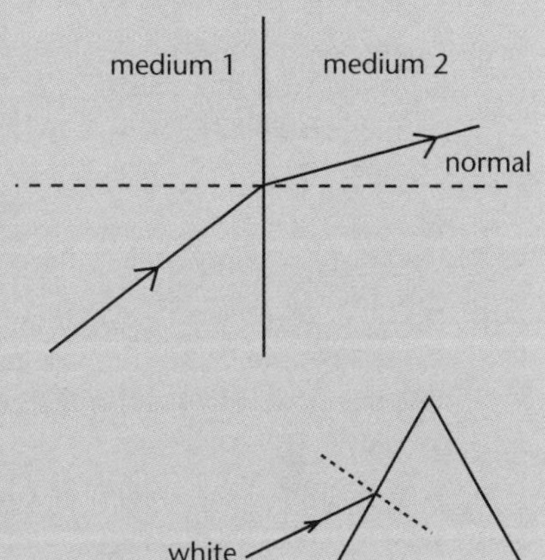

17. **a**. White light is made up of seven main colours. Copy and complete the diagram to show what happens after white light enters the prism. Clearly label the red and violet emergent rays.

white light

b. The emergent rays can be combined to give white light using a second prism. On the diagram you drew for **a**, draw in this second prism so that white light is produced from the coloured rays.

18. Complete the diagram to correctly show a ray of light passing from glass into a parallel layer of air and then into glass again.

glass air glass

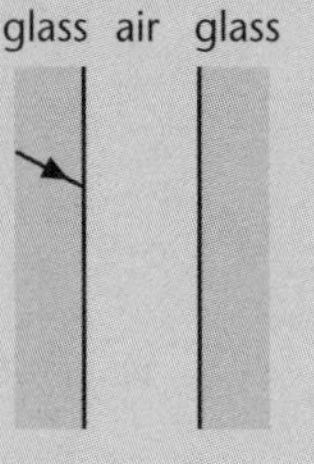

19. An observer is looking at a stone at the bottom of a tank of water. At what position does the stone appear to be?

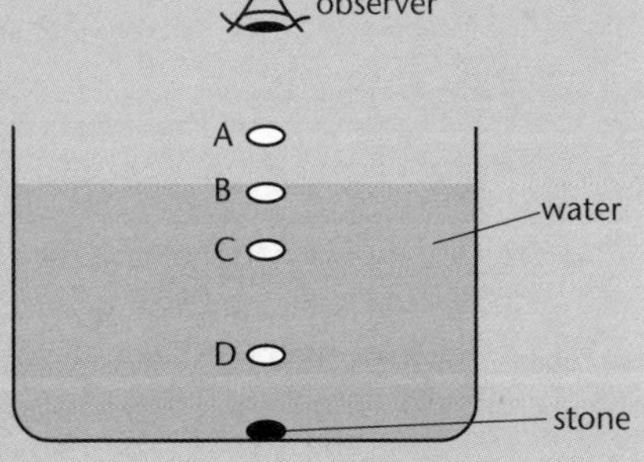

20. a. Copy and complete the diagram to show the path of the ray as it:

(i) Enters the glass block.

(ii) Leaves the glass block.

(b) What is the line called?

(c) Explain what happens at A.

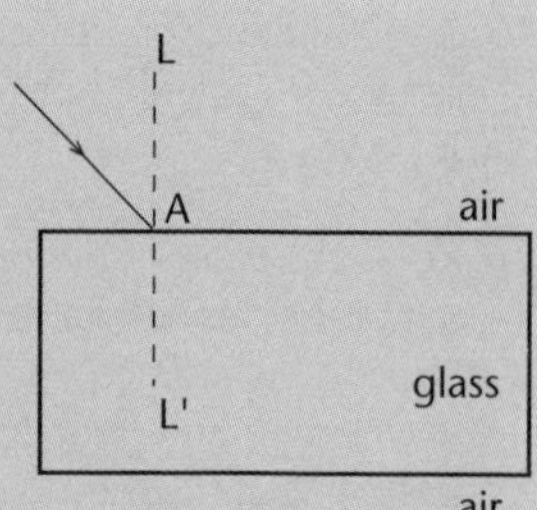

d. The word GREEN is viewed from immediately above a glass block. Copy and complete the diagram to show how the missing letters E and N would appear through the block.

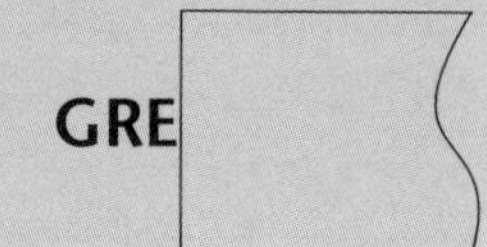

21. Jimmy was sitting on the wharf, fishing early in the morning. The water was very still and clear.

a. He looked at the water and could see a reflection of his fishing rod. The diagram below shows Jimmy sitting on the wharf with his fishing rod. An incident light ray going from the tip of the rod to the water is shown.

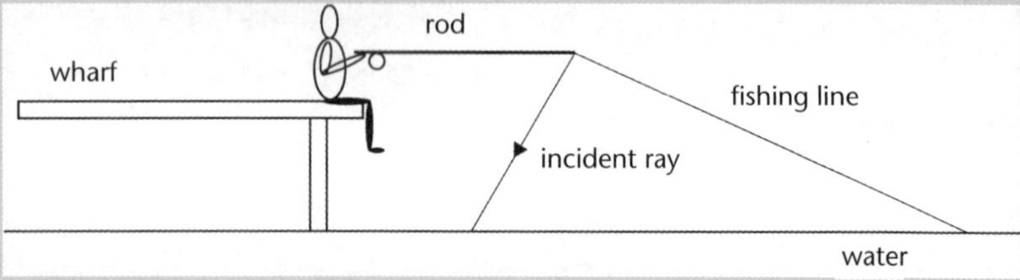

i. Draw the reflected ray on the diagram below.

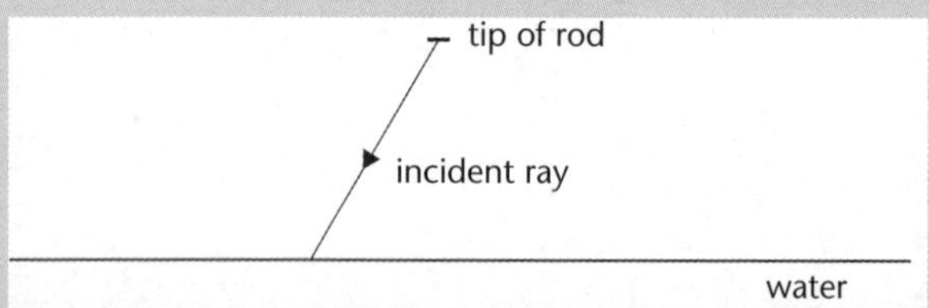

ii. On the diagram in a. i., label the angle of reflection with the letter 'r'.

iii. On the same diagram, draw another incident ray from the tip of the rod to a different point on the water. Then use both rays to show the position of the image of the tip of the rod. Label the image with the letter 'I'.

b. Jimmy sees a fish swimming below the water's surface. The diagram below shows where the fish is, but this is not where Jimmy sees it. The diagram also shows the position of Jimmy's eye.

water

i. On the diagram above, draw rays to show where Jimmy sees the image of the fish. Your diagram requires two rays to be drawn from the same point on the fish. Your diagram needs to show how these two rays enter Jimmy's eye.

ii. State why the image of the fish is virtual.

c. A second fish is swimming towards the first fish. The second fish can see the first fish, even though there is a very large rock obscuring its direct view.

i. State the name of the physical phenomenon that allows the second fish to see the first fish.

ii. Discuss the conditions necessary for this phenomenon to occur in this situation.

d. At night, Jimmy is out on the wharf playing with a powerful spotlight torch. He shines the torch into the water as shown in the diagram below. (Note that the diagram is not to scale.)

Calculate the angle of refraction of the torch beam in the water given that the refractive index of water (n_{water}) is 1.33 and the angle of incidence is 40°.

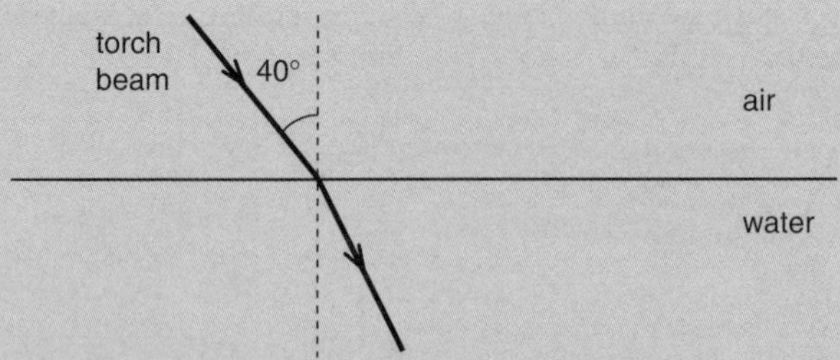

e. Jimmy rides his bike home that night. On the back of his bike there is a reflector that gives a bright red reflection when car headlights shine on it. The reflector is made up of many small prisms. Because of total internal reflection, each of these prisms acts like two reflecting surfaces (mirrors) at right angles. The diagram below shows how one ray of light is reflected.

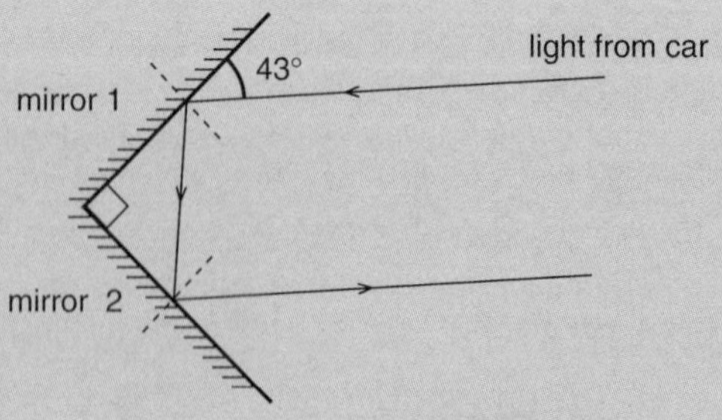

i. Calculate the angle of incidence at mirror 1.

ii. Calculate the angle of incidence at mirror 2.

Reflectors are constructed in the way explained in (e) to allow car drivers to see the bright red reflection.

iii. Explain why reflectors are constructed with right-angled reflecting surfaces. (To answer this question you may like to consider what would happen if a bike reflector were a simple, vertical reflecting surface.)

(Source: NCEA Examination paper)

Unit 12.3 Activity 3B: Multiple choice questions

1. Which of the following statements best explains the significant (main) difference between mechanical waves and electromagnetic waves?

A. Electromagnetic waves have a source and mechanical waves do not.

B. Electromagnetic waves travel faster than mechanical waves.

C. Electromagnetic waves have a higher frequency than mechanical waves.

D. Electromagnetic waves can travel through a vacuum and mechanical waves cannot.

2. Which of the following statements best explains the difference between transverse waves and longitudinal waves?
   A. Longitudinal waves have vibrations perpendicular to the direction of the motion of the wave. Transverse waves have vibrations parallel to the direction of wave motion.
   B. Longitudinal waves can travel long distances. Transverse waves cannot travel long distances.
   C. Longitudinal waves transport plenty of energy. Transverse waves do not transport plenty of energy.
   D. Longitudinal waves have vibrations parallel to the direction of the motion of the wave. Transverse waves have vibrations perpendicular to the direction of wave motion.
3. The speed of sound in air is dependent upon the temperature of the air. The dependence is expressed by the equation:
   $v = 331\frac{m}{s} + \left(0.6\frac{m}{s°C} \times T\right)$ where T is the temperature in degrees Celsius.
   What is the temperature on a cold rainy day when the speed of sound, v = 332.8 m/s?
   A. 0.3°C
   B. 3.0°C
   C. 10°C
   D. 20°C
4. Two speakers are arranged so that sound waves with the same frequency are produced and radiate through the room. An interference pattern is created (as represented in the diagram). The thick lines in the diagram represent wave crests and the thin lines represent wave troughs. Use the diagram to answer questions 4 and 5. At which of the labelled points would **constructive interference** occur?

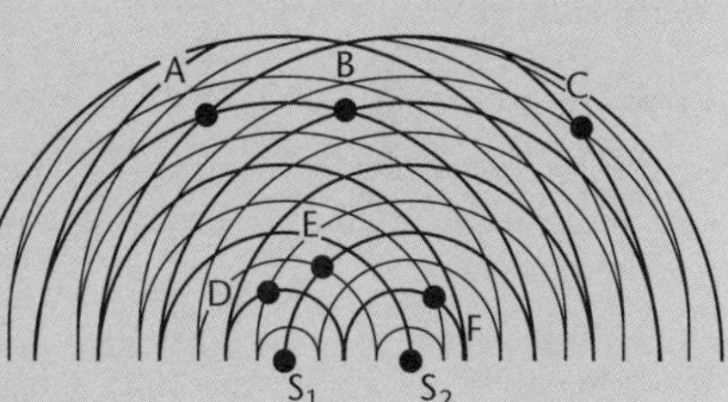

   A. B only
   B. A, B, and C
   C. D, E, and F
   D. A and B
5. How many of the six labelled points represent anti-nodes?
   A. 1
   B. 2
   C. 3
   D. 4

6. The figure at right shows waves generated by an object thrown into a still pond, viewed from above. Use this diagram to answer questions 6, 7 and 8.

Wave crests as seen from above

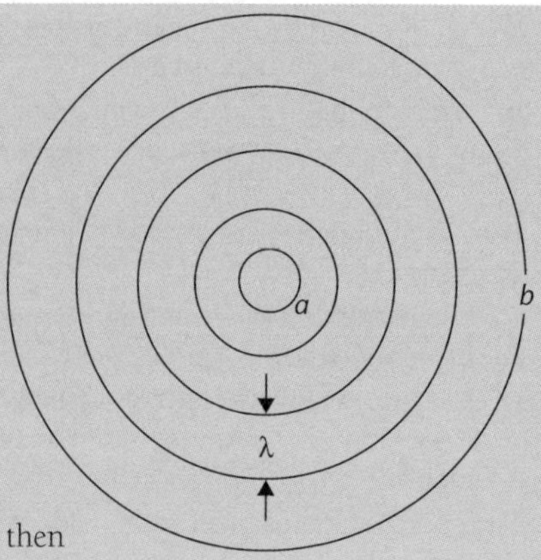

How many wavelengths are shown in the diagram?

A. 1

B. 2.5

C. 3.5

D. 5

7. In the diagram of question 6, if λ = 10 cm, then how far has the wave travelled from point *a* to point *b*?

A. 20 cm

B. 25 cm

C. 35 cm

D. 50 cm

8. In the diagram of question 6, if the velocity of the wave generated is 10 ms^{-1}, what is the period of the wave?

A. 0.01 s

B. 0.1 s

C. 10 s

D. 100 s

9. A standing wave in a 20 cm string horizontally has nodes at x = 0 cm, x = 4 cm, x = 8 cm, x = 12 cm, x = 16 cm and x = 20 cm. What is the wavelength of the standing wave?

A. 4 cm

B. 6 cm

C. 8 cm

D. 10 cm

10. Which of the following statements best explains the significant (main) difference between stationary waves and the travelling waves?

A. Stationary waves always have higher amplitude than the travelling waves.

B. Stationary waves transport energy and travelling waves do not.

C. Stationary waves have a shorter period than the travelling waves.

D. Stationary waves do not transport energy while travelling waves do.

11. Which of the following definitions best describes the term *superposition* in waves?

A. The ability of waves to travel long distance with less energy as they move through a medium.

B. The ability of waves to add their displacements and energy as they move through each other in a medium.

C. The ability of waves to move faster with less energy as they move through a medium.

D. The ability of waves to transport energy as they move through a medium.

12. Which of the following definitions best describe the terms ***constructive superposition*** in waves?

A. When two amplitudes of a wave are the same they add up.

B. When two amplitudes of a wave are opposite they subtract when added up.

C. When two amplitudes of a wave are the same they cancel each other out.

D. When two amplitudes of a wave are opposite they add up.

13. Which of the following definitions best describe the terms ***destructive superposition*** in waves?

A. When two amplitudes of a wave are the same they add up.

B. When two amplitudes of a wave are opposite they subtract when added up.

C. When two amplitudes of a wave are the same they cancel each other out.

D. When two amplitudes of a wave are opposite they add up.

14. What happens to the wavelength of a wave as the wave travels from deep water to shallow water?

A. Increases

B. Decreases

C. Stays the same

D. Doubles

15. Which of the following terms best describe different speeds of waves as the depth of the water varies?

A. Diffraction

B. Reflection

C. Refraction

D. Repulsion

Unit 12.3 Waves

Topic 4: Waves and the electromagnetic spectrum

Topic 4 deals with waves and the electromagnetic spectrum. It covers:

- Reflection and refraction of water waves.
- Superposition of waves.
- The electromagnetic spectrum.
- Diffraction.
- Standing waves.

Energy transfer

Energy can be transferred from one place to another either by the movement of matter, or by **wave motion** between the two places.

Wave motion, in general, is a process in which kinetic energy is transferred from one point to another without any transfer of matter between the points.

Example A

The waves which form when a stone is dropped in water transfer kinetic energy without an overall movement of water.

Waves

Some waves such as **mechanical waves** (eg vibrations passing through a solid, or sound waves), require a material **medium** in which to travel.

Electromagnetic waves such as radio waves and light are produced when **electrons** are made to **accelerate** or when electrons change energy levels in an **atom**. Electromagnetic waves are able to travel through a vacuum, ie they do not need a medium.

Most waves involve the regular repetition of the same motion. They are therefore called **periodic**. A **pulse** is a wave of short duration.

There are two types of wave, **longitudinal** and **transverse**.

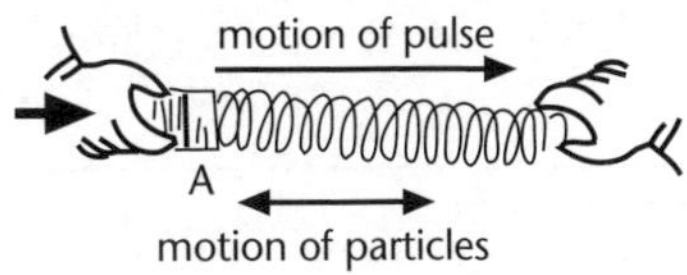

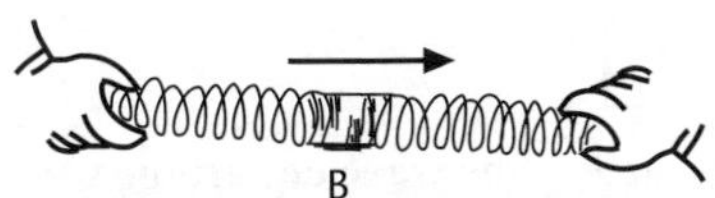

Longitudinal – the vibration is parallel to the direction of motion of the wave, eg sound waves. The diagram shows a longitudinal pulse along a stretched slinky spring.

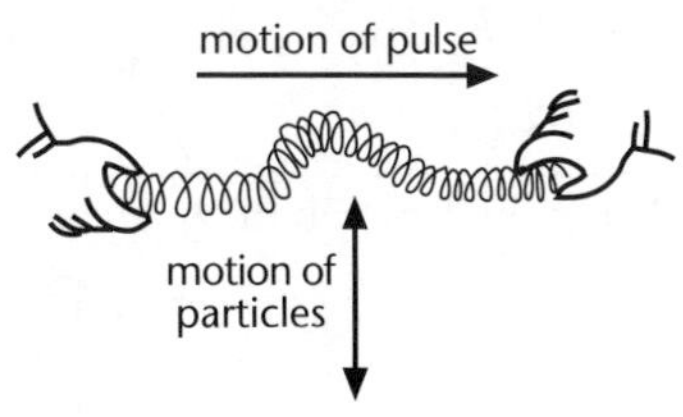

Transverse – the vibration is perpendicular to the direction of motion of the wave, eg electromagnetic waves such as light. A transverse pulse is shown moving along a stretched slinky spring.

Waves on a slinky

Longitudinal waves are often represented in a form similar to transverse waves, by plotting on a graph the particles' displacement at right angles to the direction of **propagation** (the direction in which the wave travels).

Definitions and symbols

The **amplitude**, A, of a wave is the maximum distance a particle is displaced from its rest position.

The **wavelength**, λ, is the distance between two successive corresponding positions in a wave, eg the distance between two successive **crests** (or **troughs**).

The **wave velocity**, v, is the velocity of the wave-shape.

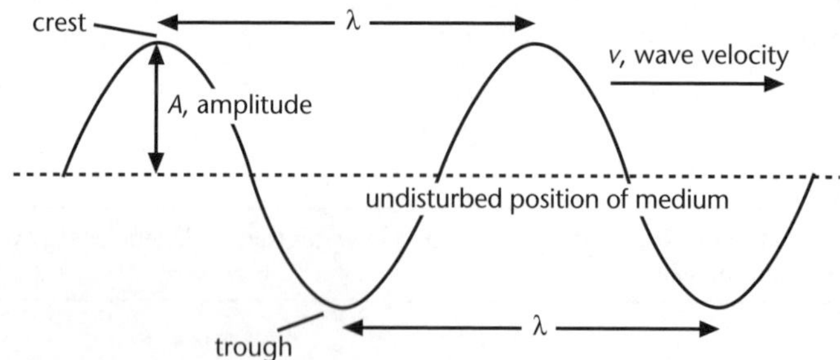

Describing waves

The number of waves passing any point each second is the **frequency** (symbol f) of the waves. The frequency is measured in **hertz** (symbol Hz).

1 Hz = 1 wave per second.

The **period**, T, is the time (in seconds) taken for one wave to pass any point.

Example B

A fisherman standing on some rocks counts 8 waves that pass him in 4 seconds.

The frequency of the waves is $\frac{0}{4}$ = 2 Hz. The time taken for one wave to pass him is $\frac{4}{0}$ = 0.5 s.

The relationship between the frequency, f, of a wave and its period, T, is:

$$T = \frac{1}{f} \text{ or } f = \frac{1}{T}$$

The wave velocity, frequency and wavelength are related by the **wave equation**:

$$v = f\lambda$$

where v is measured in m s^{-1}, f in Hz and λ in m.

Example C

a. Water waves are produced in a tank at a frequency of 20 Hz and with a wavelength of 2 cm. The speed of the wavefront is:

$v = f\lambda$

$= 20 \times 2$ [substituting]

$= 40 \text{ cm s}^{-1}$

b. Sound travels at 330 m s^{-1} in air. A note with frequency 110 Hz will have a wavelength of:

$v = f\lambda$

$\lambda = \dfrac{330}{110}$ [substituting and rearranging]

$= 3.0 \text{ m}$

Wavefronts

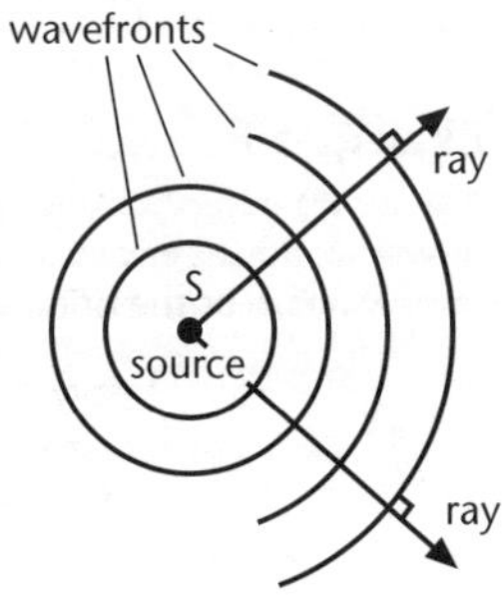

Wavefronts and rays

Consider waves generated at a **point source**, S, as shown figure to the right. This could be done by dipping a finger regularly in and out of the water of a calm pond or water tank. The waves travel outwards and form concentric circles called **wavefronts**. Wavefronts are at right angles to the direction in which the waves are travelling. A line in the direction of a wave's motion is called a **ray**.

As time goes on each wave travels further out while new wavefronts are generated and move out from the point source. At points far away from S the wavefronts become nearly straight (eg light from the sun reaches the earth in **plane** (flat) wavefronts).

Phase

Points along a wave undergoing similar motion at the same time are said to be in **phase**. Points exactly out of phase are ones moving oppositely to each other. For travelling waves, points in phase are a whole number of wavelengths apart and points exactly out of phase are an odd number of half wavelengths apart.

Example D

At the instant shown figure to the right, points A, C and E are momentarily stationary and points B, D and F are moving in the directions shown. Points A and E are in phase and points B and F are also in phase. Points A and C are exactly out of phase, as are points B and D.

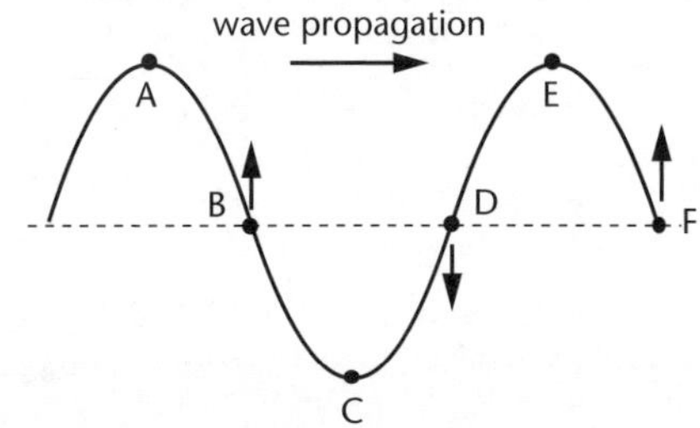

Phase differences in a wave

Phase is usually measured as an angle. 'In phase' means a phase difference of 0°. 'Exactly out of phase' means a phase difference of 180°.

Example E

Points A and B in Example D are $\frac{1}{4}$ wavelength apart and so have a phase difference of 90° ($\frac{1}{4} \times 360°$)

Reflection and transmission of pulses

When a pulse (or a wave) moves from one medium to another, some of the pulse is reflected. The type of reflection depends upon whether the pulse is moving from a slow medium to a fast medium or the other way round.

Example F

A pulse on a heavy string moves toward a lighter string (figure below). The pulse moves slower along the heavy string and faster along the lighter string. A small reflected pulse the same way up as the original pulse moves back along the heavy string.

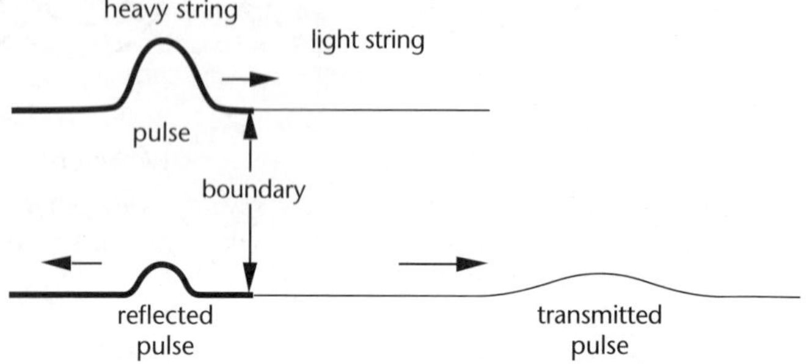

Heavy to light strings

A pulse on a light string moves toward a heavy string (figure below). A small reflected pulse upside down to the original pulse moves back along the light string.

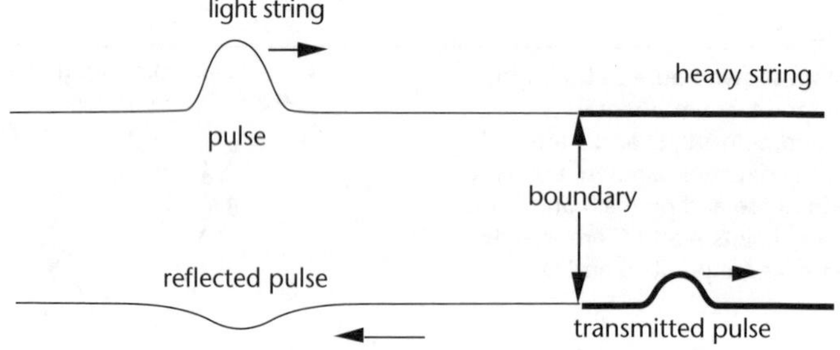

Light to heavy strings

The amplitudes of the reflected and transmitted pulses are less, showing loss of energy. The pulses in the lighter string are further from the boundary, as they are travelling faster.

Unit 12.3 Activity 4A: Pulses and waves

1. A cork is floating on water. A wave approaches it, as shown below.

Complete the diagram below by placing the cork in the correct position after the wave has passed.

2. A wave travels along a piece of rope. The source of vibration S is on the left.

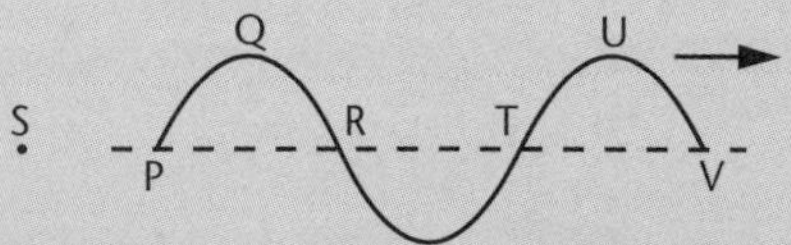

   a. State the name given to the distance between positions R and V.
   b. Which direction will position V move in at the instant shown?
3. A wave of frequency 20 hertz has a wavelength of 4.0 m. What is the wave's speed?
4. Copy and complete the table using the formula $v = f\lambda$.

	Speed ($m\ s^{-1}$)	Frequency (Hz)	Wavelength (m)
a.		20	3
b.		10	0.5
c.	330	100	
d.	3×10^8	1×10^6	
e.	3×10^8		150
f.	350		0.1

5. The best length for a radio aerial is half of the wavelength of the waves used for transmission. If the speed of electromagnetic waves is given as $3.0 \times 10^8\ m\ s^{-1}$, what length aerial is required to receive a signal on 1 480 kHz?
6. The speed of a pulse P in spring X is less than the speed of the pulse in spring Y.

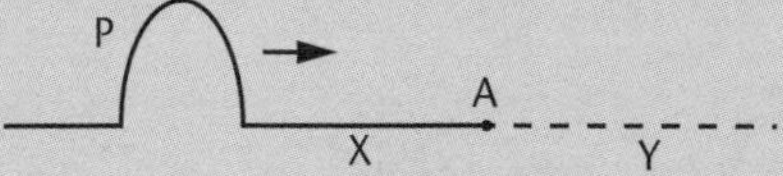

Draw a diagram that would represent the reflected and transmitted pulses.

The Wave Model of light

The **Particle Model** of light fails to explain two important facts which are known about light:

- When light passes from one medium (such as air) to another (such as glass) some light is reflected and some refracted into the second medium.

- When light passes from one medium (such as air) to an optically denser one (such as glass) the speed of light decreases.

Both of the preceding difficulties can be explained if light is considered to be a wave motion.

The **Wave Model** of light describes light as consisting of waves with a very small wavelength and travelling in straight lines from a source with a very large speed.

Water waves

Because water waves move along the surface rather than below the surface, they are known as **surface waves**. In the laboratory, a **ripple tank** is used to study waves.

The glass bottom allows images of the waves to be projected onto a screen below. These images are produced because the crests of the waves act as **converging** lenses and tend to focus the light from the lamp, while the troughs, acting as **diverging** lenses, tend to spread it out. Therefore, the crests appear on the screen as bright bands while the troughs appear dark.

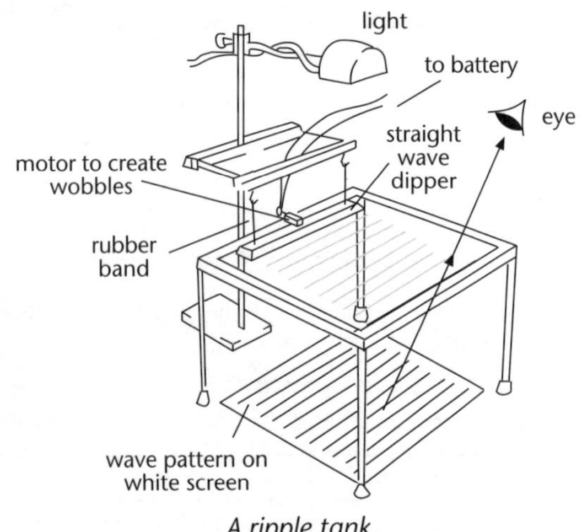

A ripple tank

Experiments done in the ripple tank show that water waves behave like light waves:

- Waves are *reflected* off barriers, obeying the same **laws of reflection** as light.
- Waves are *refracted*, travelling with different speeds depending on the depth of the water and obeying Snell's law.
- Straight wave fronts can be made to *converge* to a focus or to *diverge* from a focus using appropriately shaped sections of shallow water.
- Waves **diffract** when they pass around barriers or through gaps.

Reflected waves

Water waves in a ripple tank reflect off barriers.

Plane waves hitting flat and circular barriers are reflected as shown:

Plane wave reflections

Circular waves hitting flat and circular barriers are reflected as shown:

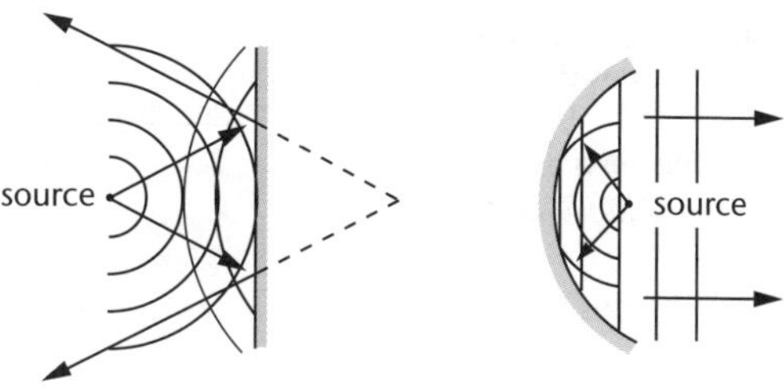

Circular wave reflections

Figure below shows the refraction of plane waves at a plane (straight) boundary between deep and shallow water.

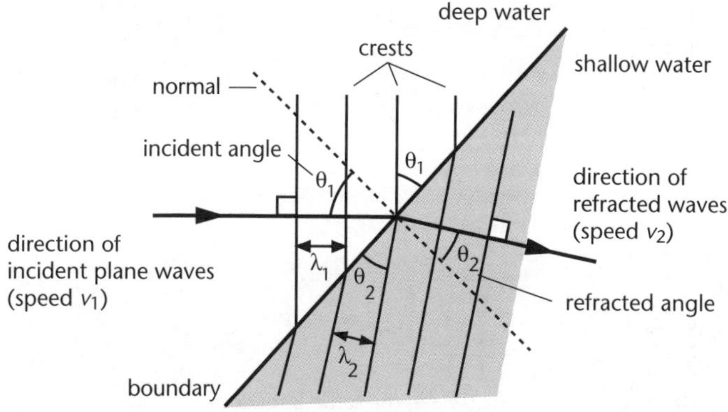

Refraction of waves

The waves' angle of incidence, θ_1, equals the angle that the incident wavefronts make with the boundary. Similarly, the waves' angle of refraction, θ_2, equals the angle that the refracted wavefronts make with the boundary.

For various angles θ_1 and θ_2: $\dfrac{\sin \theta_1}{\sin \theta_1}$ = constant (= ${}_1n_2$, the relative refractive index).

Waves travel slower in shallow water than in deep water. The ratio of the wave speeds $\dfrac{v_1}{v_2}$ is equal to the constant in the above equation; ie $\dfrac{\sin \theta_1}{\sin \theta_1} = \dfrac{v_1}{v_2}$,

and since $v = f\lambda$, this also equals the ratio of wavelengths, $\dfrac{\lambda_1}{\lambda_2}$. (The frequency of a wave does not change when it refracts.)

In summary for wave refraction:

$$ {}_1n_2 = \frac{\sin \theta_1}{\sin \theta_1} = \frac{v_1}{v_2} = \frac{\lambda_1}{\lambda_2} $$

Example G

Water waves travelling at 5.00 cm s^{-1} and with a wavelength of 2.00 cm are incident from deep water to shallow water as shown in the diagram.

a. Determine the relative refractive index.

b. Determine the speed of the waves in the shallow water.

c. Determine the wavelength of the waves in the shallow water.

d. Determine the frequency of the waves.

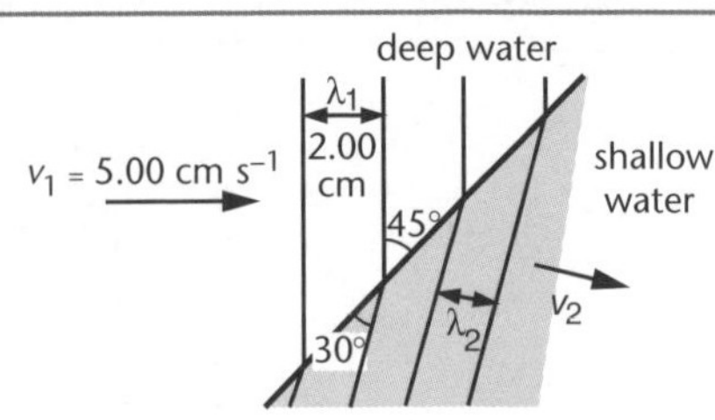

Solution:

a.
$${}_1n_2 = \frac{\sin\theta_1}{\sin\theta_2}$$
$$= \frac{\sin 45°}{\sin 30°}$$
$$= 1.41 \text{ cm}$$

b.
$${}_1n_2 = \frac{v_1}{v_2}$$
$$1.41 = \frac{5.00}{v_2}$$
$$v_2 = \frac{5.00}{1.41}$$
$$= 3.54 \text{ cm s}^{-1}$$

c.
$${}_1n_2 = \frac{\lambda_1}{\lambda_2}$$
$$1.41 = \frac{2.00}{\lambda_2}$$
$$\lambda_2 = \frac{2.00}{1.41}$$
$$= 1.41 \text{ cm}$$

d.
$$v = f\lambda$$
$$f = \frac{v_1}{\lambda_1}$$
$$= \frac{5.00}{2.00}$$
$$= 2.5 \text{ Hz}$$

$$v = f\lambda$$
$$f = \frac{v_2}{\lambda_2}$$
$$= \frac{3.54}{1.41}$$
= 2.5 Hz ie the same frequency

Unit 12.3 Activity 4B: Multiple choice questions

1. Which of the following statements best describes a light wave?
   - **A.** Travels in a straight line and is wavelength dependent which ranges from 100 nm to 400 nm in an electromagnetic spectrum
   - **B.** Travels in a straight line and is wavelength dependent which ranges from 400 nm to 700 nm in an electromagnetic spectrum
   - **C.** Travels in a straight line and is wavelength dependent which ranges from 750 nm to 1 mm in an electromagnetic spectrum
   - **D.** Travels in a straight line and is wavelength dependent which ranges from 1mm to 1 km in an electromagnetic spectrum.
2. Which of the following statements best explains why a hibiscus flower appears red when a visible light falls on it?
   - **A.** Only the wavelength of red light is absorbed while the rest are reflected
   - **B.** Only the wavelength of red light is transmitted while the rest are absorbed
   - **C.** Only the wavelength of red light is refracted while rest are reflected
   - **D.** Only the wavelength of red light is reflected while the rest are absorbed.

3. Which of the following surfaces will virtually transmit all the light when a visible light falls on it?
   A. Black surface
   B. Red surface
   C. Glass
   D. Blue surface
4. When a light ray is incident on the surface of a transparent medium, the incident angle is the angle:
   A. in between the reflected ray and the normal
   B. in between the incident ray and the surface of the medium
   C. in between the reflected ray and the surface of the medium
   D. in between the incident ray and the normal.
5. When a light ray is incident on the surface of a transparent medium, the critical angle is the angle of incidence:
   A. that when a light ray is propagated through it, the refracted ray is completely absorbed by the first medium.
   B. that when a light ray is propagated through it, the reflected ray is reflected back to the boundary of the second medium.
   C. that when a light ray is propagated through it, the refracted ray grazes the boundary of the two media.
   D. that when a light ray is propagated through it, the refracted ray is completely absorbed by the second medium.
6. When a light ray is propagated at an angle greater than the critical angle on a transparent material, what will happen to the refracted ray?
   A. total absorption by the medium itself
   B. some portion of ray absorbed while the rest exit the medium
   C. grazing the interface of the two media
   D. all refracted ray will exit the medium.
7. A ray of light is incident on a rectangular block of glass of index of refraction 1.50 at an angle of *a* as shown in the figure below. Use this diagram to answer questions 7, 8 and 9. 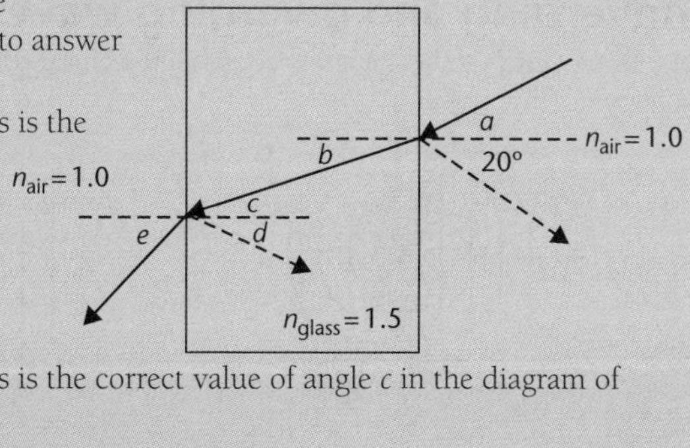
   Which of the following answers is the correct value of angle *a*?
   A. 13°
   B. 20°
   C. 25°
   D. 30°

8. Which of the following answers is the correct value of angle *c* in the diagram of question 7?
   A. 13°
   B. 20°
   C. 25°
   D. 30°

9. Which of the following answers is the correct value of angle *e* in the diagram of question 7?
   A. 13°
   B. 20°
   C. 25°
   D. 30°
10. A ray of light is incident on a rectangular block of glass of index of refraction 1.450 at an angle of θ_i measured with respect to the normal as shown in the diagram below. The thickness of the glass is 4 cm. The angle of incidence is $\theta_i = 30°$. Use this diagram to answer questions 10, 11 and 12.
   What is the value of incident angle θ_r?
   A. 10°
   B. 20°
   C. 30°
   D. 40°

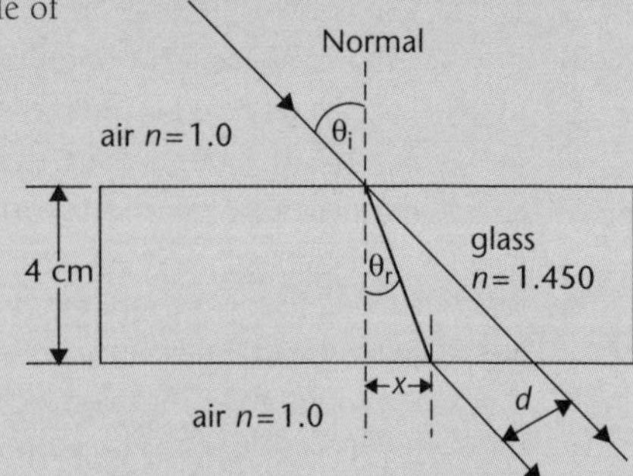

11. What is the value of *x*?
   A. 1.20 cm
   B. 1.40 cm
   C. 1.46 cm
   D. 1.50 cm
12. By how much has the ray deviated, *d*?
   A. 3.70 cm
   B. 3.72 cm
   C. 3.75 cm
   D. 3.77 cm

Converging and diverging waves

The figure below shows shaded shallow regions used to focus plane water waves and to diverge plane water waves.

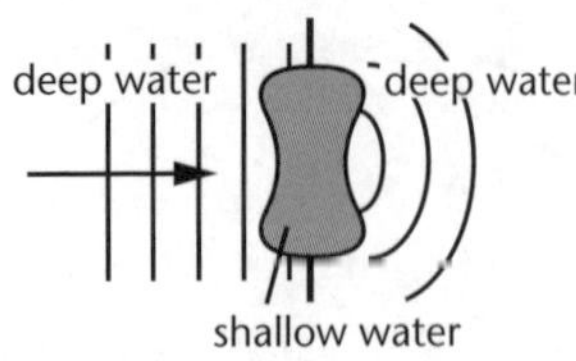

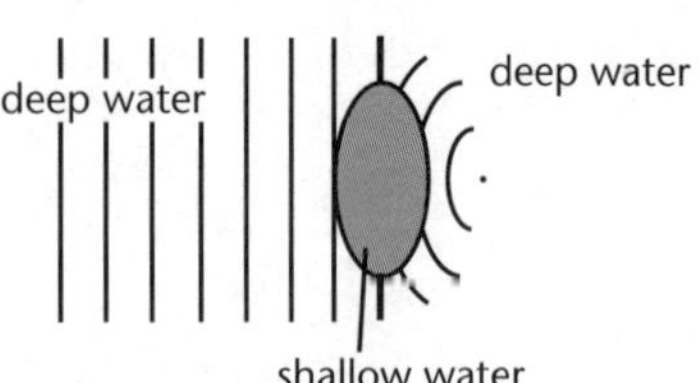

Refraction caused by varying depth

Other refraction effects of light such as **total internal reflection** can also be demonstrated using water waves.

Diffraction of waves

The *smaller* the wavelength, the less the diffraction of waves when they pass around a barrier, ie the less the angle by which the waves bend.

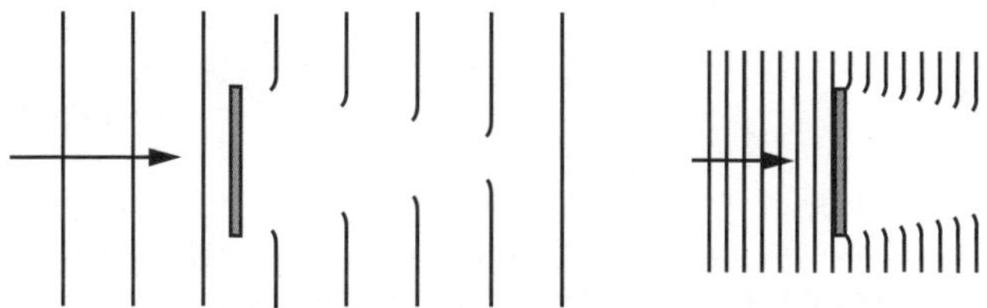

In this diffraction example the wavelength does not change.

Diffraction around obstacles

The *smaller* the wavelength compared to the size of the gap, the less the diffraction of waves through a gap.

In this diffraction example the wavelength does not change.

Diffraction through gaps

When waves pass through a gap the waves are striking the ends of two barriers. The waves which pass through the gap bend behind both barriers. The effect is most noticeable when the gap is approximately the same size as the wavelength of the waves.

Sound waves bend around barriers. A listener indoors does not have to be in line with an open window in order to hear noises coming in from the street.

Diffraction is observed for light as well as for other kinds of waves, and the effect is one of the main items of evidence for concluding that light has wave properties.

Unit 12.3 Activity 4C: Reflection, refraction and diffraction of waves

1. When ripple tank waves are projected onto a paper screen below the tank the wavelength is magnified. How can you work out the magnification produced by the ripple tank? How could you measure the wavelength of the waves?
2. Draw an accurate diagram showing the reflection of plane waves from a straight barrier lying at an angle of 30° to the direction of the incoming waves.

3. The diagram shows the change of direction experienced by water waves passing from one depth to another. Explain why the waves in depth Y have a lower velocity.

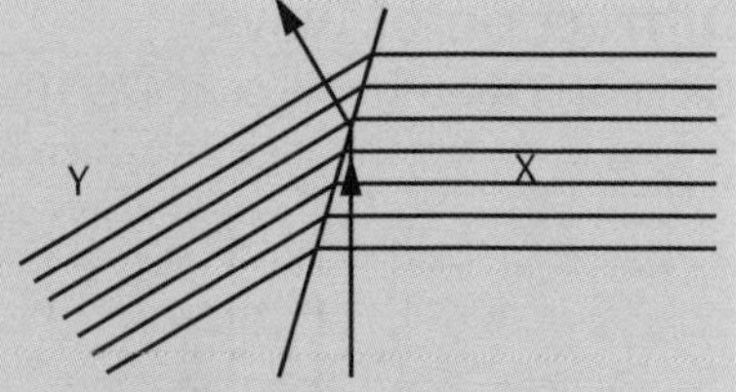

4. A water wave initially of wavelength 3.0 cm and period 0.50 s enters water of a different depth where its speed is 2.0 cm s^{-1}. Calculate the changed wavelength.
5. Waves of frequency 60 Hz pass across a boundary from deep water, where their speed is 4.0 m s^{-1}, into shallow water, where their speed is 2.0 m s^{-1} . Calculate the wavelength in each region, and draw a diagram showing the waves if the refracted waves are leaving the boundary with a refracted angle of 20°.
6. The diagram shows a set of straight waves passing through a narrow gap from shallow to deep water. Copy and complete the diagram by drawing waves in the deep water.

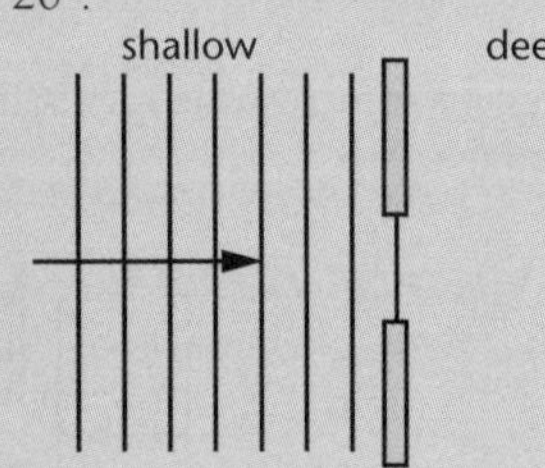

Sound waves

When a guitar string is plucked, or a tuning fork vibrates in air, **compressions** and **rarefactions** of the air are caused which spread outwards with a longitudinal wave motion.

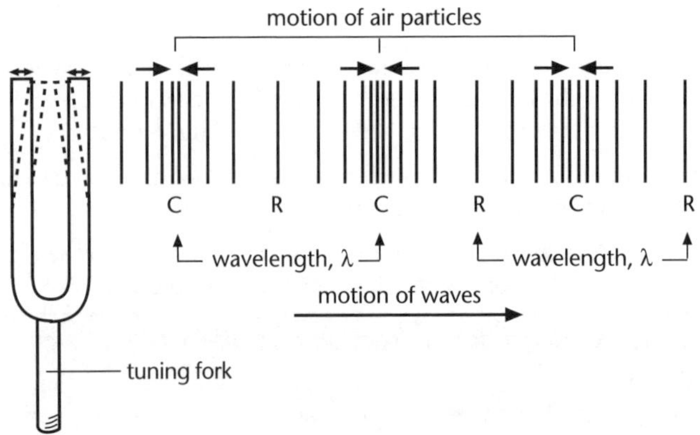

Sound waves

When sound waves reach the ear, they cause the eardrum to vibrate rapidly. The sensation of loudness depends on the **intensity** of the sound waves which reach the listener's ear. The intensity of a sound is the *rate of flow of energy*. A louder sound wave has a greater amplitude. The **pitch** of the sound is related to the frequency of the wave.

A medium such as air, water or iron is necessary if sound is to be transmitted from one place to another; *it cannot travel through a vacuum.*

Longitudinal waves, such as sound, are often represented as transverse displacement waves. This is shown on an **oscilloscope**, where a microphone changes the sound signal into an electrical signal and this is displayed on the screen as a graph of **voltage** against time.

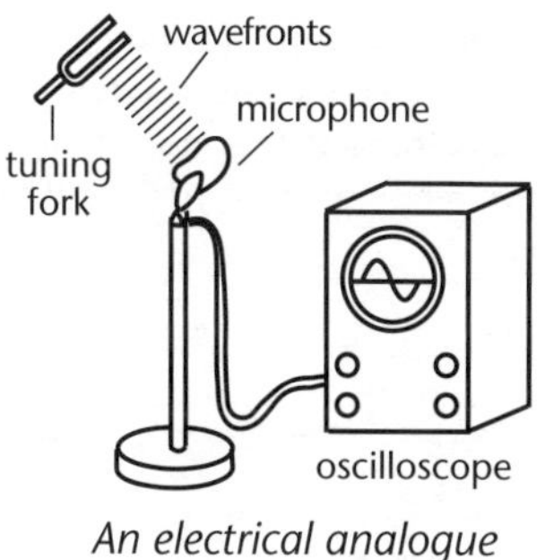

An electrical analogue of a sound wave

The speed of sound

Thunder is heard some seconds *after* the associated lightning is seen, showing that sound travels slower than light.

The speed of sound varies depending on the medium. In air at 0° C, the speed is 332 m s^{-1}. The speed of sound in water is 1 410 m s^{-1} and in iron it is 5 000 m s^{-1} at 20° C.

Superposition

Superposition is the ability of waves to **superimpose** (add their displacements and their energy) as they move through each other.

Example H

When two stones are dropped into water at different places two sets of circular wave ripples are set up. Each set can move across the surface and through the other, *as though the other waves were not present.*

When several people talk at the same time in a room, the different sounds move from place to place with no effect on each other.

Two rectangular pulses moving in opposite directions in a spring or rope can pass through each other so that each pulse remains the same afterwards. The **principle of superposition** allows the two waves to be 'added' together by adding their displacements together at each instant in time.

Constructive superposition refers to the case where two pulses are the same way up and their displacements add (lines 2 to 4).

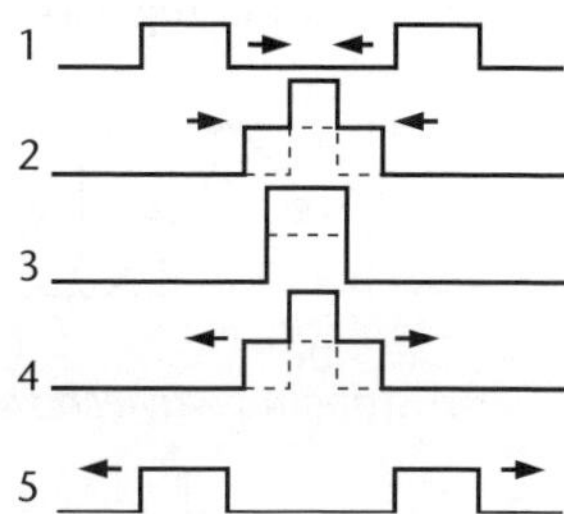

Constructive superposition

Destructive superposition refers to the case where the pulses are inverted with respect to each other and their displacements subtract to give zero displacement (lines 2 to 4).

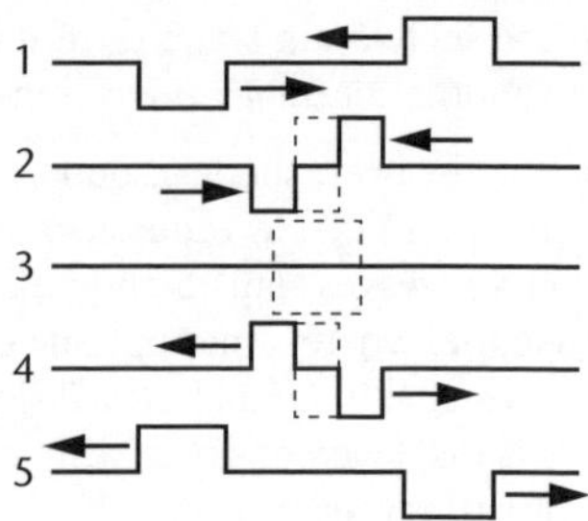

Destructive superposition

Example I

Two pulses each travelling along a string at 1.5 cm s^{-1} are shown at time = 0 s. Draw the shape of the string at the instant 2.0 seconds later.

Solution:

Each wave moves a distance:

$d = vt$

$= 1.5 \times 2.0$

$= 3.0$ cm towards the other.

The superimposed pulses give the resultant shape shown.

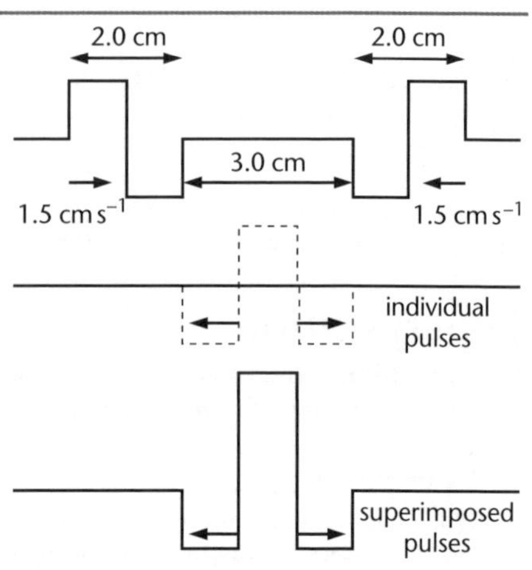

Standing waves

If two waves of the same amplitude meet, a resultant displacement of twice the amplitude occurs. At this position an **antinode** is formed. If two waves of opposite amplitude overlap, the resultant displacement is zero. At this position a **node** is formed.

If the two waves of equal amplitude, frequency and speed are produced from the ends of a hollow pipe or string, a **standing wave** can result.

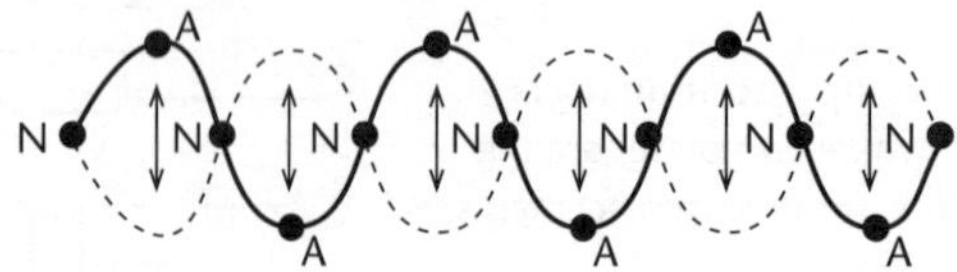

A standing wave showing antinodes (A) and nodes (N)

Ripple tank interference

The pattern of water waves produced by two point sources separated by a little distance apart is shown in the figure below. The point sources are two dippers that move in and out of the water in a ripple tank at the same time.

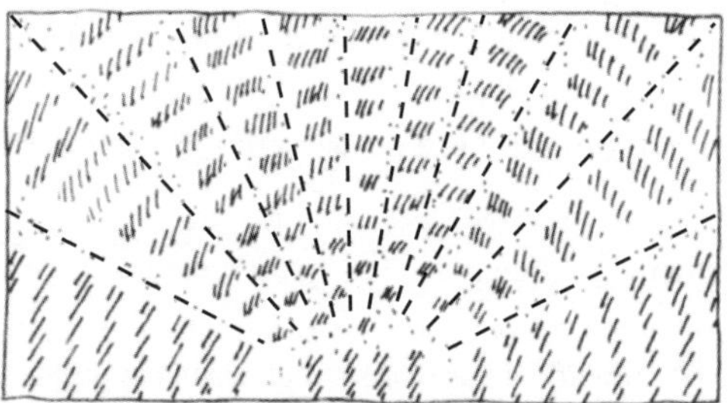

Water interference

In the above figure, darkened areas show where trough meets trough (also known as places of **constructive interference**), dashed areas represent undisturbed (calm) water (where crest meets trough) (also known as places of **destructive interference**), and places where crest meets crest are left white (also places of constructive interference).

The whole pattern on the water surface is called an **interference pattern**. As the troughs and crests move away from the sources, a continuous series of points appear, forming undisturbed lines of water. These lines are called nodal lines. Between **nodal lines** are moving double crests and moving double troughs. These are called **antinodal lines** (lines of maximum disturbance and wave energy).

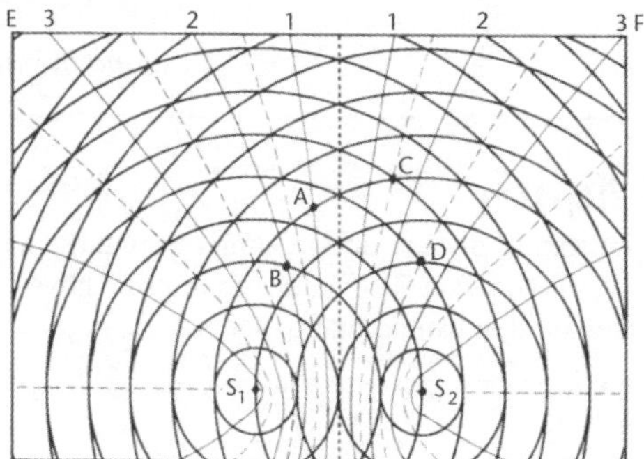

Light solid lines mark nodal lines and antinodal lines are indicated by dotted lines.
Dark solid lines mark the wavefronts (crests) produced by the two sources.

Schematic view of interference

The interference pattern in the above figure is symmetrical about the central dotted line (an antinodal line of choppy water). Point A is on nodal line 1 (the first nodal line from the centre). The path difference between lines from the two sources S_1 and S_2 to A is:

$AS_1 - AS_2 = {}^1/_2\lambda$

If A is moved to any other position along nodal line 1, the path difference would still be $^1/_2\,\lambda$. The path difference between lines from the sources to any point on nodal line 2 (point B, say) is equal to $1^1/_2\,\lambda$. In general, the path difference between lines from the sources to any point, P, on the nth nodal line (destructive interference) (n = 1, 2, 3…) is:

$$PS_1 - PS_2 = (n - {}^1/_2)\,\lambda$$

In the same way path differences can be determined for points C, D, etc, lying on the antinodal lines.

For any point, Q, on the nth antinodal line (constructive interference; n = 0, 1, 2, 3…):

$$PS_1 - PS_2 = n\,\lambda$$

Antinodal lines are numbered slightly differently since there is a central line (number 0).

Moving along a line parallel to the two sources (the line EF in the figure on the previous page), alternative areas of calm and areas of disturbance are encountered. The same effect is noticed with sound waves when moving in front of two loudspeakers playing the same note.

Example J

A person moving across in front of the speakers will hear an alternating increase and decrease in the level of sound.

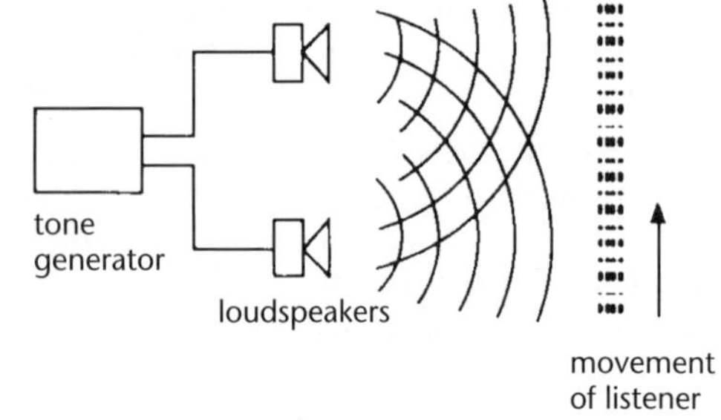

Sound interference

Diffraction of light

By holding two fingers in front of one eye and looking through the narrow slit between them, diffracted light is able to be seen. The light seen appears larger than the slit and there seems to be several **fringes** of light extending out across the slit.

Both waves and light are able to be *diffracted*. With respect to diffraction, light behaves like a wave and not a particle.

Interference of light waves – Young's experiment

In 1810, Thomas Young demonstrated that light passing through two small holes very close together diffracts and forms an interference pattern.

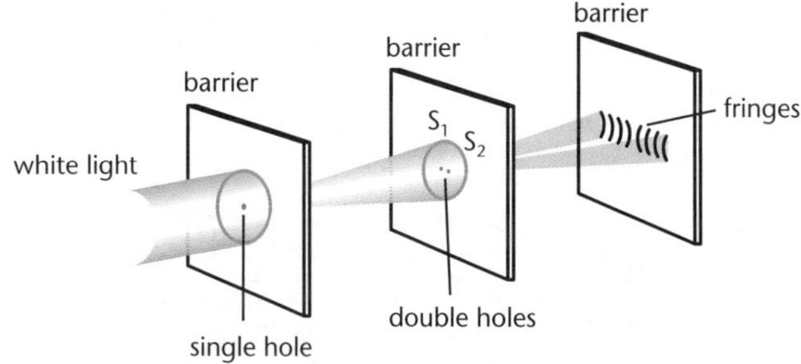

Young's experiment

In the above figure, white light passes through hole S and then through holes S_1 and S_2 on to a screen. An interference pattern is visible on the screen. The interference pattern consists of dark bands (called fringes) which are of varying widths, separated by coloured bands (consisting of red, orange, yellow, green, blue, etc).

Nowadays when Young's double-slit experiment is performed, an incandescent lamp with a long, straight filament is used and the light from it is allowed to pass through two narrow slits. The slits are parallel to the light source and very close together (about 0.1 mm apart).

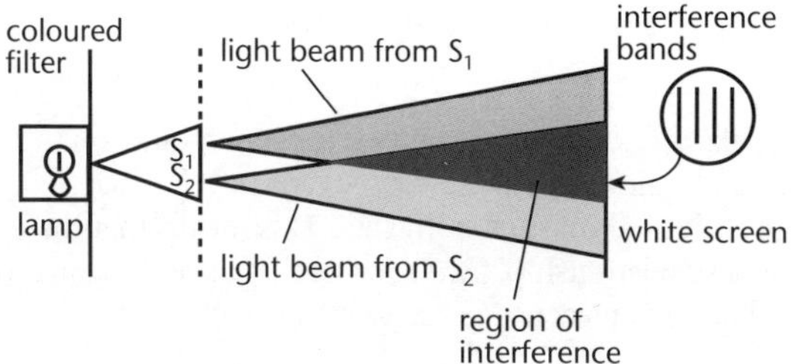

Young's experiment (today)

Instead of using white light, monochromatic light (light of one colour) is used. This is done by using a filter (such as a piece of coloured cellophane to cover the bulb). A laser which only emits one particular wavelength of light can also be used.

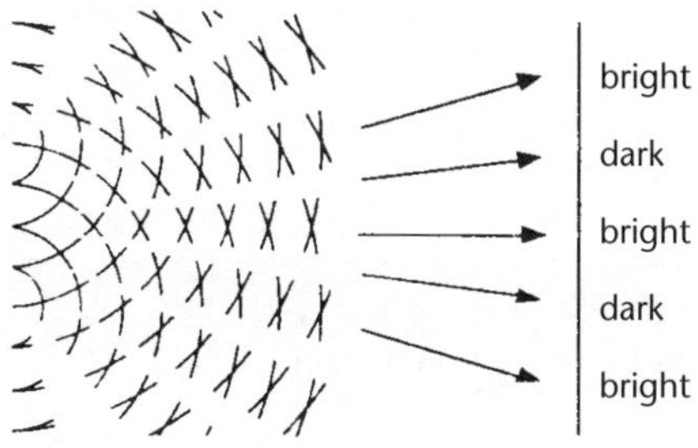

Interference fringes

When light of one colour (eg red) is used instead of white light, a much clearer series of bright (red) and black fringes is seen. The spacing of the fringes in the patterns depends on the separation of the slits, the distance from the pair of slits to the screen, and the wavelength of the light. Red light has a relatively long wavelength and the bright fringes are wider apart than any other colours.

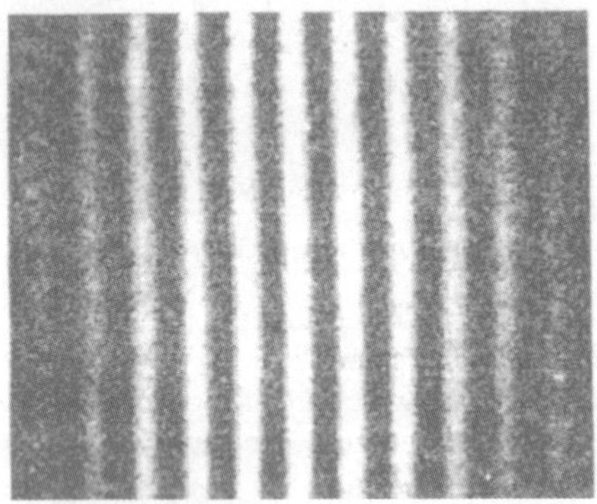

Interference pattern produced with red light.
Both the black fringes and red fringes are evenly spaced.

Interference bands

Interference bands have never been observed using light waves from two independent sources (eg two lamp filaments) or even from two separate points on the same source. This is because light is emitted randomly from a source. For interference to occur, light from the same wave must pass each time through both slits. This means that the two sources must maintain a constant phase relationship. Two sources that have the same frequency and have a phase relationship that remains constant are said to be coherent.

The electromagnetic spectrum

Visible light waves form a tiny part of the family of electromagnetic waves, which also includes radiowaves, microwaves, X-rays and gamma rays. All of these waves travel at the same speed, the speed of light, c, in a vacuum.

$c = 3.0 \times 10^8$ m s^{-1}. The waves differ in their frequency and wavelength. The wave formula $v = f\lambda$ can still be applied to electromagnetic waves.

Example K

Find the wavelength of radiowaves transmitted from an FM station with a transmitting frequency of 98.2 MHz.

Solution:

$$v = f\lambda$$

$$3.00 \times 10^8 = 98.2 \times 10^6 \times \lambda \qquad \text{[substituting]}$$

$$\lambda = \frac{3.0 \times 10^6}{98.2 \times 10^6}$$

$$= 3.05 \text{ metres}$$

The electromagnetic spectrum extends infinitely in each direction. Theoretically, there is no limit to the frequency of electromagnetic waves. They can have extremely low frequencies or extremely high frequencies.

The range of frequencies and wavelengths for each type of electromagnetic wave is not clearly specified and some ranges overlap.

Infrared waves are produced by hot objects and so these waves are often called *radiated heat*.

Ultraviolet waves are often called *ultraviolet light* even though the waves are outside the range visible to the human eye.

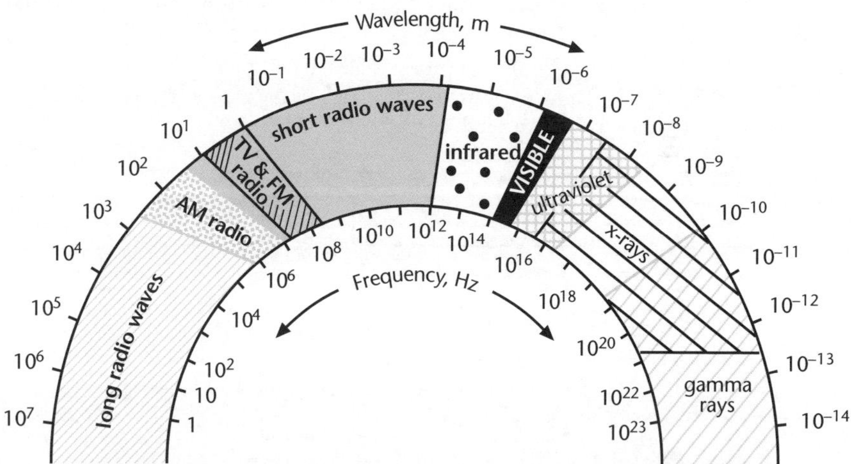

The family of electromagnetic waves is laid out according to their frequency and corresponding wavelength.

The electromagnetic spectrum

Electromagnetic waves get their name because they are made up of oscillating electric and **magnetic fields**. These two fields are positioned at right angles to each other and oscillate at right angles to the wave propagation direction.

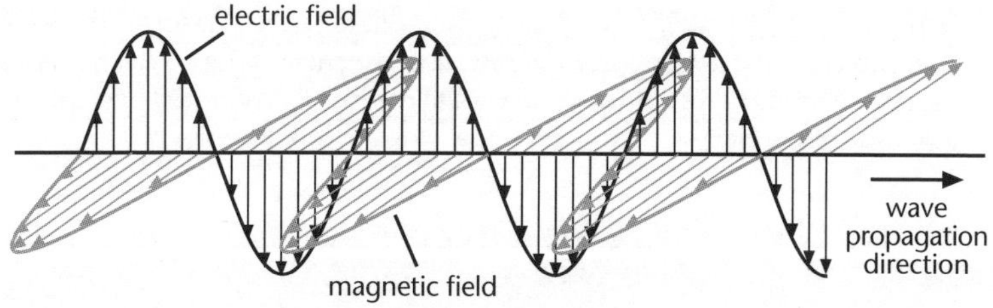

Electromagnetic waves

The energy in an electromagnetic wave is stored in the two oscillating fields.

Unit 12.3 Activity 4D: Diffraction, superposition and electromagnetic waves

1. Which one of the following waves is diffracted most by an open square window of width 0.50 m?
   A Microwaves of frequency 10 GHz (10×10^9 Hz) and speed 3.0×10^8 m s^{-1}?
   B Light waves of wavelength 650 nm?
   C Sound waves of frequency 1 050 Hz and speed 350 m s^{-1} ?
   Explain clearly the reasons for your answer.
2. Ordinary AM radio signals of frequency of about 1 MHz can usually be received behind a hill, while the reception of an FM radio signal frequency 100 MHz is often very poor and cannot be heard. Explain clearly why this happens, assuming that the speed of the waves is 3.0×10^8 m s^{-1}.
3. The diagrams show a pair of pulses approaching each other along a rope. Their speed is 1 cm s^{-1}. Draw diagrams of the shape of the rope at 2-second intervals until both pulses have passed through each other.

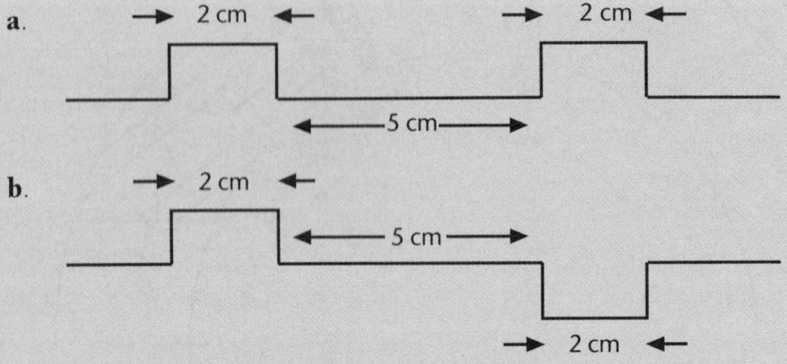

4. State what type of wave an electromagnetic wave is.
5. State the physical quality that is the same for all electromagnetic waves.
6. The frequency of an electromagnetic wave is measured to be 5.6×10^{14} Hz. According to the electromagnetic spectrum, what classification of electromagnetic wave would it be?
7. How many times is the speed of light in air faster than the speed of sound in air (at 0°C)?
8. Microwaves are used to cook food because their energy is best absorbed by water molecules which vibrate strongly at microwave frequencies. The wavelength of microwave radiation produced by a microwave oven is 12.2 cm. Calculate the corresponding frequency.

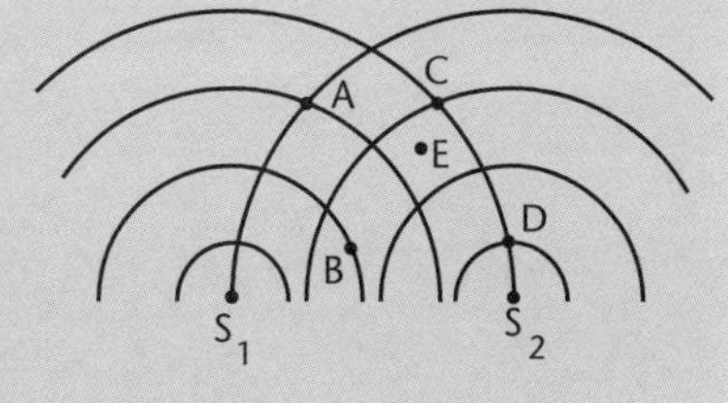

9. The diagram shows an interference pattern produced in a ripple tank, by two points S_1 and S_2 which dip in and out of the water in phase (ie together). The wavefronts are 4 mm apart.

a. For points A, B, C, D and E, find the difference in distance (in wavelengths) from points S_1 and S_2.

b. Which points are constructive and which are destructive interference points?

10. In an operatic auditorium, sound is heard very loudly in some seats, while in other positions very little sound is heard. Give the reason for this and suggest a method to reduce this effect.

11. In a double-slit interference experiment, light from a lamp passes through a colour filter that transmits red light only.

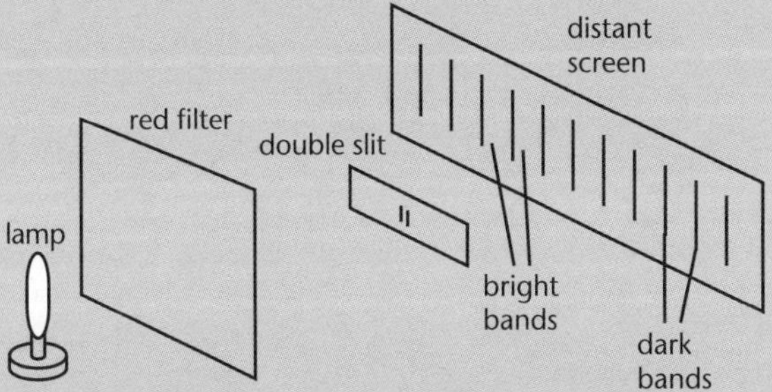

The red light passes through two narrow parallel slits which are close to each other and form alternating bright and dark bands on a distant screen. If the filter was replaced with another which transmitted only blue light (which has a smaller wavelength), explain what would happen to the spacing between the bright bands.

Topic 5: Reflection of light – review and extension

Topic 5 provides a review and extension of the reflection of light, concepts which we have already covered in the preceding Topics.

Introduction

Light is a type of **electromagnetic radiation** which our eyes are sensitive to. Electromagnetic radiation (abbreviated to em radiation) is made up of oscillating electric and magnetic fields which travel together near a speed of 3.00×10^8 m s^{-1}, called the **speed of light**, *c*. Other forms of em radiation include radio-waves and X-rays, which also travel at the speed of light.

Light is a form of energy. For each colour of light, the brighter the light the more energy it has. When light from the sun shines on us, the light energy is changed to heat energy and we feel warm.

How light travels

In general, light appears to travel in *straight lines*. This is sometimes referred to as the **straight line propagation** of light and can be illustrated in various ways.

Example A

The observations illustrated below can be explained by the fact that light travels in straight lines.

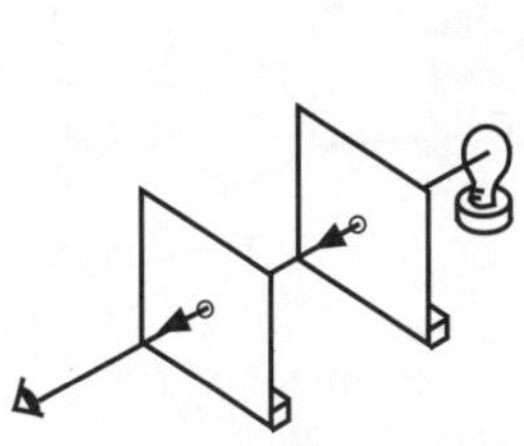

When two screens each with a small hole are placed in front of a bulb, the bulb can only be seen by an observer if the two holes are in a straight line with the bulb.

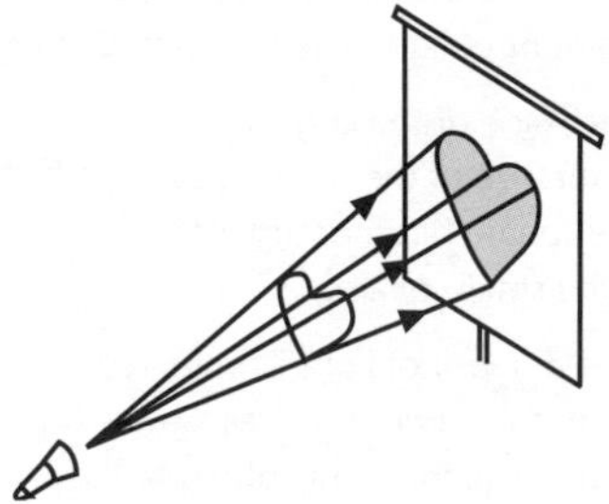

If an object is placed between a small source of light and a screen, a *sharp* shadow is seen on the screen.

Straight line propagation of light

A **pinhole camera** can be made by drilling a small hole in the centre of one end of a tin can. The other open end is covered with tissue paper to make a screen and a tight-fitting cardboard tube is attached to the tissue paper as shown.

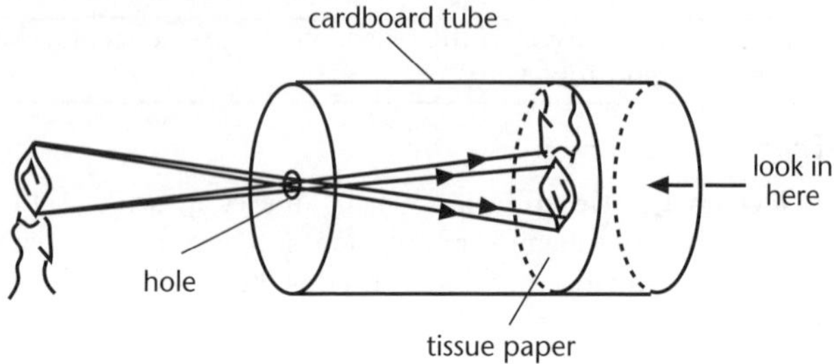

A pinhole camera

When a light source such as a candle flame is viewed with a pinhole camera the observer sees an inverted (upside down) picture of the flame on the tissue paper.

Light travels out in all directions from a light source. The *further* from the light source, the less the **illumination**. The figure below illustrates an important relationship between the distance from the light source and the illumination.

Light spreads out in all directions from a point P.

The large square is twice the distance from the light source as the small square and the same amount of light falls on both squares.

Since each side of the large square is twice the size of the small square, the large square has four times the area of the small square.

The same amount of light illuminates (shines on) four times the area, so the light on the large square is only one quarter as intense as on the small square.

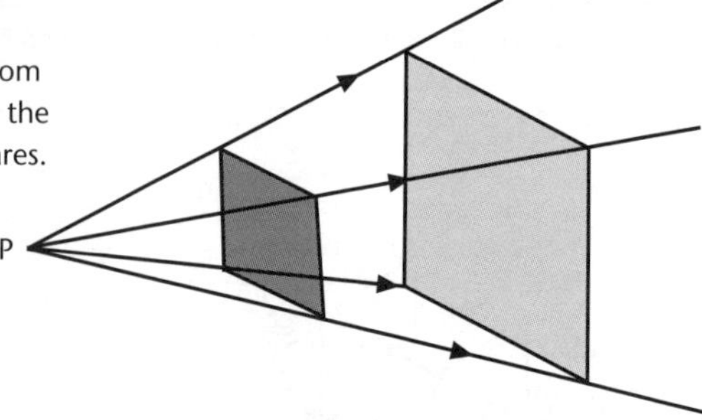

Decreasing illumination from a light source

Illumination is *inversely proportional* to the distance squared. If E is the illumination, and d the distance from the light source, the **inverse square law of illumination** states:

$$E \propto \frac{1}{d^2}$$

Light meters are used by photographers and cricket umpires to measure the illumination of light.

Reflection of light

When light reaches a mirror or polished surface it is **reflected**. This is illustrated using a **ray** of light in the figure below.

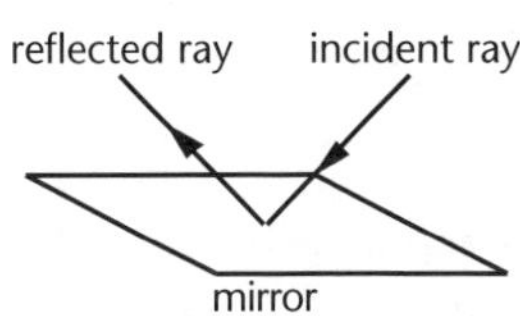

The ray reaching the surface is the incident ray and the ray reflected from the surface is the reflected ray.

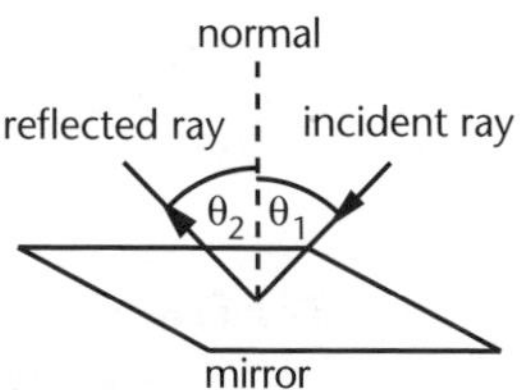

The **normal** is the line perpendicular to the surface at the point where the incident and reflected ray meet. The angle of incidence (θ_1) and **angle of reflection** (θ_2) are measured from the normal.

Reflection from a mirror

Two important results called the **laws of reflection** are:

- The incident ray, reflected ray and normal all lie in the same plane.
- The angle of incidence is equal to the angle of reflection, ie:

$$\boxed{\theta_1 = \theta_2}$$

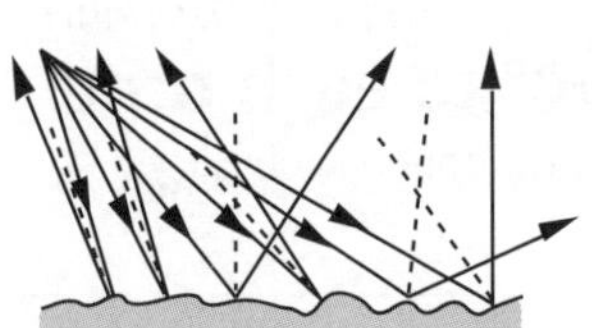

Irregular (or diffuse) reflection occurs with most surfaces. This is when light is reflected from a surface in many *different* directions.

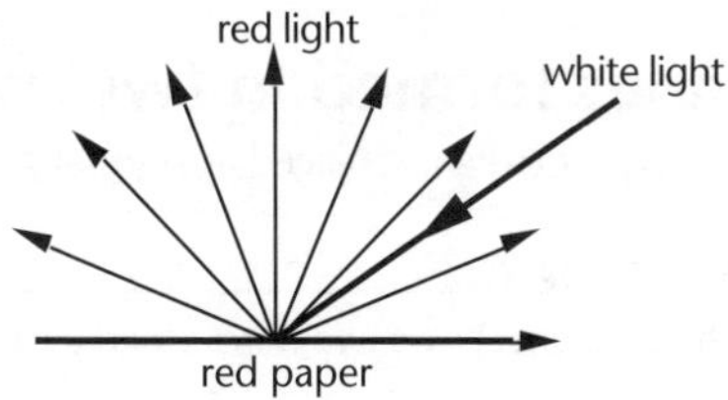

Often the reflected light has a different colour from the incident light. Thus a red reflecting surface reflects only the red light and *absorbs* the other colours.

Irregular reflection and colours

At each point on a rough surface light is reflected in such a way that the laws of reflection are followed.

The image formed by a plane mirror

When we look into a **plane** (flat) mirror, the **image** (what we see) appears to be behind the mirror. This occurs because the position of the image that is seen depends on the direction from which light enters the eye. The light entering the eye is reflected from the mirror and appears to come from behind the mirror. The human eye always 'wants' to accept rays that appear straight entering it, and so accepts the illusion of the image in the mirror.

Example B

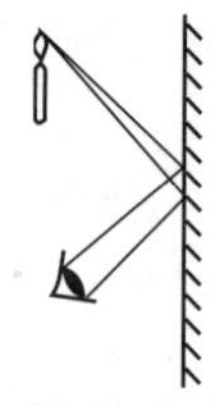

Actual light rays

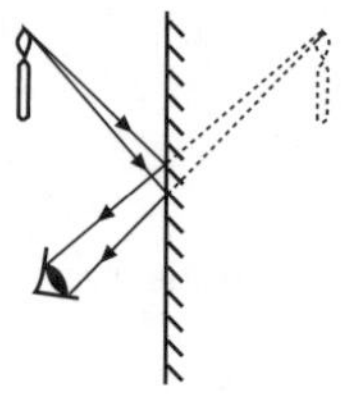

Image appears to be behind the mirror

Mirror, mirror on the wall...

Careful measurements on the diagram in Example B will show that for a plane mirror, the image is the same size as the object and is the same distance directly behind the mirror as the object is in front.

The image in a plane mirror is described as **virtual** because there is actually no light at the image position. Behind the mirror, at the image position, the image would not be able to be seen, photographed or projected onto a screen. Virtual rays from a virtual image are drawn dotted to distinguish them from actual rays of light. The image appears to be at the *intersection* of the two reflected rays.

A left hand seen in a plane mirror looks like a right hand. Such observations are referred to as **lateral inversion**.

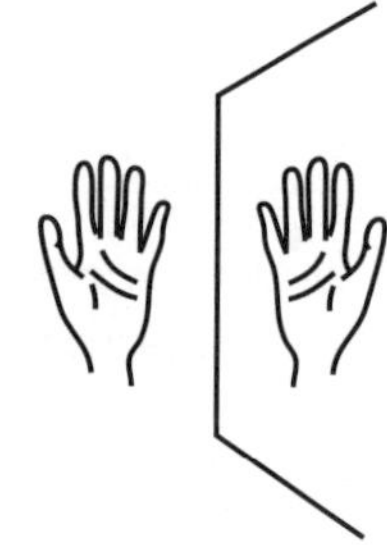

Left becomes right?

Images formed in two mirrors at 90° to each other

When two mirrors are placed at right angles, *three* images form from one object.

Example C

The diagram shows two mirrors and a candle.

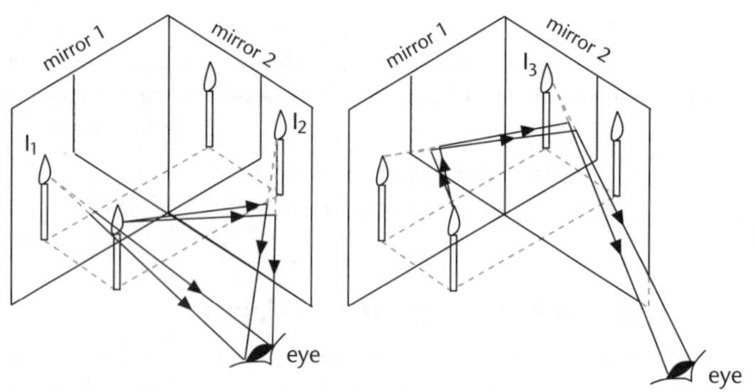

Two images of the candle (I_1 and I_2) are each formed by reflections from one mirror only.

The third image (I_3) is formed by reflections from both mirrors.

An extra image?

Parallax and binocular vision

The view from the window of a moving train or car shows trees, chimneys, hills and other objects appearing to move relative to each other. This *apparent* movement of two objects due to movement of the observer and to the objects being at different distances from the observer, is called **parallax**.

The scale of a measuring instrument must be placed near the object being measured in order to reduce errors caused by parallax. A high-quality moving coil meter uses a mirror behind the scale to help reduce **parallax error** in reading the scale.

The position of the eyes in the head of a person permit both eyes to be focused on the same object – this allows distances to be more easily estimated. This **binocular vision** is an important factor in judging distance and depth.

Unit 12.3 Activity 5A: Plane mirrors

1. A ray of light makes an angle of 25° with the normal to a plane mirror. The mirror is turned 6° so the angle of incidence becomes 31°. Through what angle is the reflected ray rotated?
2. A boy 1.5 m tall whose eyes are 1.4 m from the floor when he stands straight, can just see his complete image in a wall mirror.
   a. What is the smallest length, *l*, of the mirror that allows the boy to see his complete image? The mirror is on a wall 3.0 m from the boy.
   b. What is the height, *h*, of the mirror's bottom edge from the floor?

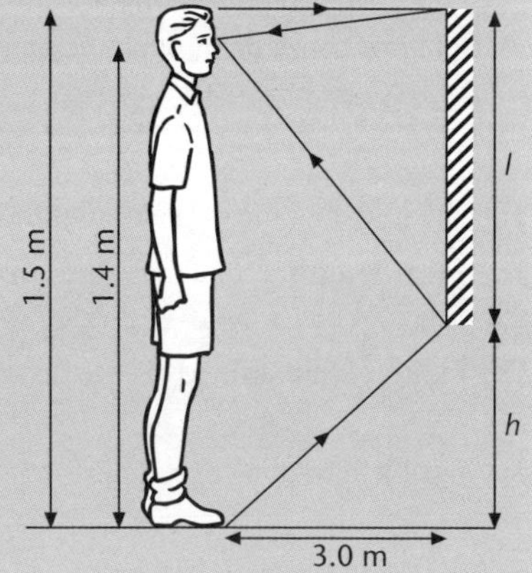

Curved mirrors

Curved mirrors can be made by taking part of the shell of a hollow glass sphere and silvering it. Silvering the glass on the *outside* gives a **converging** (or **concave**) mirror, while silvering on the *inside* gives a **diverging** (or **convex**) mirror. The **pole** (point P), **centre of curvature** (point C), **principal axis** (the line PC) and **radius of curvature**, *r*, for each type of mirror is shown in the following diagrams.

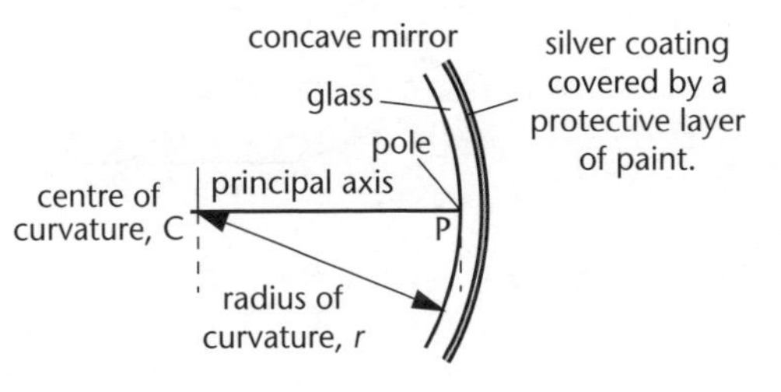

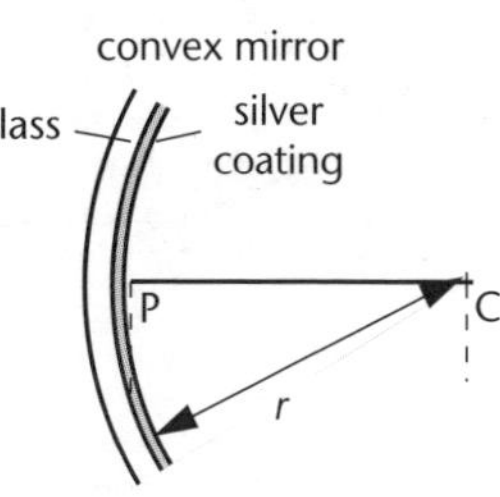

Curved mirrors

In practice, a curved mirror is only a *small* part of a sphere's surface. When this is the case, light rays *parallel* to the principal axis are reflected so that they pass through (or appear to come from) one point called the **principal focus**, *F*.

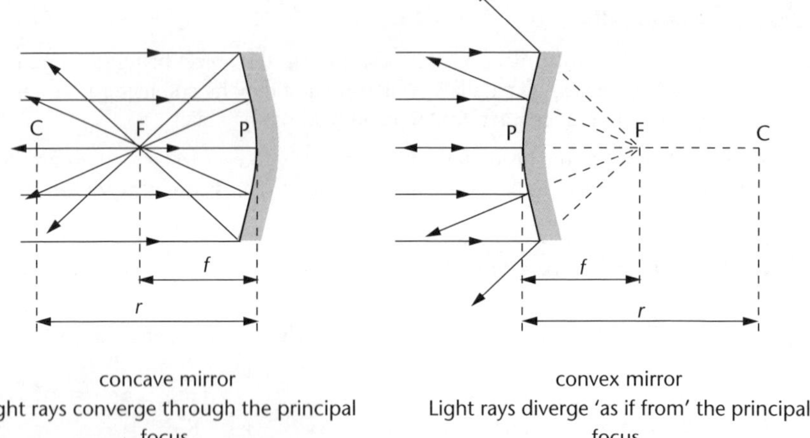

concave mirror
Light rays converge through the principal focus.

convex mirror
Light rays diverge 'as if from' the principal focus.

Different types of focus

The shading on a mirror symbol indicates the *non-reflecting* side of the mirror.

The **focal length**, *f*, of a curved mirror is the distance between the pole of the mirror and the principal focus.

The radius of curvature, *r*, is the distance between the pole of the mirror and the centre of curvature.

The focal length of a curved mirror is half the radius of curvature, ie:

$$f = \frac{r}{2}$$

Reflectors used in searchlight and headlamp reflectors have a **parabolic** shape (rather than spherical). Parabolic mirrors can bring a much wider beam which is parallel to the principal axis into sharper focus than a spherical mirror.

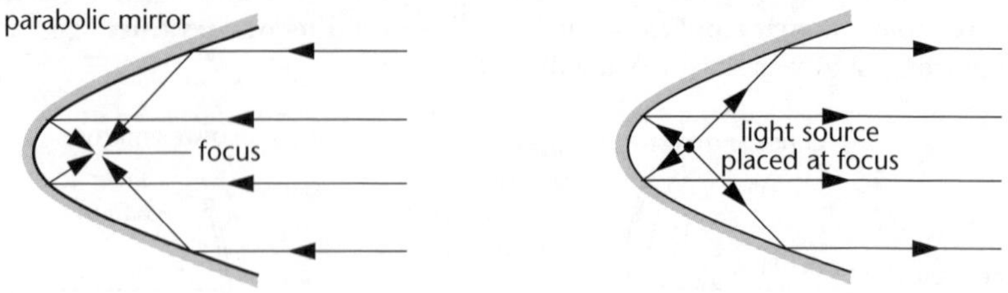

Parabolic mirrors

Images

Images can be virtual or **real**. A real image is formed from actual light rays and can be shown on a screen.

Example D

If a concave mirror is held near a lamp, an inverted image of the lamp can be formed on a sheet of white paper used as a screen. The image is *real* because it is formed with real light rays.

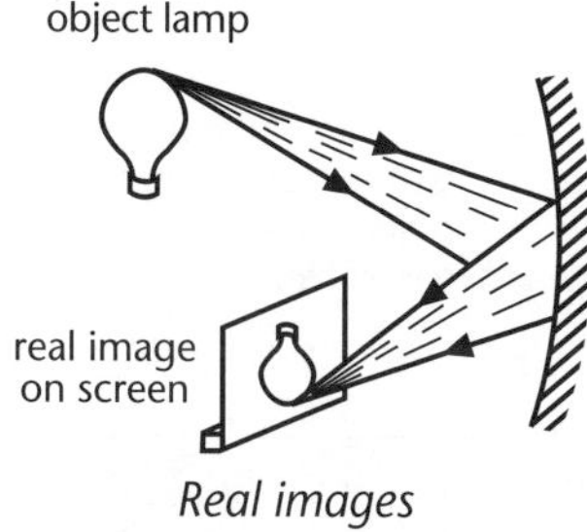

Real images

Concave mirrors can produce either real or virtual images, depending on the position of the object.

Example E

Real and virtual images

This can easily be detected with a soup spoon. Start with the concave side towards you at arm's length, and bring it closer to touch your eyebrow. Can you decide which is the real and which the virtual image, and the focal length of your spoon?

Ray diagrams

Ray diagrams are scale drawings which can be used to find details of the images formed by curved mirrors and lenses. The usual details required from a ray diagram are the *size*, *position* and **nature** of the image. The nature of the image refers to whether the image is *inverted* (or *upright*), *virtual* (or *real*) and whether it is *enlarged* (or *diminished*).

The following example illustrates how a ray diagram is drawn.

Example F

An object 2 cm high is placed 12 cm in front of a concave mirror of focal length 4 cm.

1. Draw the principal axis, mirror line, mirror symbol and principal focus. Draw the object as an arrow.

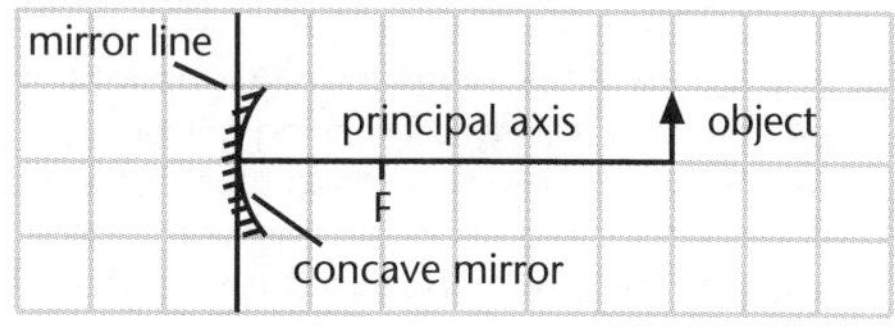

2. Draw an incident ray from the top of the object to the mirror, *parallel to the principal axis*. This ray is *reflected through the principal focus* F. Draw the reflected ray. Each ray should have an arrow drawn on it to indicate the direction of the ray's travel.

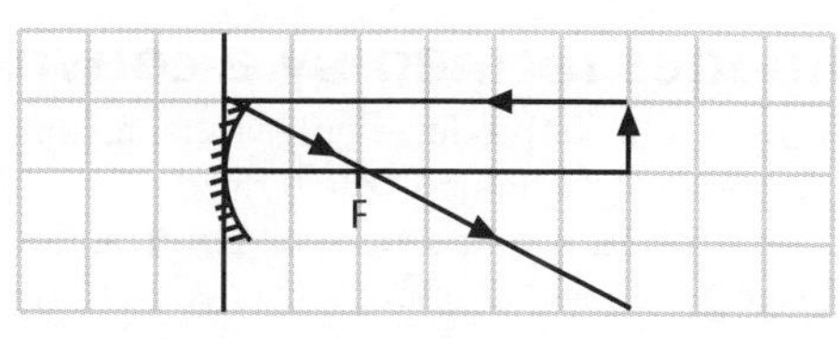

3. Draw a second incident ray from the top of the object *through the principal focus* F. This ray is *reflected from the mirror parallel to the principal axis.* Draw the reflected ray.

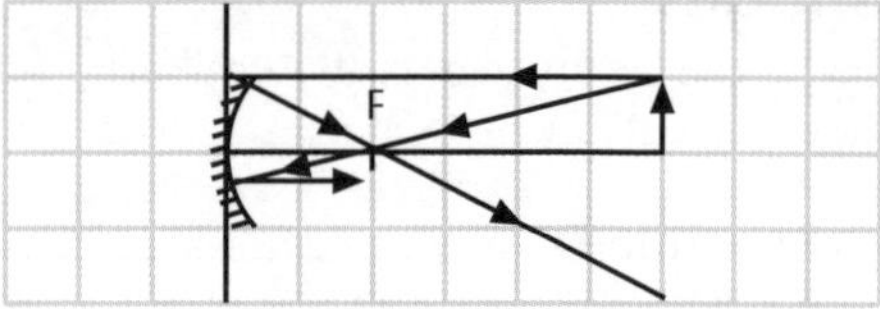

4. The top of the *image* is where the two reflected rays from the top of the object meet. Draw the image as a vertical arrow.

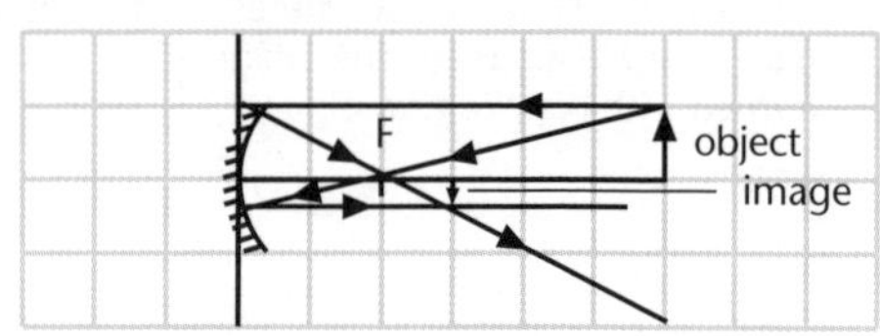

The nature of the image is described by looking at the image. (In Example F the image is inverted, diminished and real.)

It is sometimes helpful to sketch the ray diagram before drawing it. Exaggerate the vertical scale if necessary.

Reflections must occur at the mirror line to ensure a true representation of the image is formed.

Example G

An object 1.5 cm high is placed 2.5 cm in front of a concave mirror of focal length 5.0 cm.

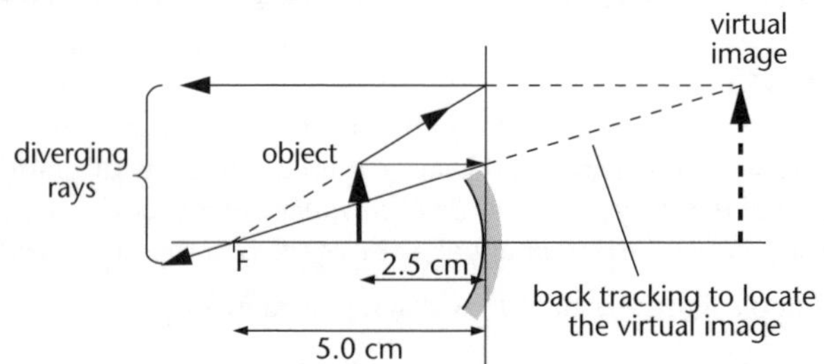

A curved mirror problem

The nature of the image formed is upright, magnified and virtual. A concave mirror can be used in this way as a shaving mirror or a cosmetic mirror to give a magnified view of a person's face. The soup spoon touching your eyebrow in Example E confirms this.

Virtual rays are drawn as dotted lines and do not have arrows drawn on them.

Images formed by a convex mirror

A convex mirror produces only virtual images which are always upright and smaller than the object. Virtual images are formed behind the mirror between the principal focus and the mirror. Turn your soup spoon over, and repeat the procedure in Example E. Notice the image does not change from inverted to upright.

Example H

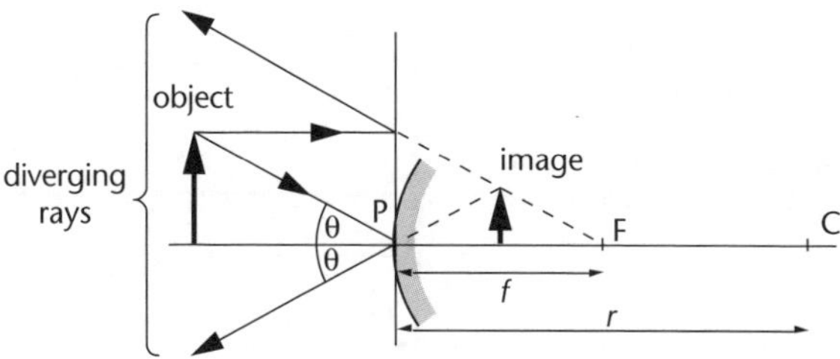

Convex mirrors

A ray incident on the mirror's pole, P, at an angle θ is reflected back at an equal angle θ.

Convex mirrors can be used as car rear-vision mirrors – because they always give an upright image and a wide field of view. The drawback with them is that they make following cars appear to be smaller and so they appear further away than they really are.

Formulas for use with spherical mirrors

It is often quicker (and usually more precise) to find details about an image using a formula rather than a ray diagram.

Both Descartes' and Newton's formulae are commonly used for both spherical mirrors and lenses.

Examples used in this chapter will use Descartes' formula, and in the next chapter only Newton's formula will be used. Students should try both with mirrors and lenses and decide which they prefer.

If D_o is the distance the *object* is in front of the mirror and D_i is the distance the image is in front of the mirror, then **Descartes' formula** gives:

$$\frac{1}{f} = \frac{1}{D_i} + \frac{1}{D_o}$$

If H_o is the height of the object, H_i is the height of the image and m is the magnification then:

$$m = \frac{H_i}{H_o} = \frac{D_i}{D_o}$$

The magnification measures the proportion by which the object's size has been enlarged or diminished.

Descartes' formula is named after René Descartes (1596–1650), a French philosopher and mathematician.

Example I

In Example F, an object 2 cm high is placed 12 cm in front of a concave mirror of focal length 4 cm. Using the following sketch and Descartes' formula, the position of the image and the image's height can be calculated.

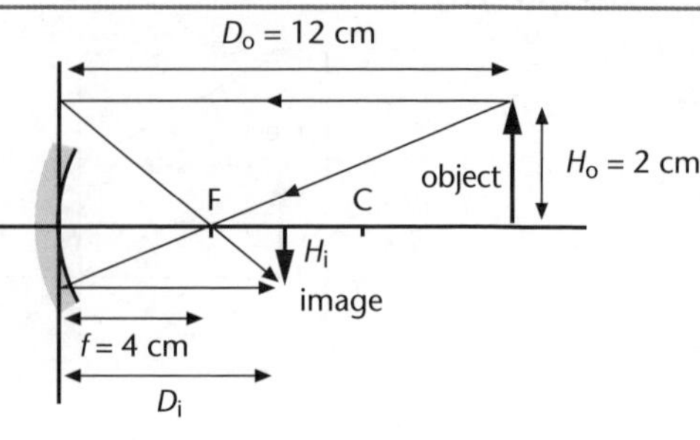

$$\frac{1}{f} = \frac{1}{D_i} + \frac{1}{D_o}$$

$$\frac{1}{4} = \frac{1}{12} + \frac{1}{D_i} \qquad \text{[substituting } D_o = 12\text{ cm, } f = 4\text{ cm]}$$

$$\frac{1}{D_i} = \frac{1}{4} - \frac{1}{12}$$

$$= \frac{2}{12}$$

$$D_i = \frac{12}{2}$$

$$= 6\text{ cm}$$

$$m = \frac{H_i}{H_o}$$

$$= \frac{D_i}{D_o}$$

$$= \frac{6}{12} \qquad \text{[substituting } D_i = 6\text{ cm and } D_o = 12\text{ cm]}$$

$= 0.5$ ie the image is half the size of the object.

ie $H_i = 0.5 \times 2$

$= 1$ cm.

In Descartes' formula *virtual distances* (distances behind a mirror) are given a negative sign. For curved mirrors, this means that:

- If D_i is behind the mirror, then it is negative.
- The image height, H_i, for a virtual image is negative.
- The focal length, f, for a convex (diverging) mirror is negative.

Example J

In Example G an object 1.5 cm high is placed 2.5 cm in front of a concave mirror of focal length 5.0 cm.

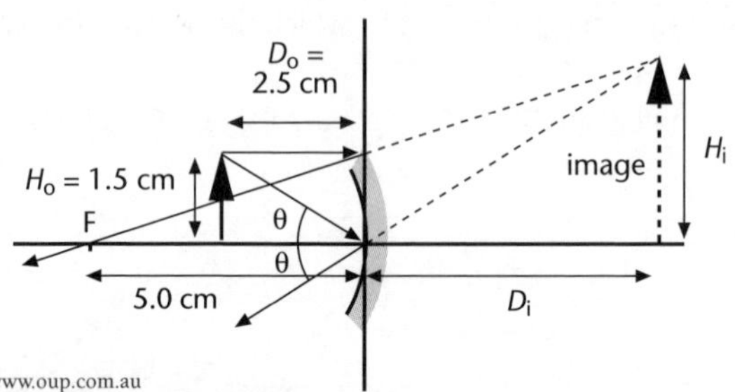

$$\frac{1}{f} = \frac{1}{D_i} + \frac{1}{D_o}$$

$$\frac{1}{5.0} = \frac{1}{2.5} + \frac{1}{D_i}$$ [substituting $D_o = 2.5$ cm and $f = 5.0$ cm]

$$\frac{1}{D_i} = \frac{-1}{5.0}$$ [rearranging and simplifying]

$D_i = -5.0$ cm, showing that the image is 5.0 cm behind the mirror.

$$m = \frac{D_i}{D_o}$$

$$m = \frac{-5.0}{2.5}$$ [substituting $D_o = 2.5$ cm, and $D_i = -5.0$ cm]

$= -2.0$ The image is twice the size of the object and is virtual (negative magnification factor).

ie $H_i = 2.0 \times 1.5$

$= 3.0$ cm.

Newton's formula is an alternative formula used to determine an image's position:

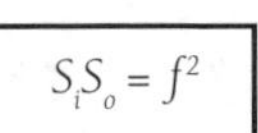

$$S_i S_o = f^2$$

S_i is the distance of the image to the principal focus, F; S_o is the distance of the object to the principal focus; and f is the focal length

The magnification, m, is:

$$m = \frac{H_i}{H_o} = \frac{f}{S_o} = \frac{S_i}{f}$$

Newton's formula has the advantage that all distances are positive. Care must be taken, however, in calculating S_i and/or S_o from the information given in a problem. A sketch ray diagram should be drawn as an aid to determining the solution.

Unit 12.3 Activity 5B: Reflection of light

1. An object is placed half-way between the focal point, F, and the pole, P, of a concave mirror. Copy the diagram and draw at least two rays to find the *position* of the image. State the *magnification* of the image.

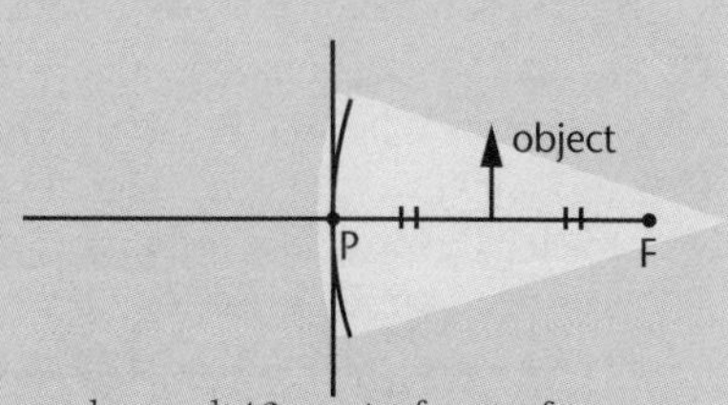

2. **a**. Find the position of the image of a candle flame located 40 cm in front of a concave spherical mirror of radius of curvature 64 cm.

b. Describe the type of image formed.

3. How far should an object be placed in front of a concave spherical mirror, with a radius of curvature 36 cm, to form a real image $\frac{1}{9}$ of its size?

4. A concave mirror forms an image of an object as shown below. The object height is 6 cm, the image height 2 cm. When the object is 12 cm from the mirror, the image is formed 4 cm in front of the mirror.

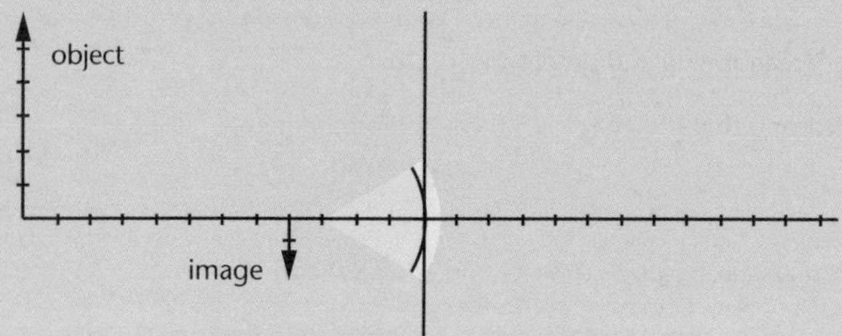

   a. Calculate the magnification of the image.
   b. State the nature of the image.
   c. Calculate the position of the focus of the mirror.
   d. Explain what would happen to the image if the object was moved 1 cm closer to the mirror.

5. An object 7.0 cm high is placed 15 cm from a convex spherical mirror of radius 45 cm.
   a. Find the position of the image. b. Find the height of the image.
6. How far must a man stand in front of a concave mirror which has a radius of curvature 120 cm, in order to see an erect image of his face four times its normal size?
7. An object 6.0 cm high is placed 30 cm in front of a convex mirror of focal length 40 cm. By using:
   a. a ray diagram b. a formula
   find the position, nature and size of the image formed.
8. The light from a distant star is collected by a concave mirror in a telescope. If the radius of curvature of the mirror is 150 cm, how far from the mirror is the image?
9. A concave shaving mirror has a radius of curvature of 30 cm. When a person's face is 10 cm from the mirror, what is the magnification of the image?
10. An object and its image in a mirror are the same height when the object is 36 cm from the mirror. What is the type of mirror is used and what is its focal length?
11. Large convex mirrors are often used in shops to improve security. Tracey is looking at herself in a convex security mirror of focal length 2.0 m. Tracey is 4.0 m away from the mirror. Tracey is 1.5 m tall.
    Draw a ray diagram in order to calculate:
    a. the position, and b. the size, of Tracey's image.
12. A 10 cm focal length concave mirror can produce a 4 cm high virtual image. The object is 2 cm high. Find the position of the object and the image.
13. It is possible to form an image with a magnification of 2 when the object is 12 cm away from the mirror. Find the focal length of the mirror if: **a**. the image is virtual and **b**. the image is real.

Unit 12.3 Waves

Topic 6: Refraction – review and extension

Topic 6 provides a review and extension of the refraction of light, and concepts that have already been covered in the preceding Topics. This Topic revises concepts to do with:

- Refraction.
- Refraction index.
- Total internal reflection.

Introduction

Refraction is the change in speed when a wave travels from one substance (called a medium) into another. The frequency, f, of the waves does not change when crossing the boundary. Applying the wave equation $v = f\lambda$ indicates a change in speed, v, will be proportional to a change in wavelength, λ, ie $v \propto \lambda$.

Example A

Imagine a column of well-trained soldiers marching along a concrete road into a muddy field. The rows of soldiers would be the wavefronts and the direction along the road the ray direction.

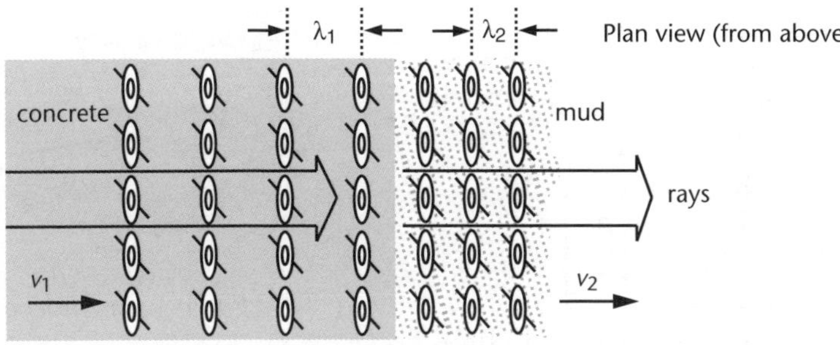

Left, right, left, right

The mud slows the soldiers down and the wavelength (λ_2) in the mud decreases from that on the concrete (λ_1).

Speeds v_2 and v_1 are in the same ratio as that of their wavelengths:

$$\frac{\lambda_1}{\lambda_2} = \frac{v_1}{v_2}$$

A more difficult concept to understand is that refraction often produces a change in direction.

Example B

Imagine the same soldiers (as in Example A) marching, but now the boundary between the concrete and the mud is at an angle to the soldiers.

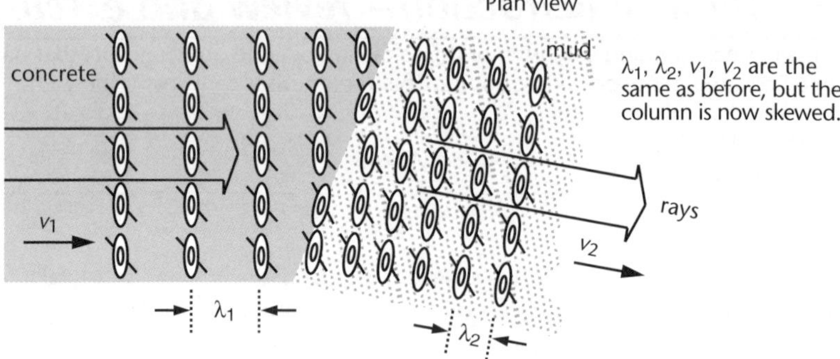

Changing direction with the mud

The effect is that the rows of soldiers skew around changing direction, as some soldiers reach the mud before others in the same row.

The change in direction as waves cross a medium is a complicated relationship that puzzled scientists for a long while, until it was found that the change in direction is related to the sines of the angles involved:

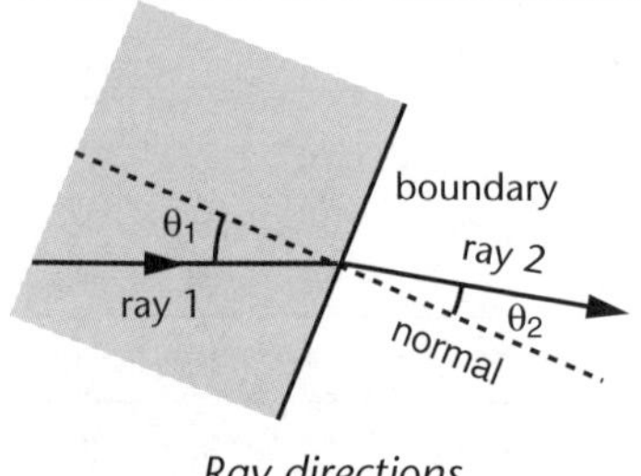

Ray directions change with refraction

$$\frac{\lambda_1}{\lambda_2} = \frac{v_1}{v_2} = \frac{\sin\theta_1}{\sin\theta_1}$$

where θ_1 is the angle of incidence (the angle in medium 1)

θ_2 is the angle of refraction (the angle in medium 2)

When light travels from one medium to another, refraction occurs. This might commonly be noticed with light travelling from air into water or glass. The change in direction is the visible feature, the change in speed definitely is not. Sound also shows refraction.

Example C

Hot air is less dense than cold air. Refraction occurs when light travels through air where different parts have different temperatures. Light reaching our eyes after travelling through a layer of air above a hot object appears to shimmer. This is because air currents cause the density of the air to vary, which, in turn, causes small and rapid changes in direction of the light. The effect is called **heat shimmer**.

Refraction

Refraction occurs when light travels from air into glass or water.

Example D

The figure to the right shows light passing at an angle from air to water.

The ray reaching the water's surface is the incident ray and the ray passing into the water is the refracted ray.

The normal is the line perpendicular to the surface at the point where the incident and refracted rays meet. The angle of incidence, θ_1, and angle of refraction, θ_2, are measured from the normal.

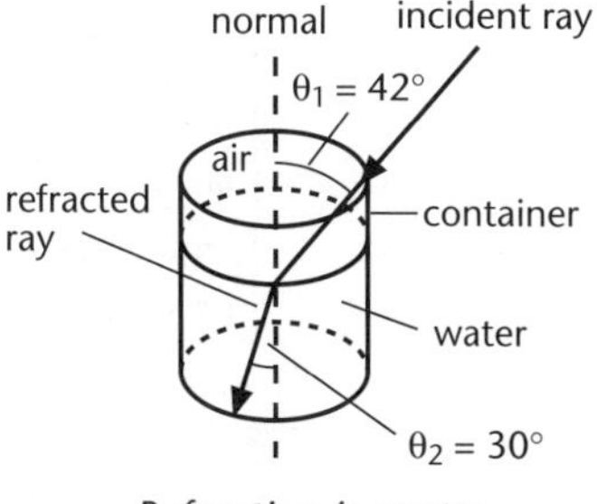

Refraction in water

The **laws of refraction** state:

- The incident ray, refracted ray and normal all lie in the same plane.
- The ratio of the sine of the incident angle ($\sin \theta_1$) to the sine of the refracted angle ($\sin \theta_2$) is a constant called the **refractive index** (symbol n);

$$n = \frac{\sin \theta_1}{\sin \theta_1}$$

Example E

In Example D the refractive index is equal to:

$$\begin{aligned} n &= \frac{\sin \theta_1}{\sin \theta_1} \\ &= \frac{\sin 42^\circ}{\sin 30^\circ} \quad \text{[substituting } \theta_1 = 42^\circ, \theta_2 = 30^\circ] \\ &= 1.33 \end{aligned}$$

The refractive index is also equal to the ratio of the speed of light through the first medium to the speed through the second; ie $n = \frac{v_1}{v_2}$, where v_1 is the speed of light in medium 1 and v_2 is the speed of light in medium 2. The refractive index, n, can be thought of as a 'slowing down' factor.

For the case of light travelling from air to water, a refractive index of 1.33 *means* that light travels 1.33 times slower in water than in air.

The refractive index mentioned above is the **relative refractive index**. For the case of light passing from glass to water, the relative refractive index is written ${}_{\text{glass}}n_{\text{water}}$. In general, for light that passes from medium 1 to medium 2, the relative refractive index, ${}_1n_2$ is:

$$ {}_1n_2 = \frac{\lambda_1}{\lambda_2} = \frac{v_1}{v_2} = \frac{\sin \theta_1}{\sin \theta_1}$$

The **absolute refractive index** of a medium, n_{medium}, is the refractive index for light passing *from a vacuum* into that medium. Light travels fastest in a vacuum but its speed changes very little when it passes through air. The absolute refractive index of a medium, then, can be taken as the refractive index for light passing from *air* into the medium. In Example D, light passes from air to water. The refractive index calculated is the absolute refractive index for water, ie $n_{\text{water}} = 1.33$.

Other substances have different absolute refractive indices.

Substance	n	Substance	n
diamond	2.42	paraffin oil	1.44
ruby	1.76	ethanol	1.36
flint glass	1.65	water	1.33
crown glass	1.52	ice	1.31
perspex	1.49	air	1.00

Table of refractive index values

Air has a refractive index of 1.00 because the speed of light in a vacuum and in air is almost the same.

The refractive index for glass is often approximated as 1.5.

For light that passes from medium 1 to medium 2, the relative refractive index ${}_1n_2$ is related to the absolute refractive indices, n_1 and n_2, by:

$${}_1n_2 = \frac{n_2}{n_1}.$$

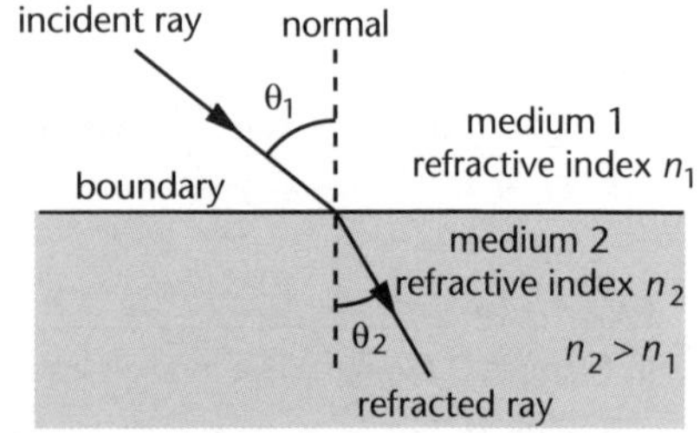

Refraction for the change medium 1 to medium 2

The complete summary is:

$${}_1n_2 = \frac{\lambda_1}{\lambda_2} = \frac{v_1}{v_2} = \frac{\sin \theta_1}{\sin \theta_1} = \frac{n_2}{n_1}$$

Combining the last two ratios (from the summary above) gives:

$$n_1 \sin \theta_1 = n_2 \sin \theta_2$$

θ_1 is the angle of incidence and θ_2 is the angle of refraction

The formula, $n_1 \sin \theta_1 = n_2 \sin \theta_2$, is known as **Snell's Law** and is another way of expressing the second law of refraction.

Example F

For a light ray passing from glass (n_{glass} = 1.52) to water (n_{water} = 1.33), the relative refractive index for the glass/water interface is:

$${}_1n_2 = \frac{n_2}{n_1}$$

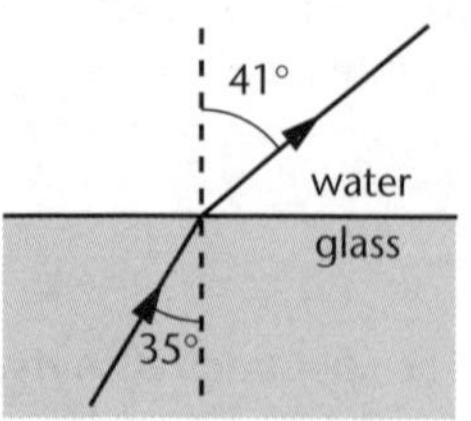

$$_{glass}n_{water} = \frac{n_{water}}{n_{glass}}$$
$$= \frac{1.33}{1.52}$$
$$= 0.875$$

The angle of refraction in water if the angle of incidence in glass is 35° is given by:

$$n_1 \sin\theta_1 = n_2 \sin\theta_2$$
$$1.52 \times \sin 35° = 1.33 \times \sin\theta_2$$
$$\sin\theta_2 = \frac{1.52 \times 0.57}{1.33}$$
$$= 0.65$$
$$\theta_2 = 41°$$

When a ray of light passes through a block of material with parallel faces, the ray emerging from the block is parallel to the incident ray but is displaced sideways by a small distance (*d*).

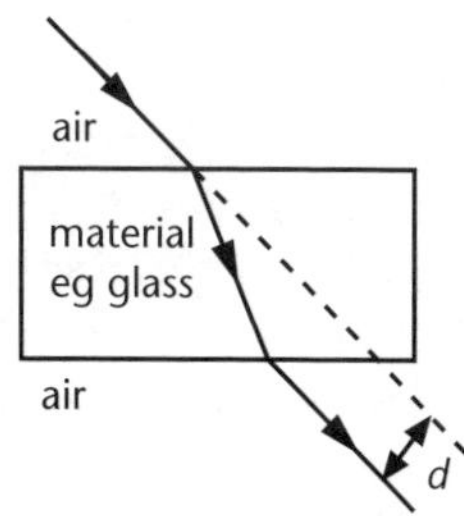

Sideways displacement

Unit 12.3 Activity 6A: Refraction

1. A beam of light strikes the surface of water at an angle of incidence of 60.0°. Some light is reflected and some is refracted. Find the directions of both the reflected and refracted rays. The index of refraction of water is $n_w = 1.33$.
2. The diagram shows a ray of light passing from glass (refractive index = 1.5) to air (refractive index = 1.0); the direction of the incident ray is shown.
   a. State the value of the angle of incidence.
   b. Calculate the angle of refraction (to the nearest degree).
3. A layer of oil ($n = 1.45$) floats on water ($n = 1.33$). A ray of light shines into the oil with an angle of incidence of 40.0°. What is the angle of refraction, θ, of the light in the water?

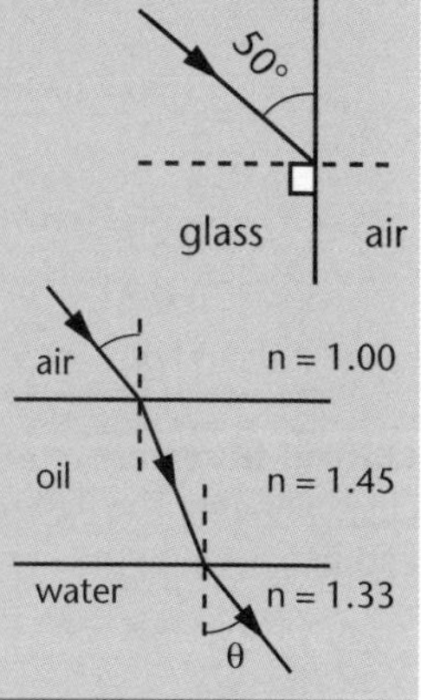

Total internal reflection

When light travels from one medium to another which is *optically less dense* (ie has a smaller refractive index than the first), the angle of incidence is *less* than the angle of refraction (ie the light bends *away* from the normal). In such situations, it is possible to have an angle of refraction of 90°.

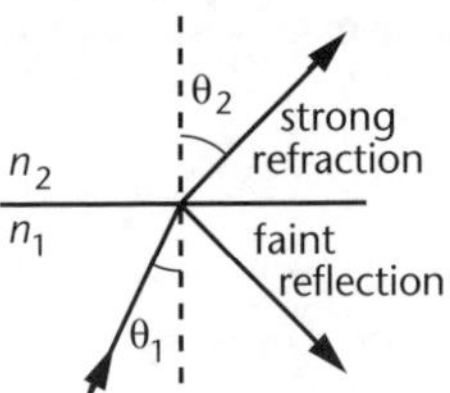

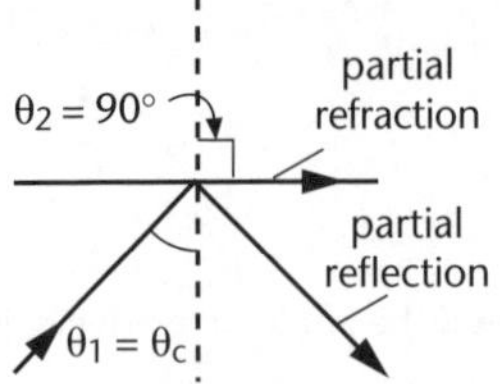

When the angle of refraction is 90° the angle of incidence is called the **critical angle** (symbol θ_c).

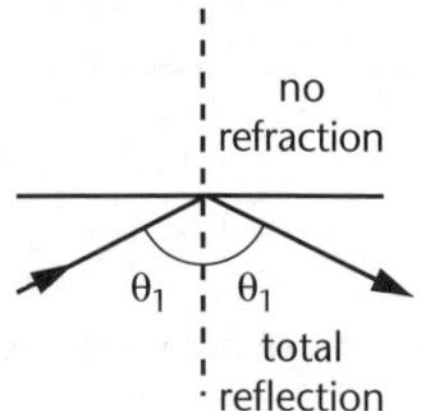

When the angle of incidence is greater than the critical angle all the light reaching the boundary is reflected and **total internal reflection** occurs.

Critical angle and total internal reflection

Total internal reflection can be seen when swimming underwater. The water's surface when seen at an angle from below appears 'mirror-like'.

The critical angle for any particular substance is easily calculated using Snell's Law.

Example G

The refractive index of glass is $n_{glass} = 1.52$, and for air, $n_{air} = 1.00$. For light passing from glass to air, the refractive index is given by Snell's Law,

$n_1\sin\theta_1 = n_2\sin\theta_2$. If θ_c is the critical angle for glass, then:

$n_1\sin\theta_c = n_2\sin\theta_2$

$$\sin\theta_c = \frac{n_2\sin\theta_2}{n_1}$$

$$= \frac{1.00 \times \sin 90°}{1.52}$$

$$= \frac{1.00}{1.52} \quad [\sin 90° = 1]$$

$$= 0.658$$

$$\theta_c = 41.1°$$

Optical fibres use total internal reflection to channel light. The light rays follow the curves and bends along the flexible cable.

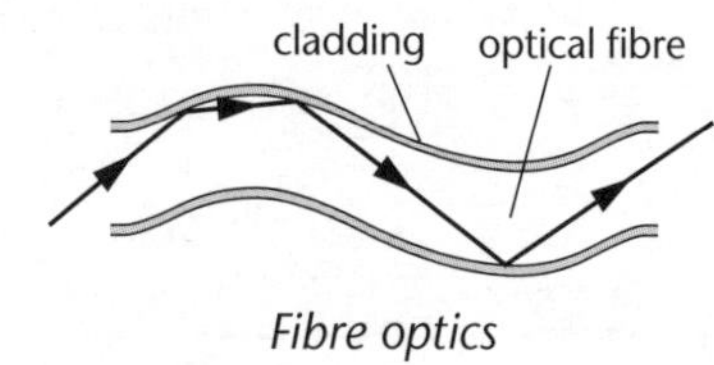

Fibre optics

Example H

An **endoscope** uses a small light and a narrow bundle of optical fibres to transmit a picture from the inside of a patient to the doctor operating the device.

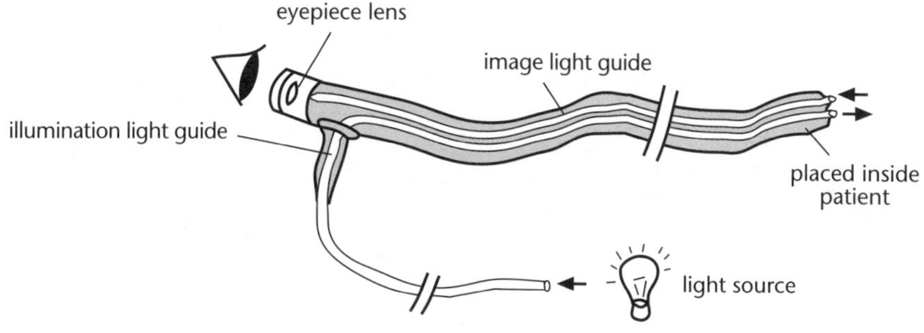

A medical examination

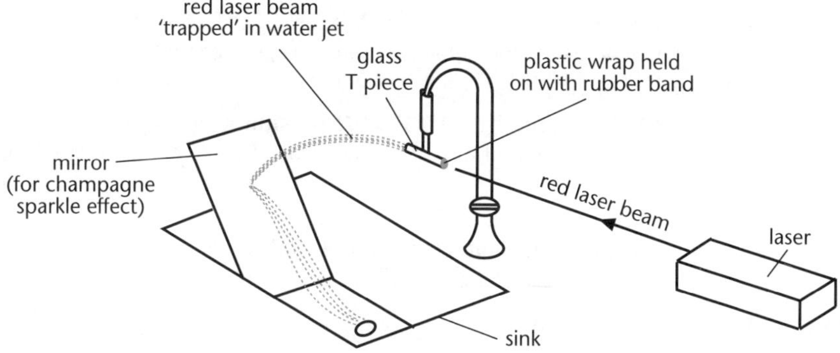

Total internal reflection can be dramatically demonstrated with a glass T piece.

Total internal reflection with a laser

Unit 12.3 Activity 6B: Critical angle

1. The refractive index for diamond is $n = 2.42$. What is the critical angle for light passing from diamond to air?
2. The critical angle for light passing from a clear substance into air is 40.5°. Calculate the index of refraction of the substance.

Dispersion – Refraction by prisms

The direction of a light beam can be changed by using a piece of glass with *non-parallel* sides, such as a **prism**. The beam of light that emerges from a prism diverges and spreads out more than the incident beam. If white light is used, the light that emerges can be seen as a **spectrum** of colours – violet, indigo, blue, green, yellow, orange and red. Red is at one end and violet at the other. The red deviates *least* and violet the *most*. You can remember this as 'violet is refracted most violently'.

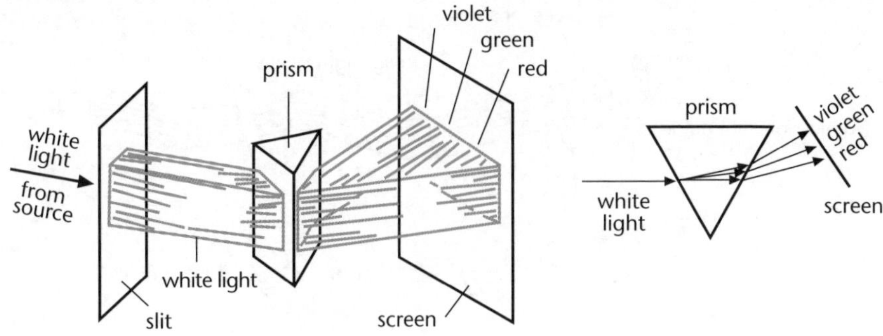

Dispersion of white light

A spectrum forms because a material such as glass has slightly different refractive indices for each of the different colours (eg the refractive index of a glass commonly used in lenses varies from 1.532 for violet light to 1.513 for red light).

The spreading out of light into a spectrum is called **dispersion**. Dispersion due to refraction occurs at the first boundary from air to glass *and* at the second boundary from glass to air.

Colour	Approximate wavelength (nm)
violet	400–430
blue	430–490
green	490–570
yellow	570–590
orange	590–640
red	640–700

Different wavelengths of colours

It must be remembered that visible light is only a small part of the whole spectrum of electromagnetic waves (em waves).

Total internal reflection in prisms

Prisms can be used as high-quality mirrors to reflect light in optical instruments such as cameras, binoculars and submarine periscopes.

Example I

A submarine periscope uses two 45° right-angled glass prisms. A submarine periscope is a combined periscope and telescope, but for simplicity in the diagram the lenses have been omitted.

Little dispersion occurs since the rays enter and leave each prism at right angles. The light reflects inside the prism by total internal reflection. The incident angle on the surface inside the prism is 45°, which is greater than the critical angle for total internal reflection.

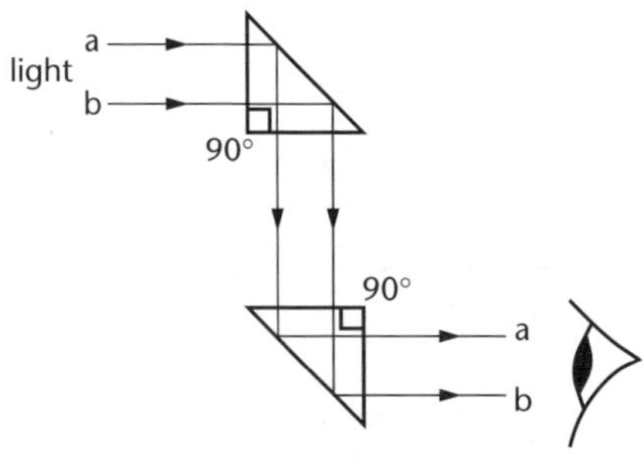

A periscope using prisms

Unit 12.3 Activity 6C: Prisms

1. Red and blue light rays pass into a glass prism from the air as shown in the diagram. Copy the diagram to show what happens to the refracted rays as they pass into the prism and out at other surface. (Show the red and blue rays clearly.)

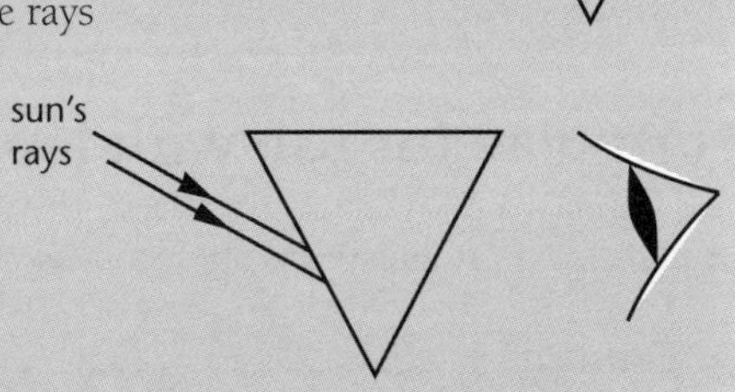

2. A narrow beam of parallel rays from the sun falls as shown in the diagram on to a glass prism. With the eye in the position shown, the light seen is yellow. If the eye is moved slightly up the page, the light seen will change colour. State what colour will be seen next.
3. A ray of light strikes the surface of a 60.0° glass prism at an angle of incidence of 40.0°. If the refractive index of the prism is 1.5, what is the angular deviation (the angle between the incident ray and the final ray that emerges from the prism) of the light ray caused by the prism?

Lenses

A simple **lens** is usually a piece of glass with spherical surfaces. The **principal axis** of a lens is the line joining the centres of curvature of its surfaces.

Converging lenses (convex lenses) cause parallel light rays passing through them to *converge* (come to a point). A ray of light, parallel to the principal axis, will converge to a point, F, on the axis *behind* the lens, called the **principal focus**.

Diverging lenses (concave lenses) cause parallel light rays passing through them to diverge (spread outward). In a diverging lens, the rays spread out *as if diverging from the principal* focus, F, in front of the lens.

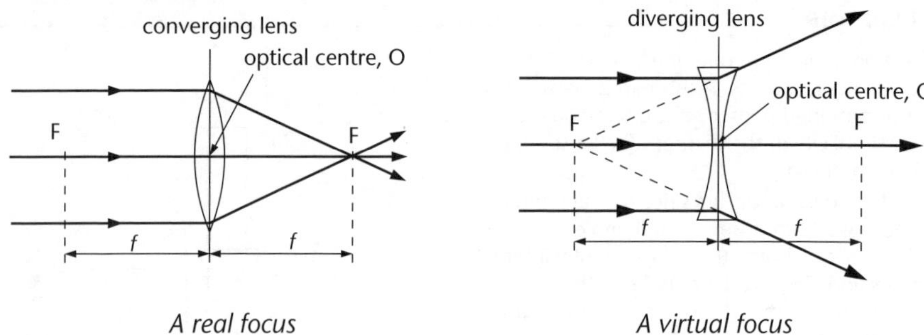

A real focus *A virtual focus*

Parallel rays *near* the principal axis which pass through a converging lens converge to a *point*.

Rays passing through the centre of a lens (called the **optical centre** or **pole**) are *not deviated* and go through the centre of a lens in a straight line.

Because light can pass through a lens from either side, each lens has *two* principal focuses, one on each side. They are at equal distances from the centre of the lens.

The **focal length**, f, of a lens is the distance between the optical centre and a principal focus.

Formulae for use with lenses

The formulae used with mirrors (Descartes' formula, Newton's formula and the magnification formulae) also apply to lenses.

Example J

An object 2.0 cm high is placed 6.0 cm in front of a converging (convex) lens of focal length 4.0 cm.

a. Draw a sketch ray diagram.
b. Determine the nature of the image.
c. Use Newton's formula to determine the position and height of the image.

Solution:

a. and **b.**

The ray diagram shows that the image is real, inverted and magnified.

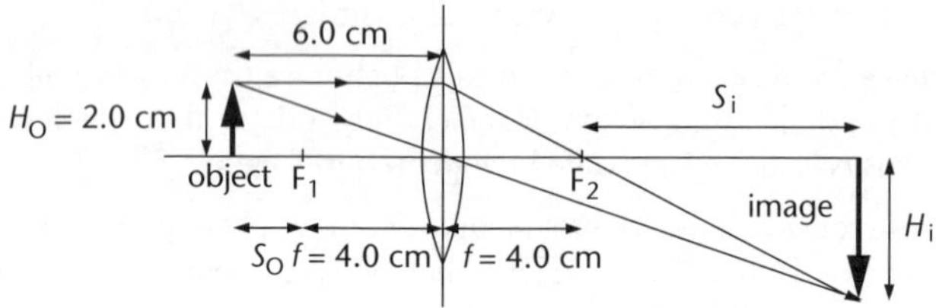

c. $S_i S_o = f^2$

$S_i \times 2.0 = 4.0^2$ [substituting $S_o = 6.0 - 4.0 = 2.0$ cm]

$S_i = \frac{16}{2.0}$

$= 8.0$ cm

ie image is 8.0 + 4.0 = 12.0 cm behind the lens.

$m = \frac{S_i}{f}$

$= \frac{8.0}{4.0}$

$= 2.0$ confirming that the image is twice the size of the object.

$H_i = 2.0 \times 2.0$

$= 4.0$ cm

Using Newton's Law for a convex lens, S_o is the distance from the object to the near focus, F_1, and S_i is the distance from the image to the far focus, F_2. This would not be apparent unless a sketch ray diagram is drawn.

If a concave lens is used, S_o is the distance from the object to the far focus, F_2, and S_i is the distance from the image to the near focus, F_1.

Unit 12.3 Activity 6D: Lenses

1. In the diagram, light falls onto two optical components represented by the open boxes.

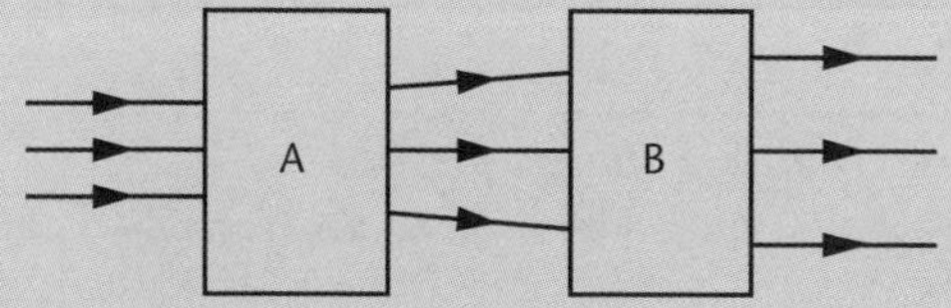

State the name of the components A and B.

2. The diagram represents a convex lens and f is the focal length.

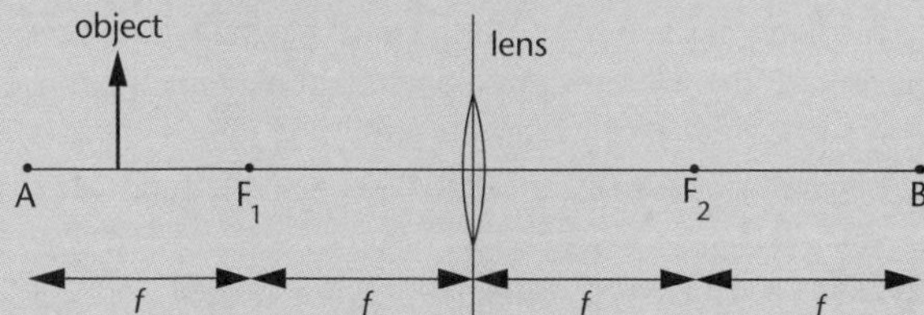

   a. Explain what happens to the image as the object moves between A and F_1.
   b. At what position should the object be placed so that the image is the same size as the object?

3. By ray diagram or otherwise, find the position and magnification of the image formed by a convex lens of focal length 100.0 cm, when the object is:
   a. 150.0 cm from the lens.
   b. 75.0 cm from the lens.

4. In what two positions relative to the object will a converging lens of focal length 7.5 cm form images of a lamp on a screen which is 40.0 cm from the object?
5. **a**. What type of lens can form a real image which is the $\frac{1}{3}$ height of an object placed 9.0 cm from the lens?
   b. Determine the focal length of the lens.
6. At night a girl uses a convex lens of focal length 20.0 cm to cast a focused image of a full moon on a piece of white cardboard. The moon has a diameter of 3.5×10^6 m and is 3.8×10^8 m from the Earth.
   a. How far is the cardboard from the lens?
   b. What is the diameter of the moon's image?
   c. What is the magnification of this image?
7. An object is placed 40.0 cm in front of a convex lens of focal length 8.0 cm. A plane mirror is 30.0 cm behind the lens – as shown in the diagram.

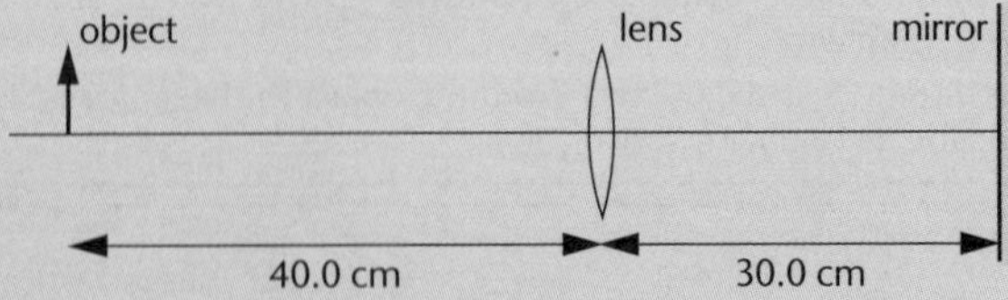

Find the position of all the images formed by this set up.

Unit 12.3 Activity 6E: Multiple choice questions

1. A concave mirror has a 50 cm radius of curvature. If an object of height 2.5 cm is placed 35 cm away from the mirror, where is the image located, what is its size and what is the nature of the image?
   A. 87.5 cm at the back of the mirror, upside down and magnified 2.5 times.
   B. 87.5 cm in front of the mirror, upright and magnified 2.5 times.
   C. 87.5 cm at the back of the mirror, upright and magnified 2.5 times.
   D. 87.5 cm in front of the mirror, upside down and magnified 2.5 times.
2. A convex mirror has a 30 cm radius of curvature. If the object is placed 18 cm away from the mirror, where is the image located, what is its size and what is the nature of the image?
   A. 8.18 cm in front of the mirror, upright and reduced by a factor of 0.45.
   B. 8.18 cm at the back of the mirror, upright and reduced by a factor of 0.45.
   C. 8.18 cm at the back of the mirror, upside down and reduced by a factor of 0.45.
   D. 8.18 cm in front of the mirror, upside down and reduced by a factor of 0.45.

3. An image formed by a concave mirror is virtual, magnified, upright and on the other side of the mirror. If d_0 is the object distance, f is the focal length and R is the radius of curvature. This implies that the object distance d_0 is:

A. $d_0 = f$
B. $d_0 > f$
C. $d_0 = R$
D. $d_0 < f$

4. An image formed by a concave mirror is upside down, unmagnified and is located at the same distance as the object distance. If d_0 is the object distance, f is the focal length and R is the radius of curvature. This implies that the object distance d_0 is:

A. $d_0 = f$
B. $d_0 > f$
C. $d_0 = R$
D. $d_0 < f$

Unit 12.3 Waves

Enrichment Topic: Vision and the camera

This Enrichment Topic deals with vision and the camera, and the application of concepts which have already been examined in this Unit. It covers:

- The eye.
- Short-sightedness.
- Long-sightedness.
- The camera.
- The relationship between physics and technology.

The **eye** and the camera use convex lenses to focus light rays.

The eye

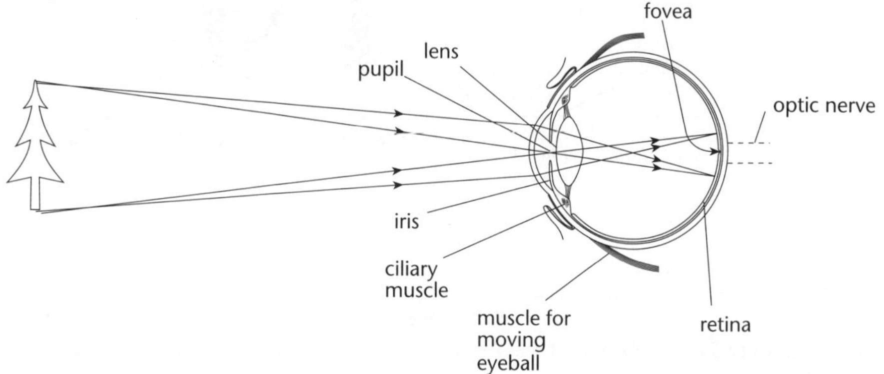

The human eye.

The **cornea** is the transparent outer cover of the eye. It protects the eye. The cornea and **aqueous humour** refract light slightly.

The **iris** controls the amount of light that enters the pupil. The pupil is the hole at the centre of the iris.

The convex lens focuses light onto the **retina**. The **ciliary muscles** contract or relax to change the shape and focal length of the lens. When the muscles relax the lens is short and fat.

The retina consists of cells called **rods** and **cones** that detect light and change it to electrical signals. The image on the retina is upside down. The **optic nerve** conducts the electrical signals to the brain. The **blind spot** is the start of the optic nerve and is not sensitive to light.

The electrical signals are conducted to the back of the brain. The brain uses stored information from past experiences to decode the signal. Sometimes the brain misinterprets the information. This causes **optical illusions**.

See also 'cross-section of the human eye' in Topic 2: Sensitivity and co-ordination in animals in Unit 11.5 of *Biology Grade 11* in the *Save Buk* series.

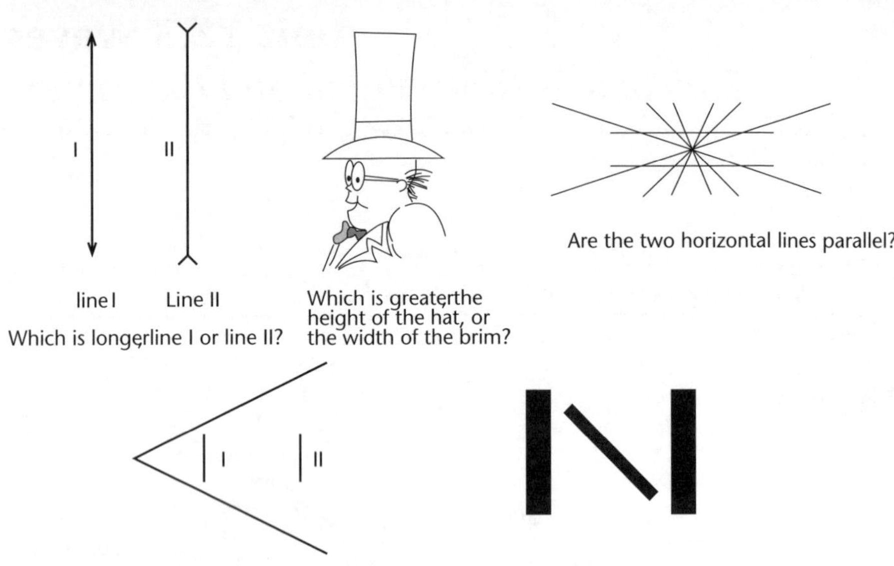

Optical illusions.

A person with normal sight can focus clearly on very distant objects (ie the far point of their vision is at infinity) as well as on very close objects (ie the near point is about 25 cm from the eye). In each case, light rays are focused on the retina to produce a clear image. Humans have two eyes (**binocular vision**) to enable distances to be judged accurately.

Defects of vision

Long-sightedness

A person who is long-sighted can see distant objects clearly but close objects appear blurry. Long-sightedness occurs because light rays that enter the eye are focused 'behind' the retina. The reason for this is that either the eyeball is too short or that the ciliary muscles are unable to relax enough to reduce the focal length of the lens sufficiently. Long-sightedness can be corrected by wearing glasses with convex (converging) lenses.

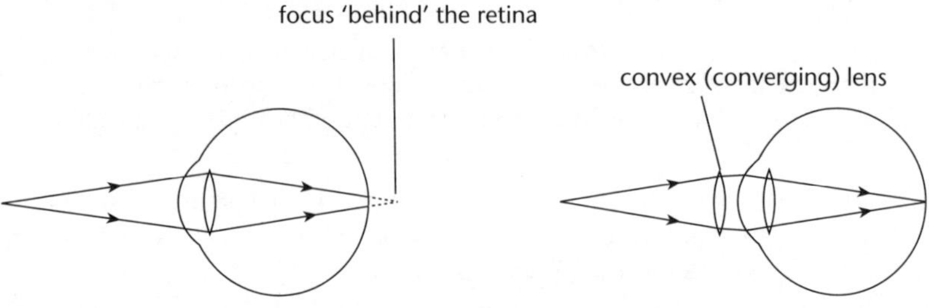

Correction of long-sightedness using a convex lens.

Short-sightedness

A person who is short-sighted can see close objects clearly but distant objects appear blurry. This happens because the light rays that enter the eye are focused in front of the retina. The reason for this is that either the eyeball is too long or the ciliary muscles are unable to contract enough to sufficiently decrease the focal length of the lens. Short-sightedness can be corrected by wearing glasses with concave (diverging) lenses.

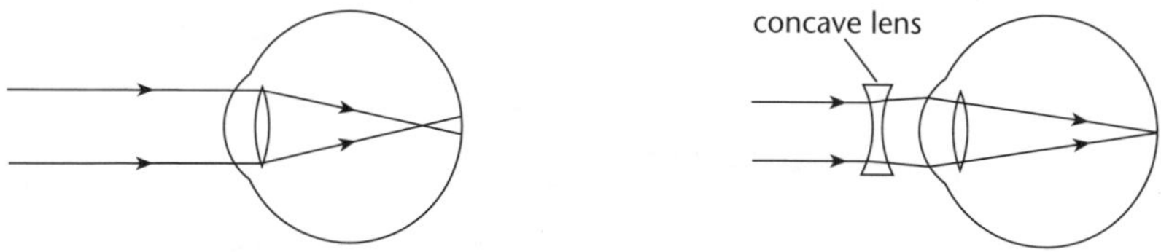

Correction of short-sightedness using a concave lens.

The camera

The simplest camera is the pinhole camera.

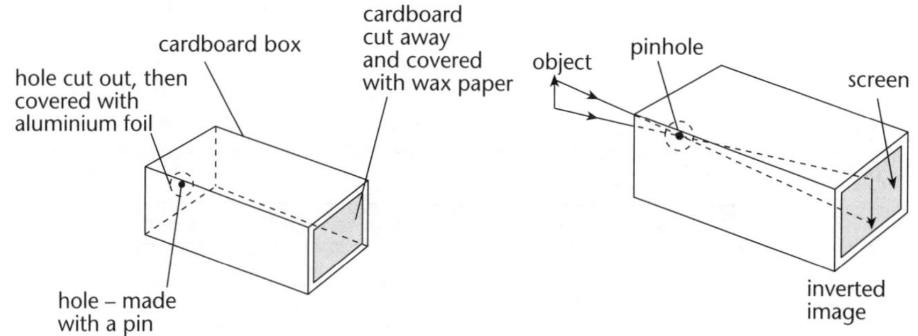

A simple pinhole camera.

The image of a pinhole camera is always upside down. If the hole is small the image will be sharp but dim. A larger hole produces a bright but blurry image. If a line is drawn through the pinhole perpendicular to both object and image, it can be shown (from similar triangles) that the magnification, M, of the object is:

$$M = \frac{\text{Height of image}}{\text{Height of object}} = \frac{\text{Distance of image from pinhole}}{\text{Distance of object from pinhole}}$$

Many people use compact cameras for photography. Keen photographers and professionals tend to use SLR (single lens reflex) 'stills' cameras – they provide more control over the quality of the image produced for a photograph.

Example A

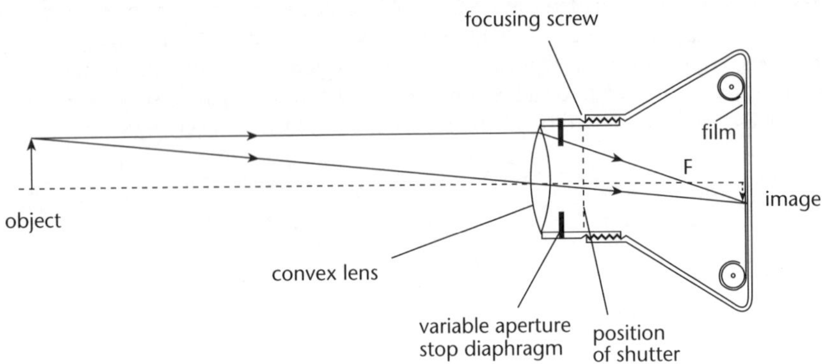

An SLR camera.

Light rays (reflected) from an object enter the camera through the **aperture**.

The **diaphragm** is used to control the size of the aperture. The size of the aperture is measured by the **f number**. A setting of f 11 means that the aperture is of the focal length and is used to take photos in bright light. To take a photograph when the light is poor, a larger aperture, such as f 4, is needed. (The f number also affects the clarity of a photo. As the f number increases, the amount of depth of the view in sharp focus increases as well.)

The **shutter speed** determines the time the aperture remains open when a photograph is taken. For fast-moving objects, a high shutter speed, such as $\frac{1}{500}$ s, is needed.

The range of shutter speeds and apertures for a basic SLR camera are shown in the table below.

f numbers (aperture settings)	Shutter trimes (seconds)
f/22	1/500
f/16	1/250
f/11	1/125
f/8	1/60
f/5.6	1/30
f/4	1/15
f/2.8	1/8
f/2	1/2

Apertures and shutter speeds.

The lens focuses the light rays onto the film at the back of the camera. The image is upside down. On the film a chemical **reaction** changes light energy into a negative image. The ASA rating indicates the sensitivity of the film. Typical films used for outdoor photography have ASA ratings of 100, 200 or 400.

There are many similarities between the eye and the camera. The list in the table below shows the corresponding parts.

The eye	The camera
Eyelid	Shutter
Pupil	Aperture
Iris	Diaphragm
Convex lens	Convex lens
Retina	Film

Corresponding parts of the eye and the camera.

Unit 12.3 Enrichment Topic Activity A: The eye and cameras

1. What is the part of the human eye that controls the amount of light that enters the eye?

2. James has made a pinhole camera and uses it to make a photograph of a 25 m high tree. The pinhole camera is 16 cm long. How far away from the tree is the pinhole camera if the image of the tree is 6 cm high?

3. An SLR camera was set up to take photos on a dull day in the early evening. What adjustments would have to be made to use the camera at midday on a sunny day?

- **A**. The shutter time should be increased.
- **B**. The camera should be moved closer to the subject.
- **C**. The size of the aperture should be reduced.
- **D**. A more sensitive film should be used.

4. Explain the function of each of the following adjustable controls on a camera:

- **a**. The shutter.
- **b**. The aperture.
- **c**. The focusing screw.

5. Write a paragraph explaining the similarities and differences between the human eye and a 'stills' camera.

6. Which of the following do the human eye and camera have in common?

- **A**. A screen where the image is formed.
- **B**. A blind spot.
- **C**. A lens of variable focal length.
- **D**. A fixed aperture.

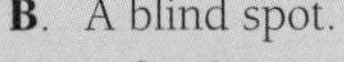

7. The diagram shows a cross-section of the human eye.

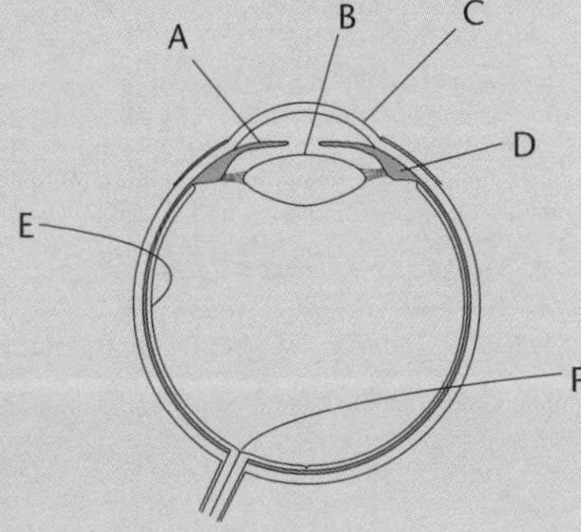

- **a**. Which letter refers to:
  - **i**. The retina.
  - **ii**. The cornea.
  - **iii**. The blind spot.
- **b**. A person enters a dark corridor after leaving a brightly lit room. Explain:
  - **i**. The effect their movement into the dark corridor will have on the pupil of their eye.
  - **ii**. How the amount of light entering their eye will change.

8. The diagram shows how a human eye views distant objects:

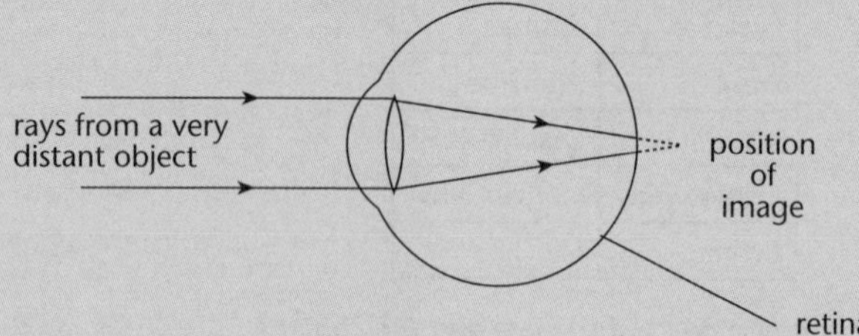

State whether the eye is long-sighted or short-sighted and how this can be corrected.

9. An image of a lamp is formed on the screen of the pinhole camera shown.

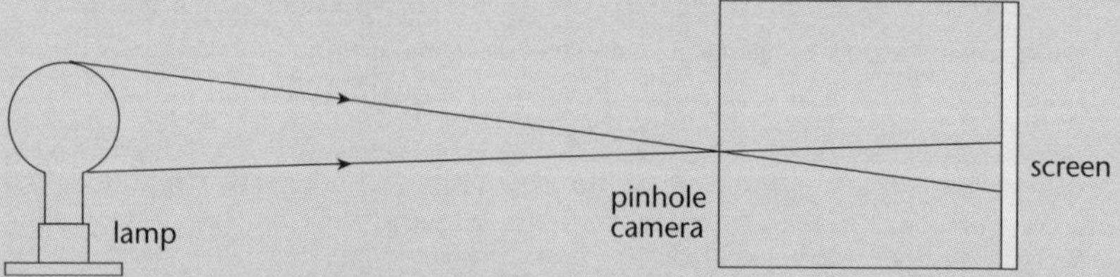

Which of the following changes, made with no other changes, would make the image larger?

A. Moving the pinhole camera nearer to the lamp.

B. Moving the pinhole camera away from the lamp.

C. Increasing the diameter of the pinhole.

D. Placing a concave lens over the pinhole.

E. Making the camera longer.

10. Kim is investigating the images formed when light passes through a small pinhole onto a screen.

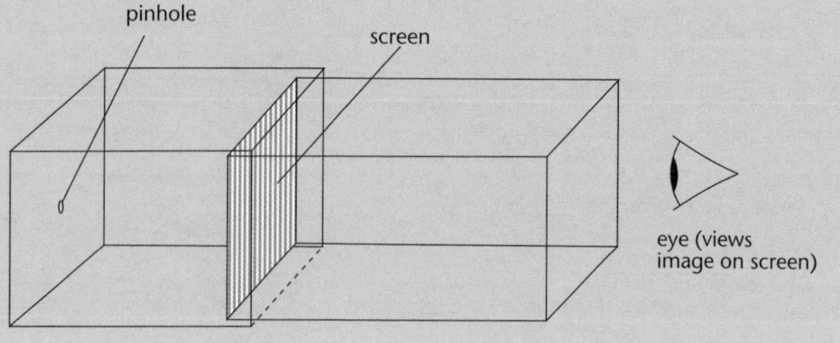

a. Copy the diagram below and draw two rays from the object to the screen, and draw the image accurately on the screen.

object

b. Is the image you have drawn:

- **i**. Real or virtual?
- **ii**. Upright or inverted?
- **iii**. Magnified or reduced?

c. Kim makes some changes to her equipment.

- **i**. The screen is moved closer to the pinhole. What happens to the image as a result of this?
- **ii**. The camera is moved closer to the object. What happens to the image as a result of this?
- **iii**. The pinhole is made larger. How does the image change?
- **iv**. Three pinholes are made instead of one. Draw the appearance of the screen.

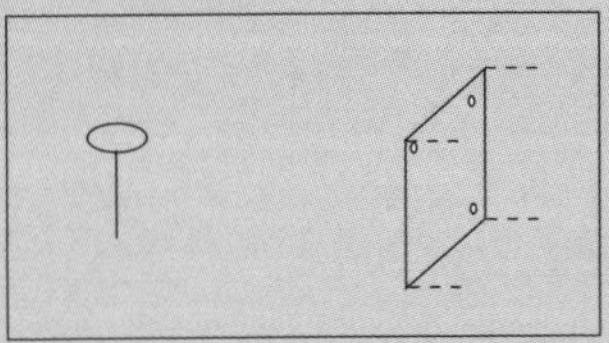

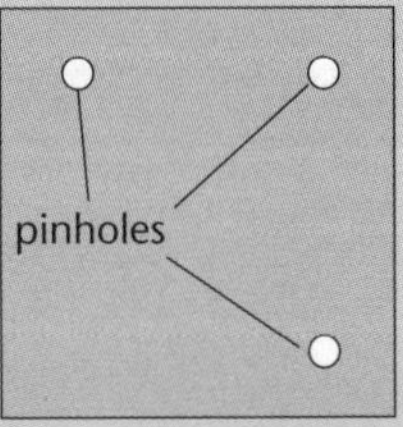

How the pinholes are arranged

11. The human eye can be compared with a camera. Copy and complete the table below which compares the functions of the parts of a camera with the parts of a human eye.

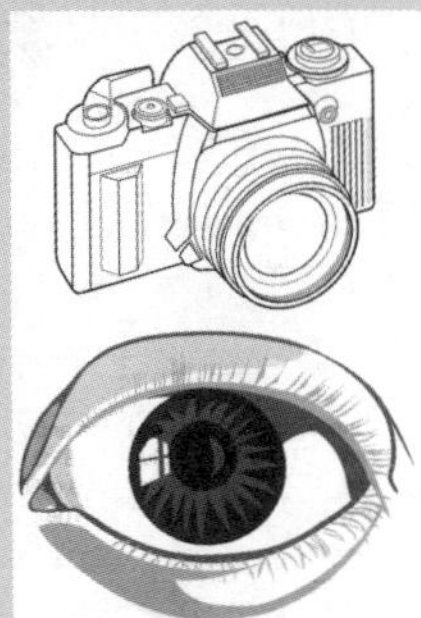

Parts of camera	Parts of eye	Function
i	Eyelid	To let in or shut out light
Lens	ii	iii
Diaphragm	iv	To control the amount of light which enters
Film	v	vi

12. Diagram A below shows how light rays which enter the eye of a person who is short-sighted focus in a spot in front of the retina. Diagram B shows how, for a person who is long-sighted, light rays which enter the eye focus in a spot behind the retina.

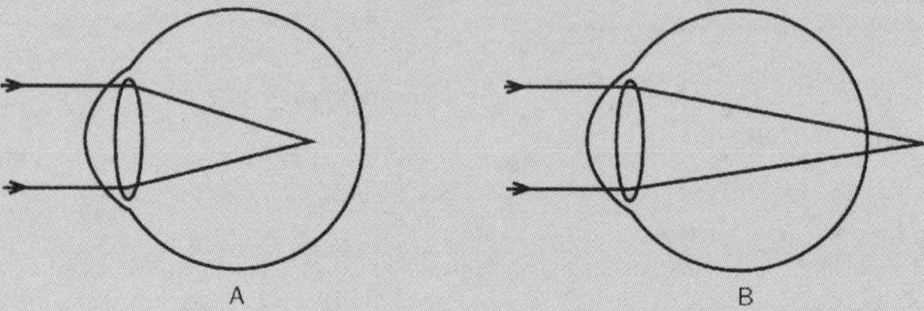

a. What type of lenses can correct the vision of the person who is short-sighted?
b. What type of lenses can correct the vision of the person who is long-sighted?

Unit 12.3 Enrichment Topic Activity B: Multiple choice questions

1. Which part of the eye focuses the light rays on to the retina?
- **A**. cornea
- **B**. iris
- **C**. aqueous humor
- **D**. lens.

2. The amount of light entering the pupil is controlled by the:
- **A**. cornea
- **B**. iris
- **C**. ciliary muscles
- **D**. cones and rods.

3. The image of a distant large object formed at the retina is:
- **A**. upright and reduced in size
- **B**. upright and magnified in size
- **C**. upside down and unmagnified in size
- **D**. upside down and reduced in size

4. Too much light reaching the blind spot may cause which of the following to the eye?
- **A**. a distant object can be easily seen
- **B**. the size of the image formed at the retina is magnified
- **C**. the eye suffers damage leading to complete blindness
- **D**. the size of the image formed at the retina is reduced.

5. Which of the following statements best explains the difference between long-sightedness and short-sightedness?
- **A**. A long-sighted person can see closer objects but distant objects appear blurry. A short-sighted person can see distant objects and closer objects appear blurry.
- **B**. A long-sighted person can see distant objects but closer objects appear blurry. A short-sighted person can see closer objects and distant objects appear blurry.

C. A long-sighted person can see closer objects but distant objects appear blurry. A short-sighted person can see closer objects and distant objects appear clearly.

D. A long-sighted person can see distant objects and closer objects appear clearly. A short-sighted person can see closer objects and distant objects appear clearly.

6. Nethlyn wants to buy a Kodak film of 36 exposures at Morobe City Pharmacy. Her camera is an SLR with a shutter speed of 1/250 seconds at f8. What film rating should she buy?

A. 50 ASA Kodak film

B. 100 ASA Kodak film

C. 200 ASA Kodak film

D. 400 ASA Kodak film

7. Beulah wants to buy a Fuji film at Morobe City Pharmacy. Her camera is an SLR with a shutter speed of 1/250 seconds at f11. What film rating should she buy?

A. 50 ASA Fuji film

B. 100 ASA Fuji film

C. 200 ASA Fugi film

D. 400 ASA Fuji film

Unit 12.4 Electromagnetism

Topic 1: Magnetic fields

This Topic deals with magnetic fields. It covers:

- Magnetic fields.
- Field lines.
- Magnetic fields around a bar magnet, the Earth, a current-carrying wire, and a solenoid.
- Applications of electromagnetism – the electromagnet, the electric bell, the relay, the reed switch, the loudspeaker.

Magnetic fields and field lines

A **magnetic field** is a region in space where magnetic forces can be detected. It can be depicted using **magnetic field lines**. The direction of the field is shown by an arrow and is the direction a small **North pole** would move if placed in the field. The density of the field lines shows the strength of the magnetic field.

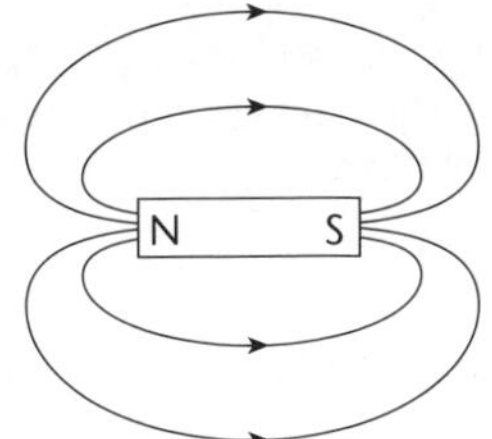

Magnetic field around a bar magnet.

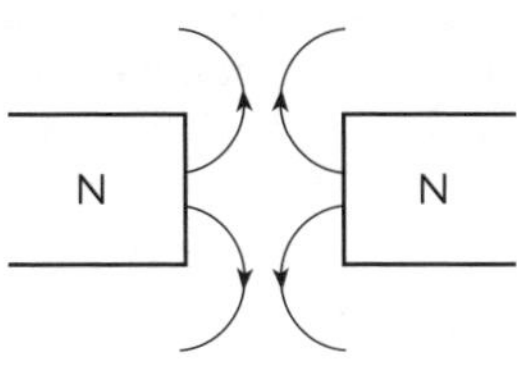

Two like poles.

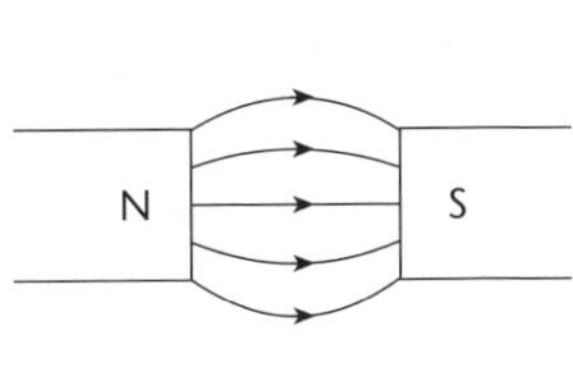

Two unlike poles.

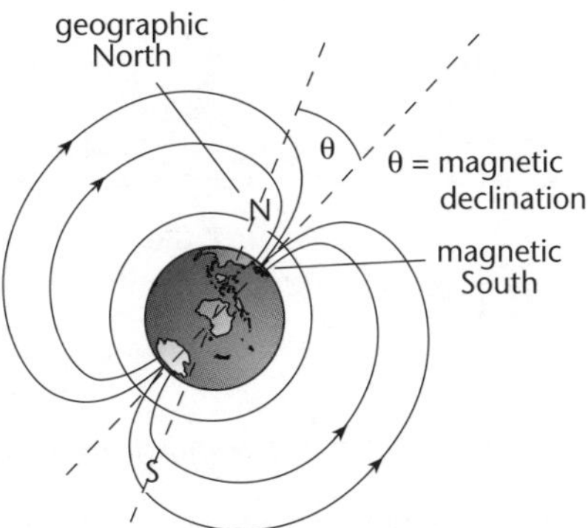

The magnetic field around the Earth.

When two magnetic poles are placed near each other, the resultant magnetic field exerts a force on the poles. The direction of this force may be determined by the rule:

'Like poles repel, unlike poles attract'.

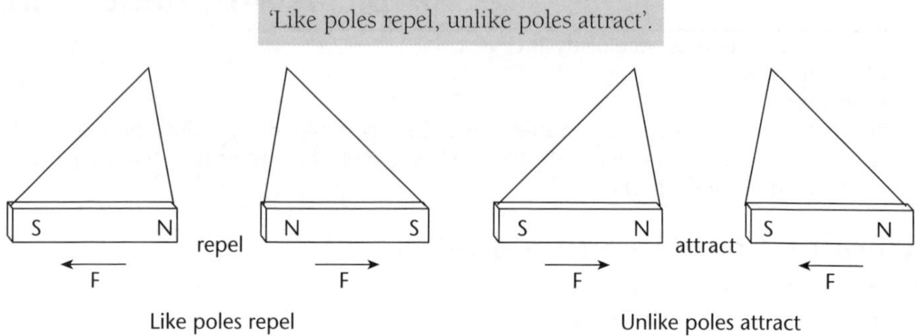

Magnetic repulsion and attraction.

A magnetic field is invisible but the field can be observed by its effects upon sprinkled iron filings or by using a **compass**. A compass consists of a small magnetic needle mounted on a pivot so the needle is able to rotate freely in a horizontal plane.

Magnetic field around a bar magnet

The diagram below shows the magnetic field around a bar **magnet**.

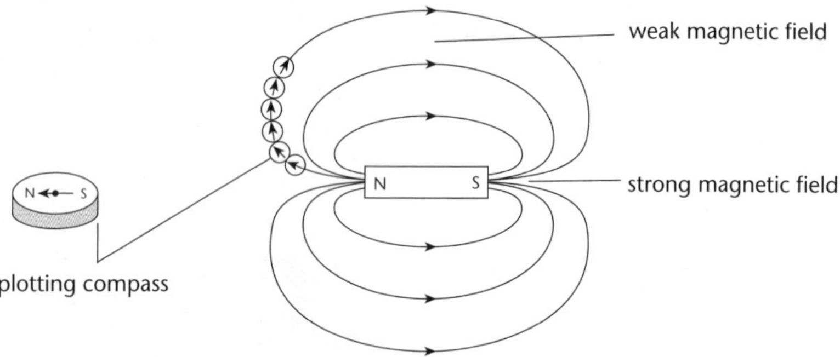

The magnetic field around a bar magnet.

Atoms act like minute magnets. In an unmagnetised bar these minute magnets point in random directions.

Arrows represent the North poles of atomic magnets

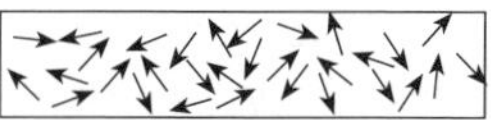

Unmagnetised bar – atomic magnets point in random directions

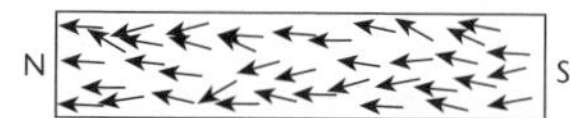

Partly magnetised bar – all atomic magnets point the same way – weak magnetic poles have formed at the ends

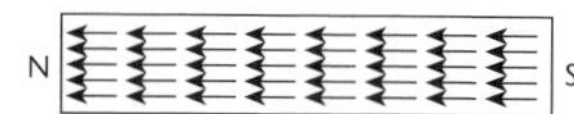

Completely magnetised bar – atomic magnets are in perfect alignment – strong magnetic poles have formed at the ends

Atomic micromagnets align to form a magnetic field.

In a magnet, the little atomic magnets are aligned and their magnetic fields combine to form a strong magnetic field around the bar magnet. If the magnet is broken in half, the halves retain their magnetism.

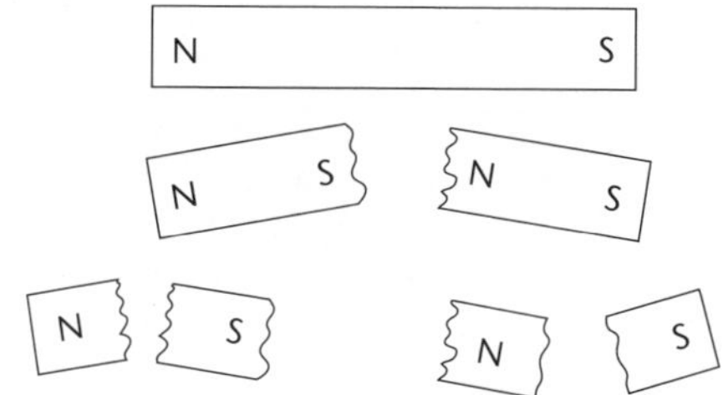

Breaking a bar magnet in half then in half again.

Magnets need to be handled and stored carefully to protect their magnetism. Banging or heating a magnet can destroy its magnetism. Bar magnets should be stored in pairs – North pole to **South pole** – with **sleepers** placed across their ends (this helps preserve their magnetism by aligning the magnetic fields around the two magnets).

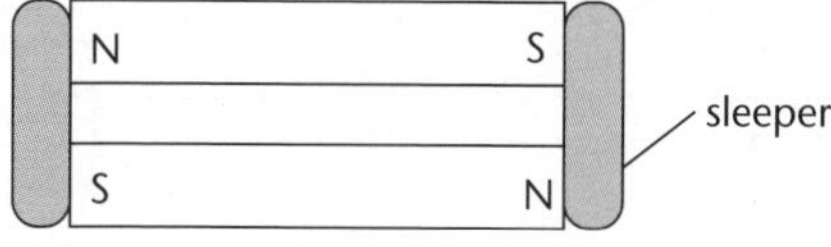

Storing magnets.

The magnetic field around the Earth

The North pole (arrowhead) of a compass needle points towards the magnetic South pole of the Earth. This is located near the Earth's geographic North pole. The angle between magnetic South and geographic North is called the **magnetic deviation**.

The magnetic field around a current-carrying wire

In 1820, Hans Christian Oersted discovered that a current-carrying wire is surrounded by a magnetic field. The direction of the magnetic field can be determined by using the **right hand grasp rule**:

> 'Grasp the wire with your right hand so that your thumb points in the direction of the current – your fingers will curl in the direction of the magnetic field'.

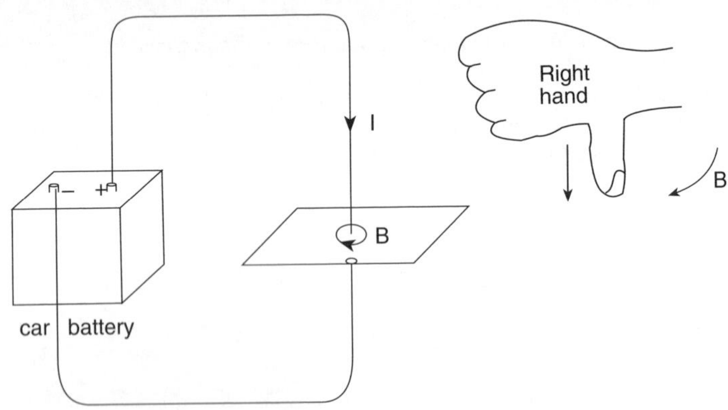

The right hand grasp rule.

The magnetic field around a current-carrying wire can be represented by using magnetic field line diagrams. The direction of the field is indicated by the arrows.

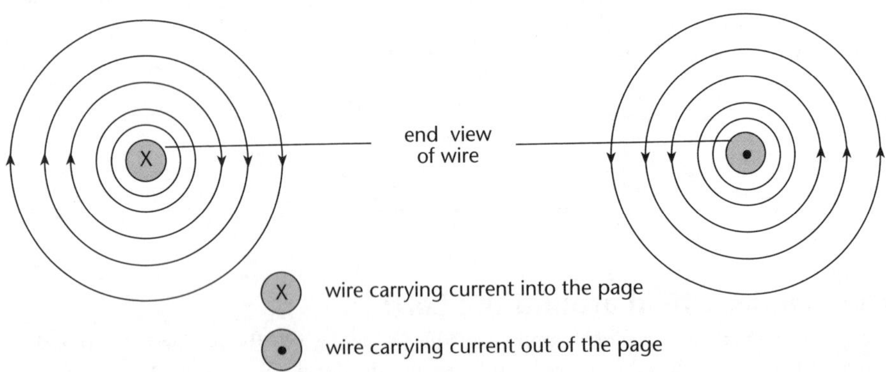

The magnetic field around a current-carrying wire.

The magnetic field around a solenoid

A **solenoid** is a coil of wire. Coiling a length of wire has the effect of concentrating the magnetic field so a *strong* magnetic field exists around the individual coils. The magnetic field around a solenoid is similar to the field around a bar magnet.

The direction of the magnetic field around a solenoid can be determined by using the **right hand solenoid rule**:

'Grasp the solenoid with your right hand so that your fingers curl in the direction of the current. Your thumb will point towards the magnetic North pole.'

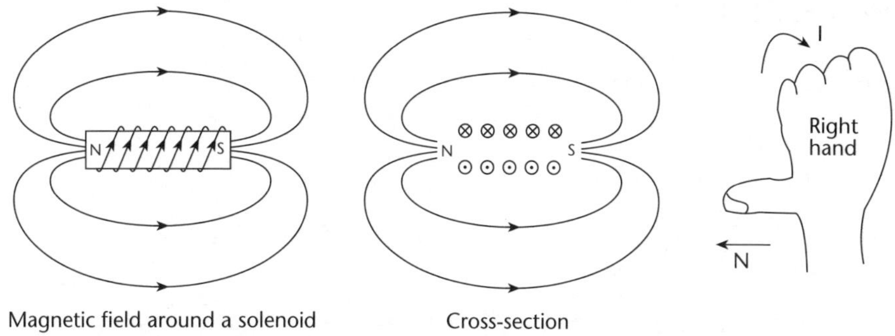

The right hand solenoid rule.

Applications of electromagnetism

The electromagnet

An **electromagnet** consists of a coil of wire wound around a soft iron core.

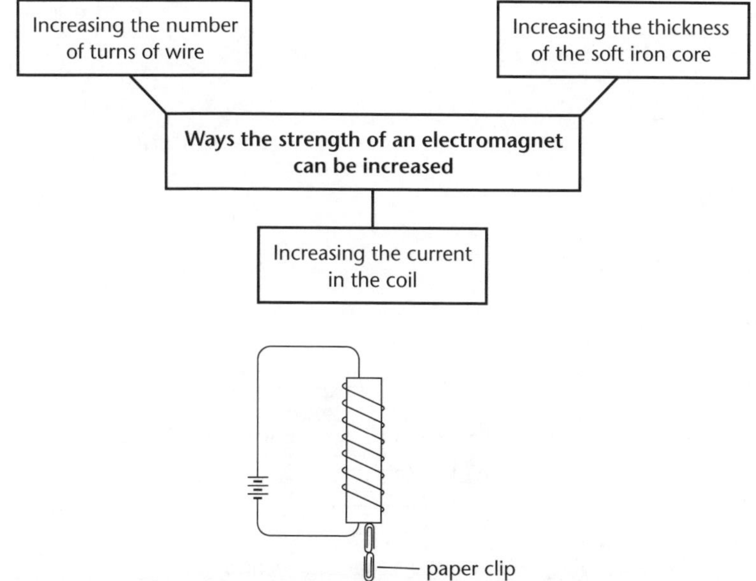

An electromagnet picking up paper clips.

The magnetic field around an electromagnet can be turned on or off by a switch and its strength can be controlled by varying the current in the solenoid.

Electromagnets are thus useful for applications such as picking up car bodies at a wrecking yard.

The electric bell

When the button is pushed the electromagnet magnetises the soft iron cores, which attracts the armature. The armature is connected to the hammer which rings the bell. When the bell is rung the **circuit** is broken and the electromagnet loses its magnetism. The armature returns to its original position, the spring presses against the contact screw and the circuit is closed again. This process repeats for as long as the button is pushed.

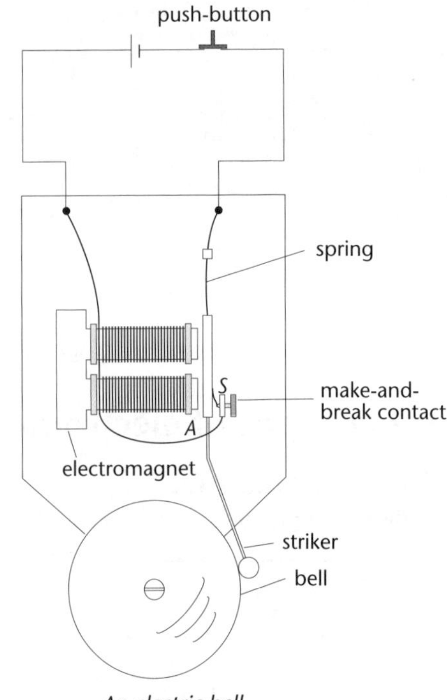

An electric bell.

The relay

A **relay** is an electromagnetic switch. When a current flows through the coil the soft iron core is magnetised and attracts the L-shaped armature. The armature is pivoted so that it pushes against the electrical contacts – this closes the switch and can be used to control a **secondary circuit**.

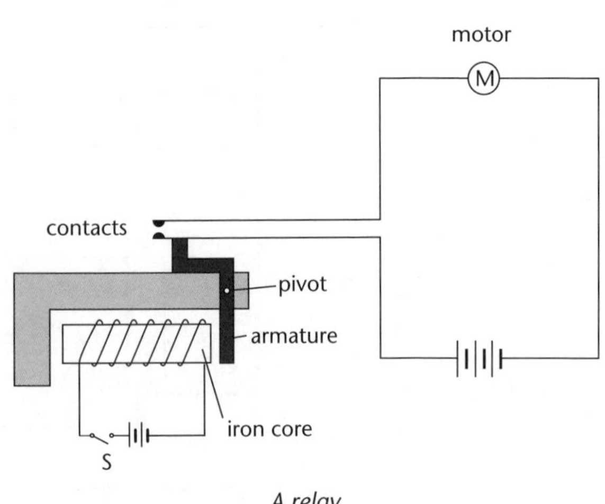

A relay.

In diagram above the relay is used to switch on an electric motor. If an ordinary switch was used, the high current drawn by the motor would cause sparking in the switch, making the motor unsafe to operate.

The reed switch

When a current flows in the coil the reeds (thin strips of metal) are magnetised and attract each other, closing the switch in the process.

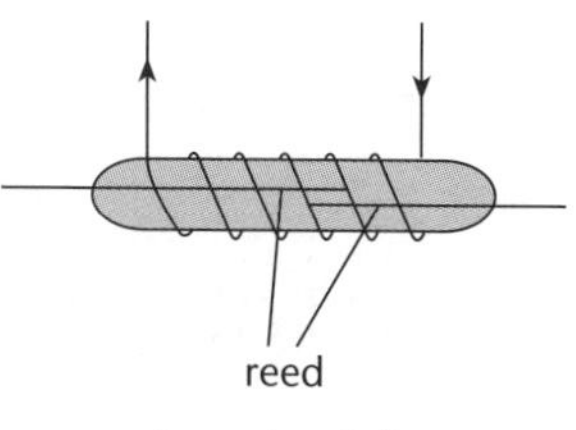

A reed switch.

The loudspeaker

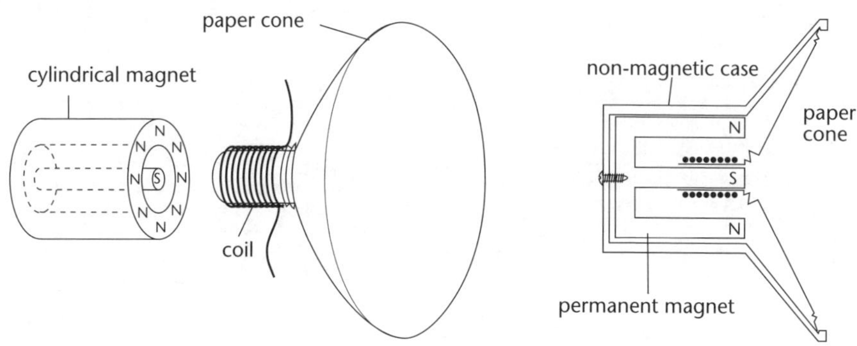

The moving-coil loudspeaker.

When a current flows in the coil of the loudspeaker a magnetic field is set up around it. (The right hand solenoid rule can be used to find the positions of the North and South poles.) Depending on the direction of the current, the magnetic field can attract or repel the permanent magnet. The coil is connected to the cone of the loudspeaker. When the coil moves, the cone moves in or out and the air particles in front of it are compressed or spread out, producing sound.

Unit 12.4 Activity 1A: Magnetic fields

1. A bar magnet is broken in two and the two halves are separated. Which of the following statements is true?

- **A.** Both halves will be unmagnetised.
- **B.** The left half will have a North pole and the right half will have a South pole.
- **C.** Each half will have a North and a South pole.
- **D.** Each half will have two like poles.

N S

2. Draw the magnetic field around each of the following:

- **a.** Bar magnet. N S
- **b.** Two like poles. N N

c. Two opposite poles.

N S

d. A wire with current going into the page.

X

e. A solenoid cross-section.

⊗ ⊗ ⊗ ⊗ ⊗
N S
⊙ ⊙ ⊙ ⊙ ⊙

3. A compass is placed above a wire that carries a current in a direction from North to South. In what direction will the compass needle point?
4. A compass is placed below a wire and the compass needle points towards West. What is the direction of the current in the wire?
5. Draw a diagram to show the magnetic field around a wire that carries a current 'out of the page'.
6. A coil of wire is suspended near a bar magnet, as shown in the diagram. Describe what happens when the switch is closed.

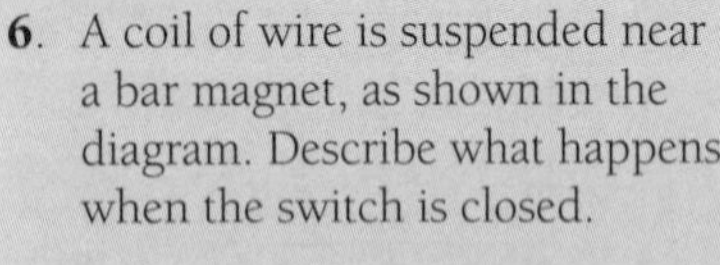

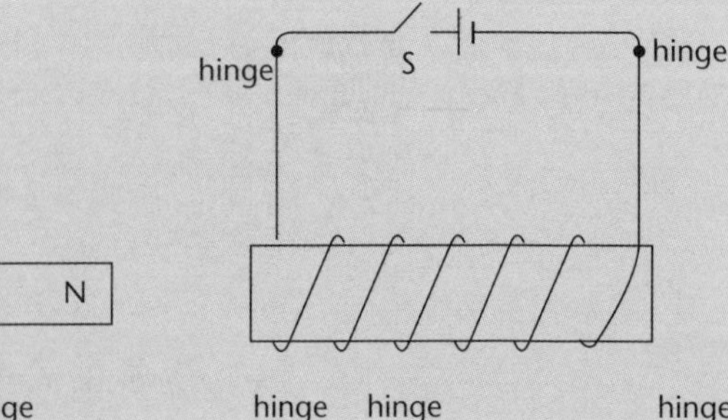

7. Two solenoids are suspended end-to-end, as shown in the diagram. Describe what happens if:
   a. The currents in each of the solenoids flow in the same direction.
   b. The currents in the solenoids flow in opposite directions.

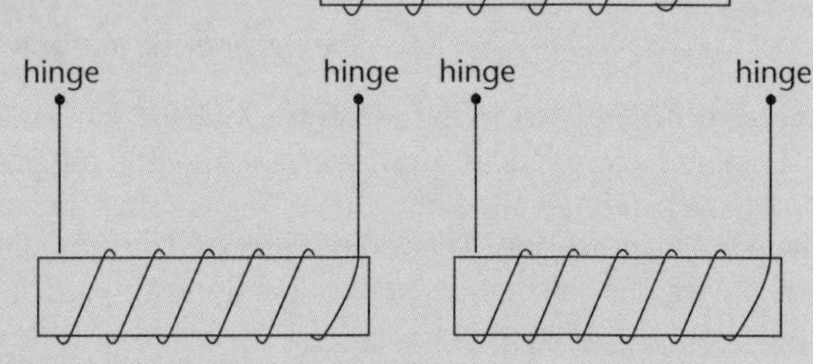

8. Explain why the insertion of a soft iron core increases the strength of an electromagnet.
9. What is the most suitable material for the core of an electromagnet?
10. The relay in the diagram controls an electric motor. When switch S is closed, it switches on the secondary circuit by remote control.
    a. Explain how the relay works.
    b. Why is a relay used to switch on the motor?
    c. Draw a relay in which one electromagnet switches on two remote circuits such as that in the diagram.

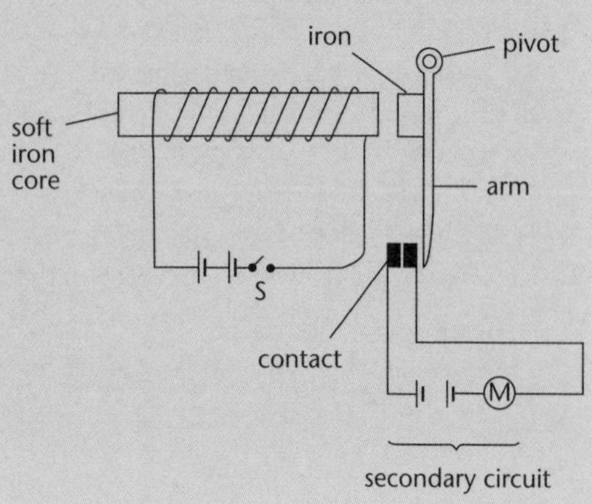

11. The diagram shows an electrically operated model railway signal.

a. Which way does the iron bar move when the current is switched on?

b. Sketch the iron and the solenoid, adding the **lines of force** linking the solenoid and the iron.

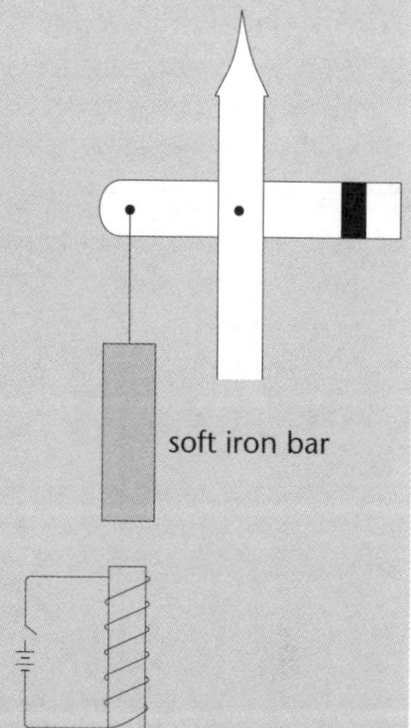

12. The following questions refer to a loudspeaker.

a. Explain why an alternating current makes a loudspeaker cone move back and forth.

b. What factors does the volume of the sound produced by the loudspeaker depend upon?

c. The terms 'squawker', 'woofer' and 'tweeter' refer to different types of loudspeaker. Find out their meanings.

13. Copy the diagram below then draw the magnetic field lines for each of the following, clearly indicating the direction of the field using arrows.

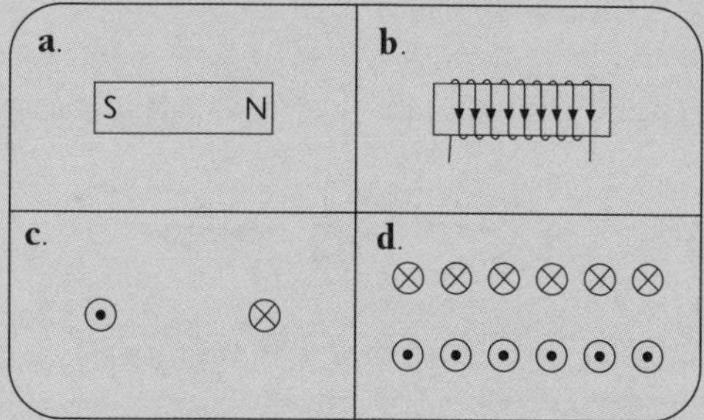

14. Draw the magnetic field around the following solenoid.

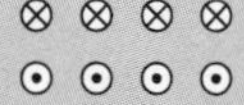

15. A long straight wire goes through the centre of a board and carries conventional current into the plane of the diagram. Draw a diagram that shows the magnetic field due to this current.

16. A coil carrying a current is placed in between two compasses, as shown. Draw arrows to show the deflection of compasses X and Y.

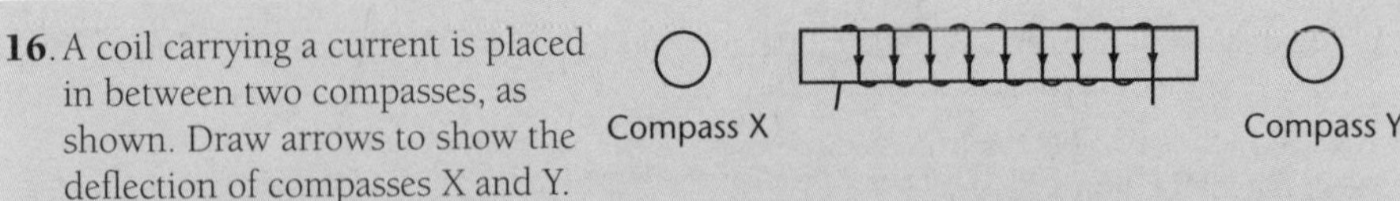

17. Draw the magnetic field lines for each of the following situations. Clearly indicate the direction of the field using arrows.

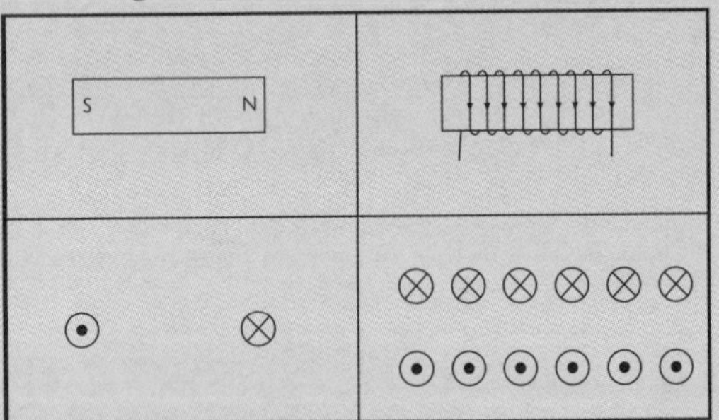

(Source: NCEA Examination paper)

18. Two magnets are placed near each other. For each of the situations shown, describe how the magnets will move.

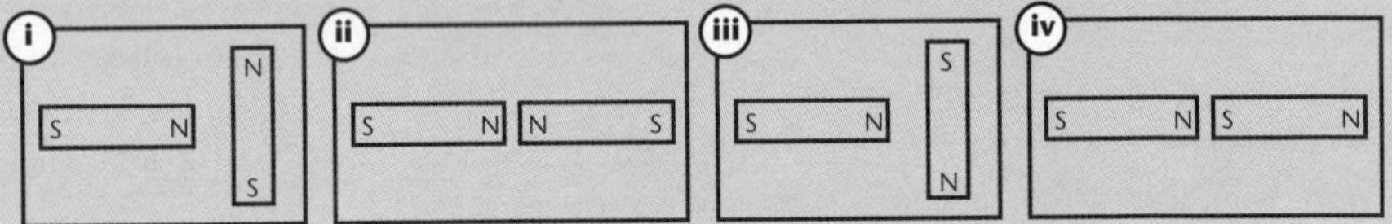

(Source: NCEA Examination paper)

19. Adam's shop sells novelties. One is a pair of 'magic rings'. These are made of a heavy grey metal. Here is some information about these rings.

When the rings are placed flat beside each other, they repel strongly

When Adam holds either of them near a particular metal block, the metal block moves towards the ring.

a. Adam thinks that the rings are magnets. Assuming this is true, where are the north and south poles on these rings?

b. Which of the following metals could the metal block be made of?

copper	brass	iron	zinc

(Source: NCEA Examination paper)

Unit 12.4 Activity 1B: Multiple choice questions

1. Which of the following is not a source of magnetic fields?

A. electric current through a wire
B. compass needle
C. the Earth
D. lodestones

2. The diagram below shows the field lines of a bar magnet pointing from north to south. If a compass is placed at point *X* as shown in the diagram, in which direction will the compass needle point?

A. *a*
B. *b*
C. *c*
D. *d*

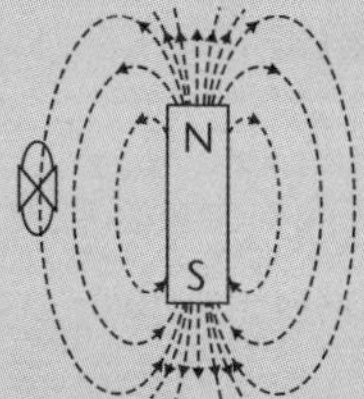

Bar Magnet

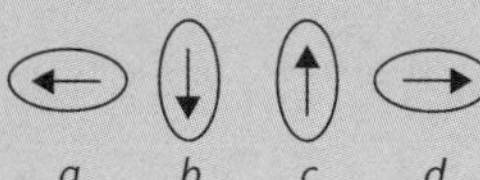

3. What happens when a completely magnetised bar magnet is broken into halves, quarters and eighths?

A. it loses its polarities of north and south poles
B. its polarities of north and south poles interchange
C. it completely loses its magnetisation
D. it retains its magnetism

4. The north pole of a compass needle points towards the magnetic south pole of the Earth which is located near the Earth's geographic north pole. The angle between magnetic south and geographic north is called:

A. magnetic angle
B. magnetic randomisation
C. magnetic deviation
D. magnetic evaluation

5. The direction of the magnetic field around a current-carrying wire can be determined by the right hand rule. Using the right hand rule, the fingers will curl in the direction of the magnetic field produced. In which direction does the thumb point?
   - **A**. parallel to the direction of the current
   - **B**. antiparallel to the direction of the current
   - **C**. parallel to the direction of the magnetic field produced
   - **D**. antiparallel to the direction of the magnetic field produced.
6. The direction of the force on a current carrying conduct in an external magnetic field can be determined by a right hand rule. In the diagram below the force is upwards and the current is to the left as shown in between opposite ends of bar magnets *A* and *B* which provide an external magnetic field. Which of the following statements about their polarities is correct?
   - **A**. south end is *B* and north end is *A*
   - **B**. north end is *B* and south end is *A*
   - **C**. both polarities *A* and *B* are north
   - **D**. both polarities *A* and *B* are south

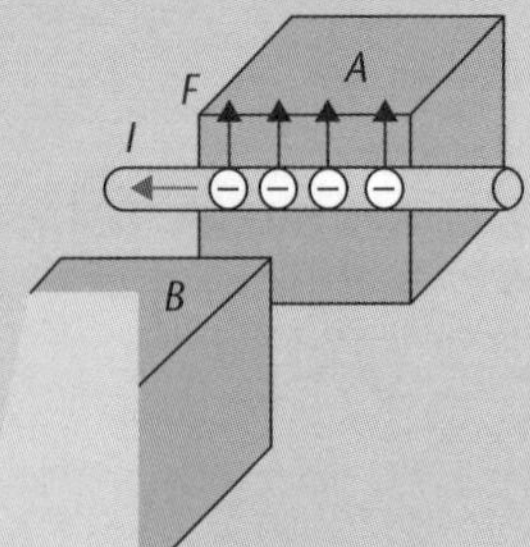

7. What would happen if the current in the diagram of question 6 was reversed?
   - **A**. south end is *B* and north end is *A*, the force is upwards
   - **B**. south end is *A* and north end is *B*, the force is upwards
   - **C**. south end is *B* and north end is *A*, the force is downwards
   - **D**. south end is *A* and north end is *B*, the force is downwards
8. Which of the following methods will not increase the strength of a magnetic field in a solenoid?
   - **A**. increase the current in the coil
   - **B**. increase the thickness of the soft iron core
   - **C**. remove the soft iron core
   - **D**. increase the number of turns of wire
9. The generation of electric voltage is done by moving a wire in between the magnetic field, hence disturbing the field, which causes electromagnetic induction (emf). If a wire is forced to move perpendicular to the magnetic field as shown in the diagram below, which direction will be the force acting on the charges in the wire?
   - **A**. upwards along the wire
   - **B**. downwards along the wire
   - **C**. in the direction of the moving wire
   - **D**. in the direction of the magnetic field

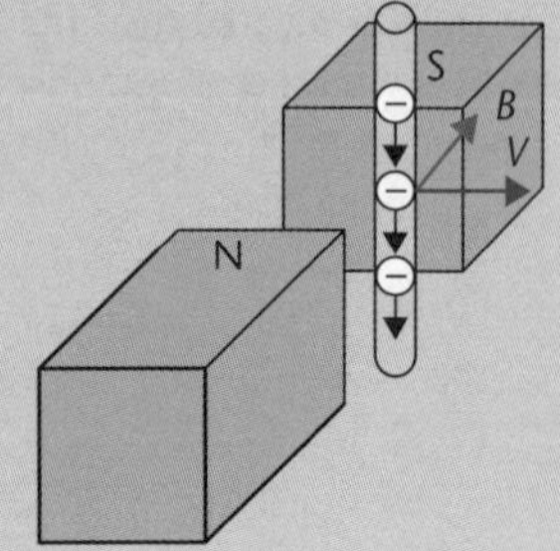

10. Why is iron metal the most suitable material that can be used for soft iron core?

A. it does not demagnetise quickly in a short period of time as the current through the coil is switched on and off

B. it takes plenty of time to become magnetised as the current through the coil is switch on and off

C. it can be magnetised and demagnetised in a short period of time as the current through the coil is switch on and off

D. it is a good conductor of electricity hence it is suitable to use as a core.

11. The following diagrams show that the amount of compass needle deflection is proportional to the amount of current flowing through the wire.

Which diagram shows a maximum current through the wire? There can be more than one answer.

A. diagram (i)

B. diagram (ii)

C. diagram (iii)

D. diagram (iv)

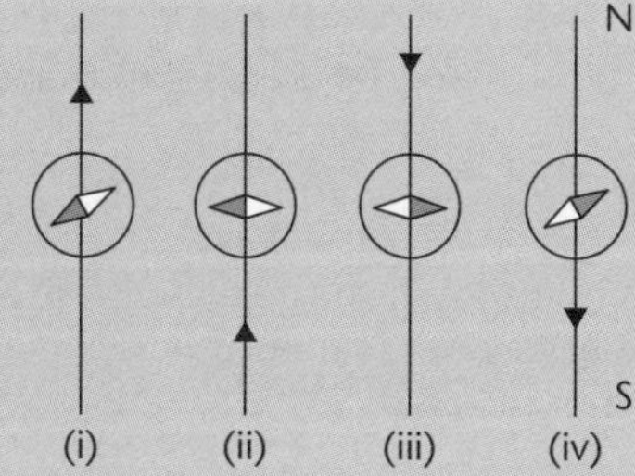

12. The diagram below shows the positions of compass needles placed around the two coils.

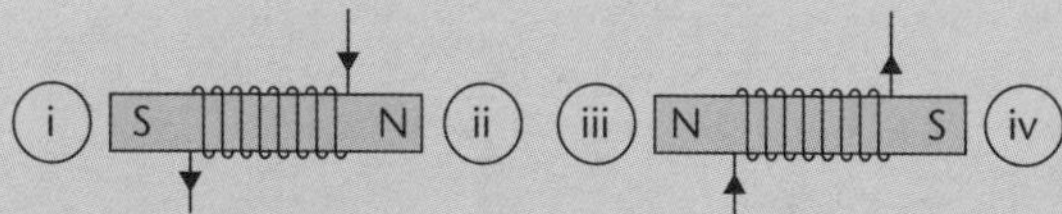

Which of the following diagrams showing compass needles would correspond to the above arrangement of coils?

A. diagram 1

B. diagram 2

C. diagram 3

D. diagram 4

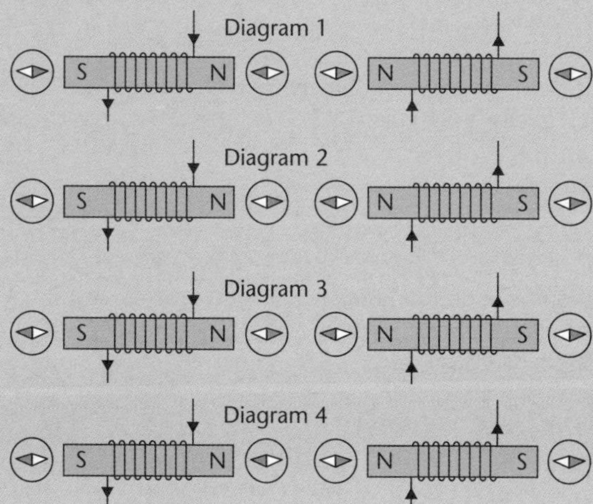

13. The diagram below shows positions of compass needles placed around the two coils.

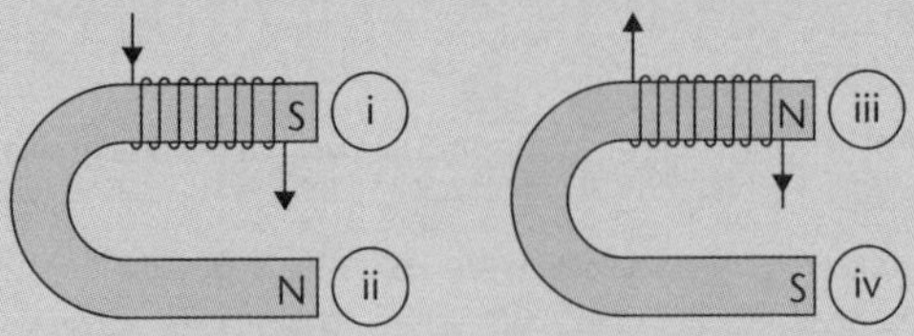

Which of the following arrangements corresponds to the above arrangement of coils?

A. arrangement 1
B. arrangement 2
C. arrangement 3
D. arrangement 4

arrangement 1

arrangement 2

arrangement 3

arrangement 4

14. The figure below is an electromagnet made by a student in a physics laboratory. The purpose of the electromagnet was to investigate the strength of the magnetic field as the number of turns is increased. With 12 turns it can pick up three pins of mass 5 grams each.

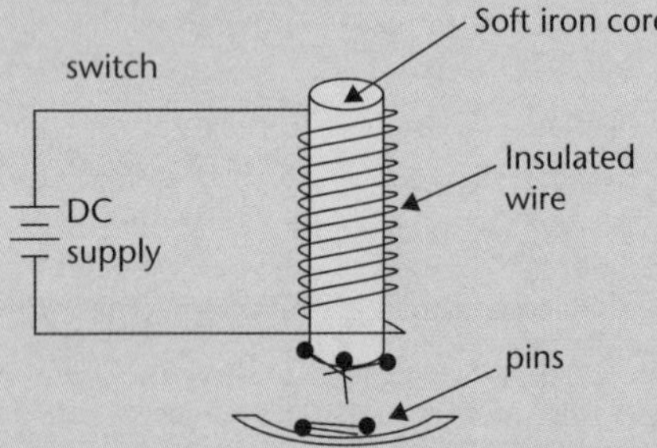

What is the magnitude of the magnetic force produced by the coil? Take acceleration due to gravity g = 10 ms^{-2}.

A. 0.015 N
B. 0.15 N
C. 1.5 N
D. 15 N

15. In question 14 above, how many turns of the coil would be required to pick up all five pins, assuming their mass is the same, the soft iron core is not changed and the DC supply is constant?

A. 14
B. 16
C. 18
D. 20

Unit 12.4 Electromagnetism

Topic 2: Magnetism

Topic 2 continues the study of electromagnetism by looking at magnetic fields and forces due to current. It covers:

- Magnetic fields.
- Force on a current in a magnetic field.
- Force on a moving charge in a magnetic field.

Revision

Magnets play a large part in our lives.

Example A

Magnets are used in the sealing strip around a refrigerator door, in motors to produce motion and in TV tubes to deflect electron beams and produce a picture.

Some of the properties of magnets (eg that a pivoted bar magnet acts like a compass needle and always points the same way), are well known.

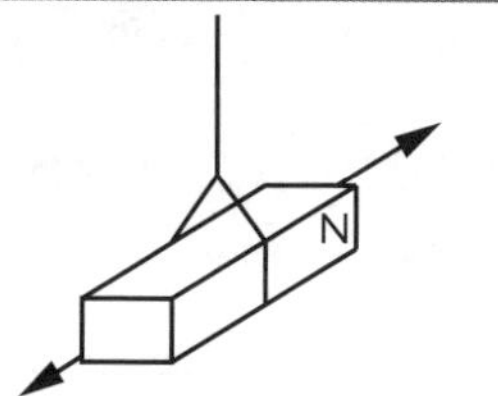

A compass needle

The compass needle always points in the same direction because the earth has a **magnetic field** around itself. One end of a bar magnet (called a **north pole**) points to the earth's geographical north pole and the other end (called the **south pole**) points toward the earth's geographical south pole.

Experiments with bar magnets show that:

Unlike poles attract each other and like poles repel.

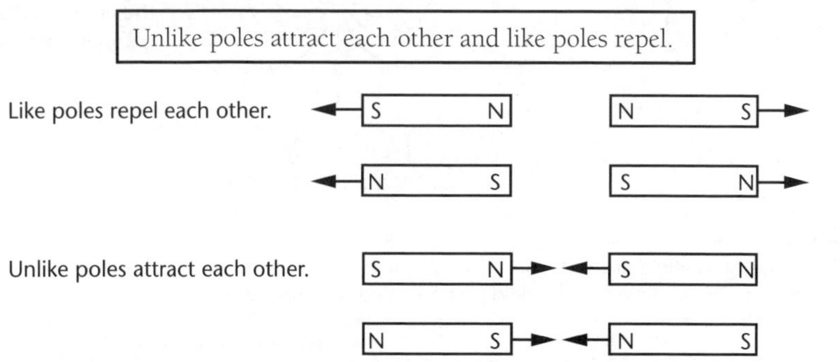

A magnet will attract some types of unmagnetised metals such as iron, nickel and cobalt and **alloys** containing them. This is because the magnet *induces* magnetism in the metals.

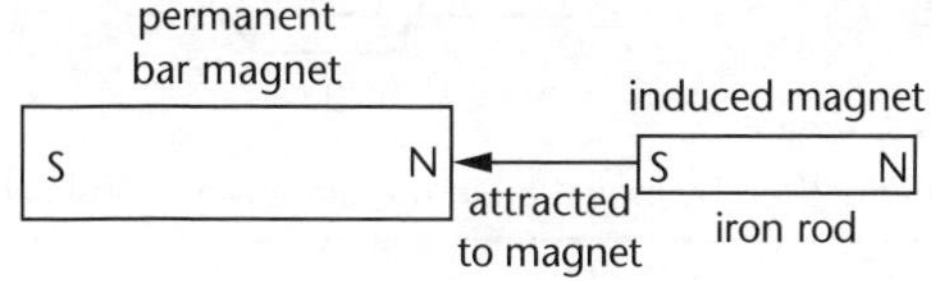

Magnetic induction

Example B

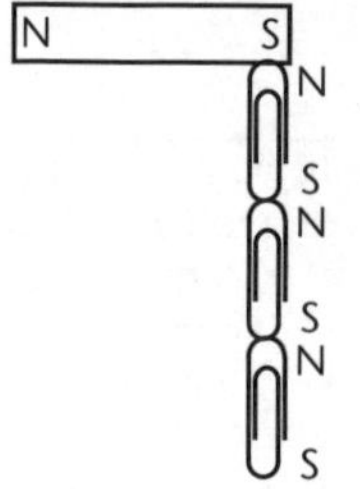

A bar magnet is able to pick up a string of paper clips. The paper clips are attracted to the magnet and to each other because in the presence of the bar magnet, the paper clips become magnets by **induction**.

The induced magnetism becomes weaker further from the bar magnet.

To test whether an object is permanently magnetised, it must display repulsion by a known magnet. Attraction is not a reliable test because some non-magnets will be attracted to the known magnet due to induction.

Steel behaves in a similar way to iron. It is not attracted as strongly but it does not lose all its magnetism when the bar magnet is removed. Thus steel is best for permanent magnets and soft iron for temporary magnets.

Magnetic fields

A **magnetic field** exists in the space around a magnetised object because other magnetic objects experience a force near the first magnetised object. Magnetic field lines show the *shape* and *strength* of the magnetic field.

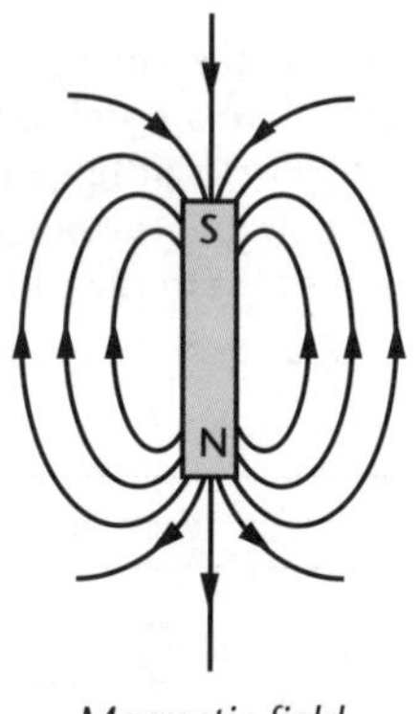

Magnetic field around a bar magnet

The magnetic field is *strongest* where the *field lines are closest together*. This is near the ends of a bar magnet.

Example C

The shape of the field around a bar magnet can be shown with iron filings or small compasses. The direction of the field is shown with a compass.

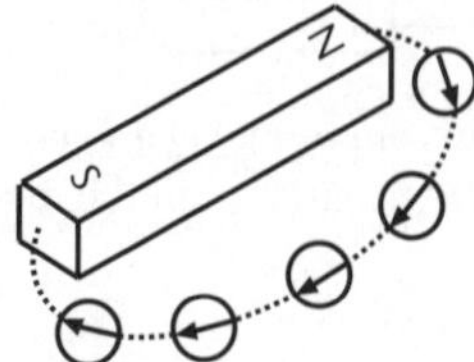

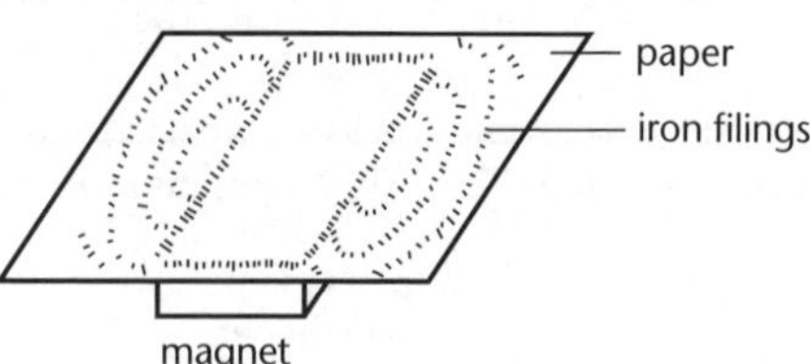

The iron filings are magnetised by induction in the same way as the paper clips.

The Earth's magnetic field, caused by electric currents circulating within the earth, is not strong enough to be able to induce enough magnetism in iron filings to see its pattern.

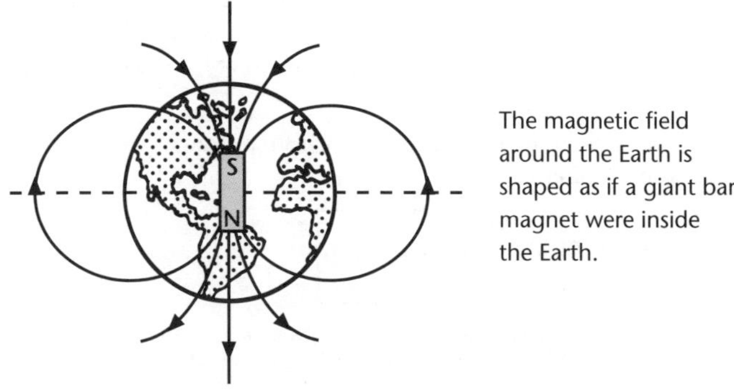

The Earth's magnetic field

The magnetic field around a current

When an electric current flows in a wire, a magnetic field is set up around the wire. The field lines form concentric circles around the wire.

Several wires used together each with a small current act like one wire with a large current and produce a greater magnetic effect. Small compasses show the shape and direction of the field.

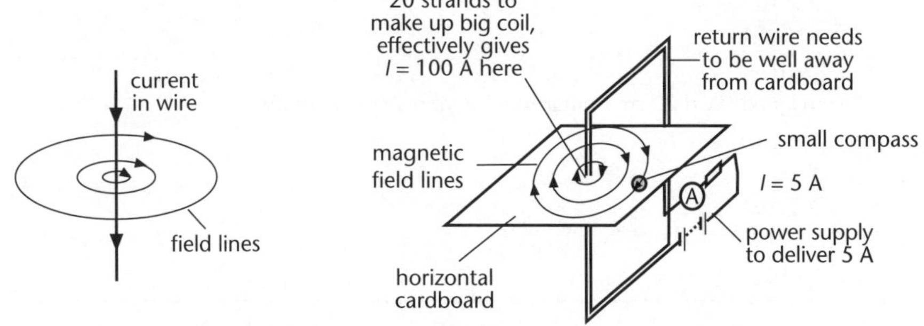

Demonstrating the magnetic field around a wire

The direction of the field can be remembered by the **right-hand grip rule**. The current carrying wire is 'grasped' with the right hand so that the thumb points in the direction of the **conventional current**. The fingers curl around the wire in the direction of the magnetic field.

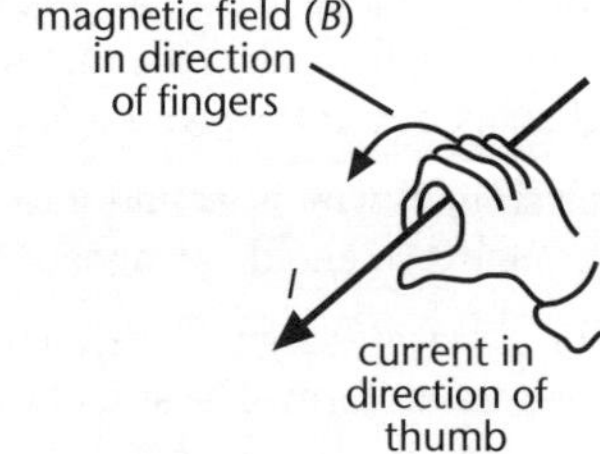

Right-hand grip rule

The size of a magnetic field is called the **magnetic field**, B. The unit for magnetic field strength is **tesla**, T.

The size of the Earth's magnetic field is about 60 μT (6×10^{-5} T) in New Zealand. The strongest magnetic fields possible in most school laboratories are about 0.5 T. A field strength of over 1 T can be achieved with massive water-cooled electromagnets, using soft iron cores or using superconducting magnets.

The size of the magnetic field, B, a distance, d, out from a straight wire carrying a current, I, is calculated from the formula:

$$B = \frac{ki}{d}$$

where k is the proportionality constant in the relationship, $k = 2 \times 10^{-7}$ N A^{-2}.

Example D

a. The magnetic field 10 cm from a wire carrying a current of 2 A is:

$$B = \frac{ki}{d}$$

$$= \frac{2 \times 10^{-7} \times 2}{0.1} \quad [10 \text{ cm} = 0.1 \text{ m}]$$

$$= 4 \times 10^{-6} \text{ T}$$

b. The magnetic field 10 cm from a wire carrying a current of 4 A is:

$$B = \frac{ki}{d}$$

$$= \frac{2 \times 10^{-7} \times 4}{0.1}$$

$$= 8 \times 10^{-6} \text{ T}$$

c. The magnetic field 20 cm from a wire carrying a current of 2 A is:

$$B = \frac{ki}{d}$$

$$= \frac{2 \times 10^{-7} \times 2}{0.2} \quad [20 \text{ cm} = 0.2 \text{ m}]$$

$$= 2 \times 10^{-6} \text{ T}$$

Example D shows that the magnetic field strength B, is *proportional to the current* I, in a wire and *inversely proportional to the distance* d, out from the wire.

The magnetic field around a current carrying wire is useful in many situations. Electromagnets, electric motors and moving coil meters all use the magnetic field produced when a current flows through a wire.

If a straight wire is wound into a circular coil, a **solenoid** forms. When a current flows through a solenoid, a magnetic field with a similar shape to a bar magnet forms.

The strongest magnetic field is inside the solenoid and near its centre; the magnetic field lines are uniform. The size of the magnetic field in and around a solenoid can be increased by increasing the current that flows through the wires of the solenoid.

The right-hand grip rule is used in the reverse **sense** to find the direction of the field lines inside a solenoid.

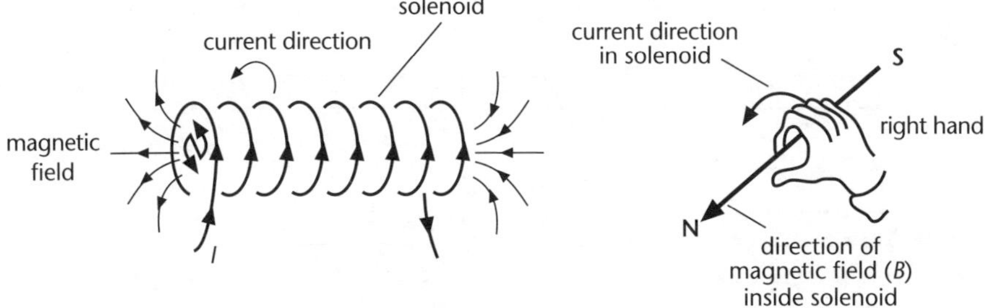

The end of the solenoid at which the field lines emerge is the north pole and the end at which the field lines enter is the south pole. Use NewS to help get field directions:

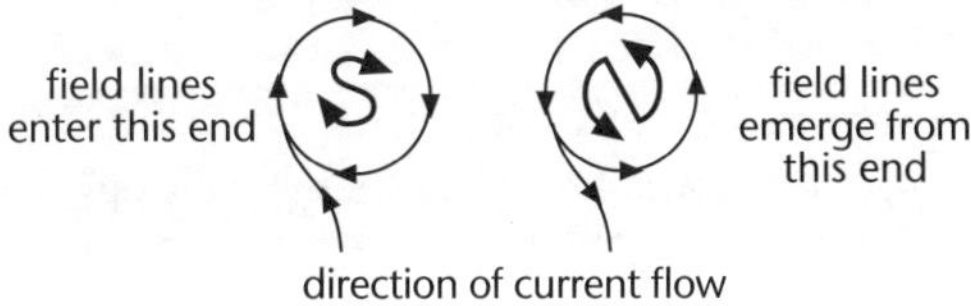

MRI body scanners are basically large solenoids in which strong magnetic fields can be produced. Atoms in a person's body emit small electromagnetic waves (radio frequency) under the magnetic field's influence. Detectors pick up these radio signals and a computer processes them to give a cross-sectional picture of the body on a TV screen.

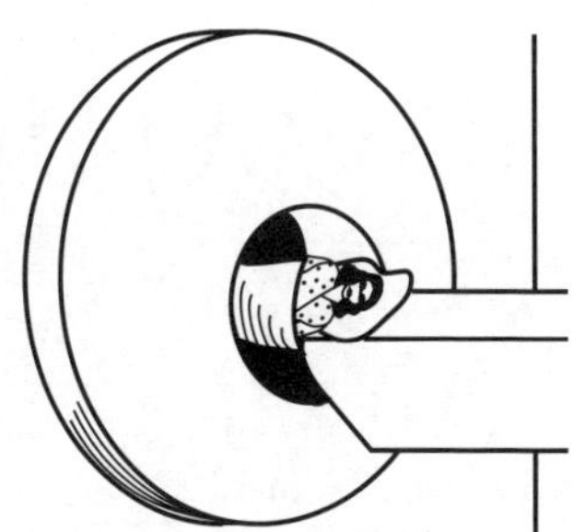

Powerful magnets (body scanners) in hospitals

A solenoid can be used as a simple **electromagnet** by inserting iron inside it. The iron core increases the magnetic field by becoming magnetised. The magnetic field can be strong enough to be able to pick up car bodies in a scrap yard. The advantage of an electromagnet is that it can be switched on and off as required, and weight-for-weight can be made stronger than permanent magnets.

Unit 12.4 Activity 2A: Magnets

1. Copy the diagrams and draw the magnetic field patterns around the magnets (assumed identical). Mark in positions of zero magnetic field with an 'X' in **b** and **c** only.

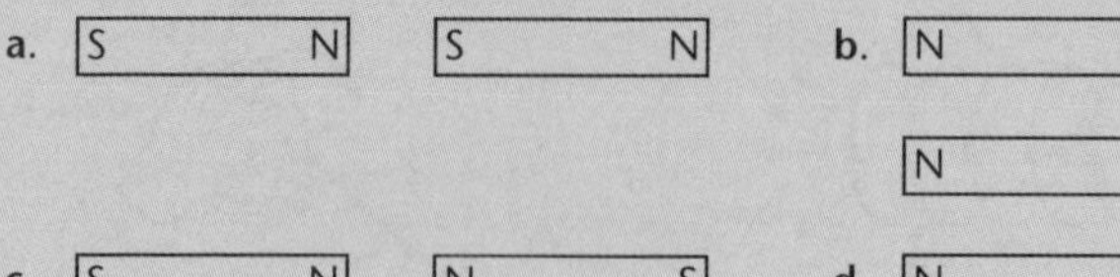

2. When not in use, bar magnets are kept in pairs with soft iron 'keepers' placed across their ends. Copy the diagram and draw the magnetic field pattern inside the keepers.

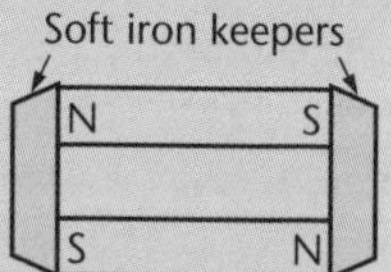

3. Currents flow through the following solenoids as shown. Copy them down and draw in the magnetic field pattern, indicating the field direction and the polarity of each end.

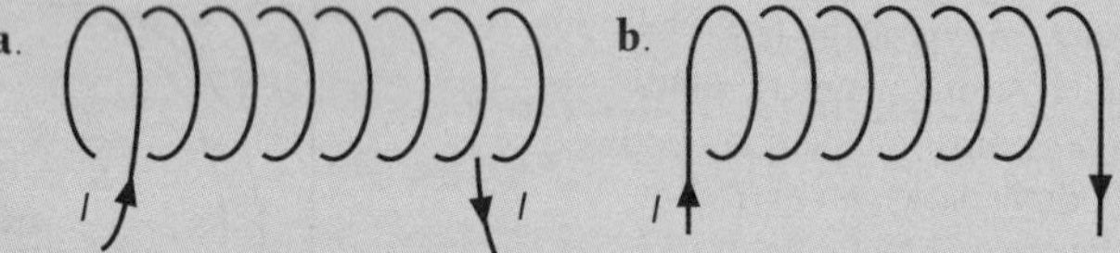

4. An improved electromagnet is constructed, as shown.
   a. Which end (A or B) is the north pole of the electromagnet?
   b. Give reasons why this type of electromagnet would be more powerful than a solenoid-type electromagnet.
5. Would the improved electromagnet still work if alternating current of frequency 50 Hz is passed through it? Explain your answer.

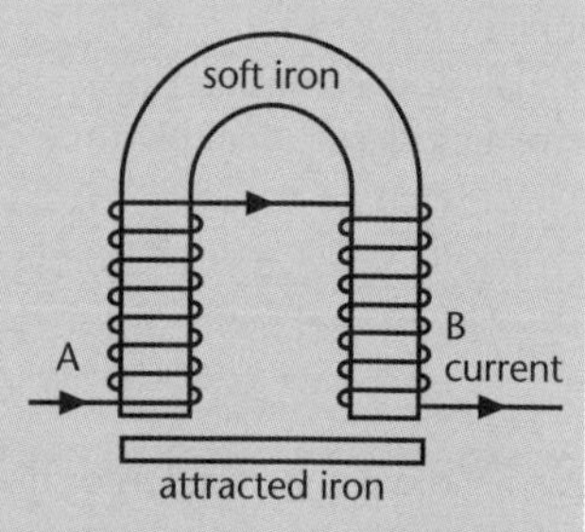

The force on a current in a magnetic field

If a current carrying wire is placed close to another magnet (permanent or electromagnet), the two magnetic fields will interact. As a result, the wire experiences a force.

A '**current balance**' can be used to study the force produced on a current carrying wire in a magnetic field.

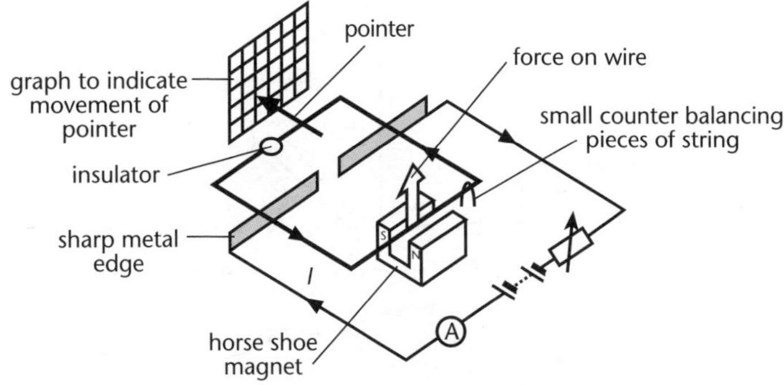

A typical current balance

The relationship between the force on the wire, F, the strength of the magnetic field, B, the current in the wire, I, and length of wire in the field, l, is:

$$F = BIl$$

If the current carrying wire is arranged so that it is *parallel* to the magnetic field, there is no force on the wire.

More accurately, the force on a current carrying wire in a field depends on the **component** of the magnetic field, which is perpendicular to the current.

If θ is the angle between the current and the field, the size of the force is:

$$F = BIl \sin\theta$$

When $\theta = 0°$, the magnetic field is parallel to the current-carrying wire. Since $\sin 0° = 0$, the force is zero ($F = BIl \sin \theta = 0$).

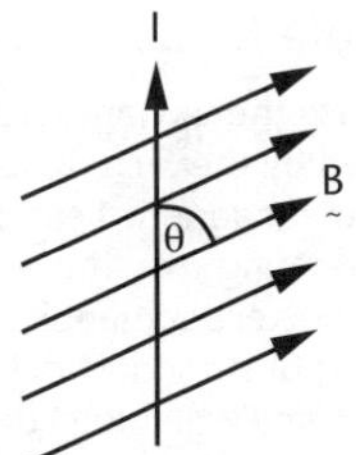

Magnetic field and current

When $\theta = 90°$, the magnetic field is at right angles to the current.

Since $\sin 90° = 1$, the force is at maximum when the current is perpendicular to the magnetic field ($F = BIl \sin \theta = BIl$).

The direction of the force on the current carrying wire is given by the **right-hand slap rule**.

If the conductor is a coil with number of turns (n), with length, l, as the circumference of a single turn, then the force experienced by the coil is given by the formula $F = nBIl \sin \theta$.

If the thumb points in the direction of the conventional current in the wire and the fingers in the direction of the field, the direction of the 'slap' from the palm of the hand is the direction of the force on the wire.

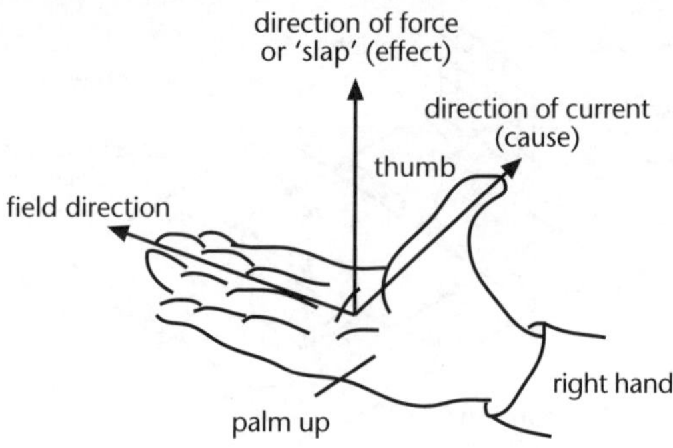

Right-hand slap rule

This is the motor version of the right-hand slap rule, getting motion from current. Consider the current in the direction of the thumb as the cause resulting in the effect, motion, in the slap direction of the palm. The right hand slap rule is reversible, just like the right-hand grip rule. When the cause in the thumb direction is motion, the effect is current in the slap direction – this is the generator version of the right-hand slap rule.

$F = BIl$ can be rearranged to give $B = \frac{F}{Il}$. The equation $B = \frac{F}{Il}$ shows the unit for magnetic field strength, T, is equivalent to N A^{-1} m^{-1}.

Example E

To balance the upward force produced by the 5 cm length of wire in the magnetic field, pieces of string are hung over the length of wire. In one experiment it was found that 0.5 g of string just balanced the upward force when the current in the wire was 2 A.

What is the size of the magnetic field between the magnet's poles?

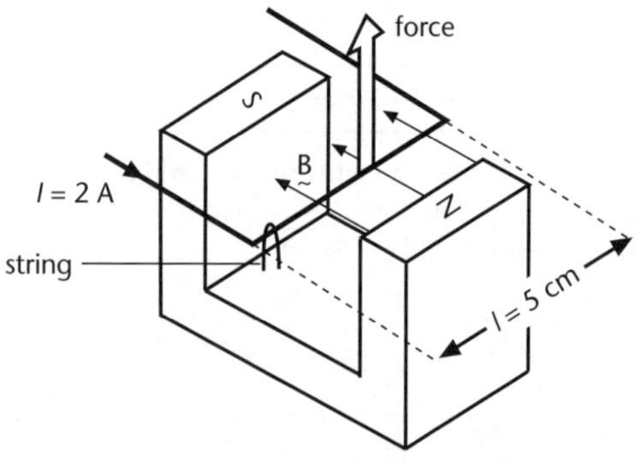

Solution:

The weight of the piece of string is:

$F_w = mg$

$= 5 \times 10^{-4} \times 10$ [0.5 g = 5×10^{-4} kg]

$= 5 \times 10^{-3}$ N

Using $F = BIl$:

$B = \frac{F}{Il}$

$= \frac{5 \times 10^{-3}}{2 \times 0.05}$ [5 cm = 0.05 m]

$= 5 \times 10^{-2}$ T or 50 mT

A current-carrying wire placed at right angles to a magnetic field experiences a force because the magnetic field around the wire interacts with the other magnetic field.

Where two field lines in *opposite* directions interact, they have a *cancelling effect* and the size of the field is reduced in that region (see diagram below).

Where both fields lines are in the same direction, the field lines *add together* strengthening the field (see diagram below). Overall the fields become distorted and a force is experienced by the wire so that the wire moves toward the region where the field is weakened.

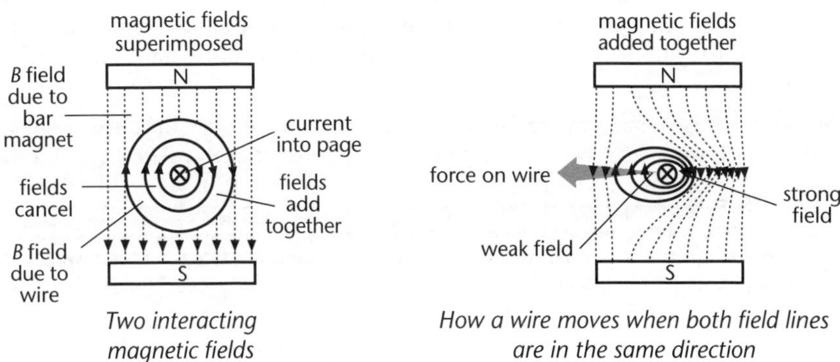

Two interacting magnetic fields

How a wire moves when both field lines are in the same direction

Force between two parallel wires

If two long parallel straight conductors separated by a distance d carrying currents I_1 and I_2, the direction of the current's flow will then determine the direction of the force experienced by each conductor. The direction of the force is determined by applying the right hand grip rule as shown below.

Case 1: Currents in the same direction. Applying the right hand rule, we can determine the direction of each force experienced by each wire.

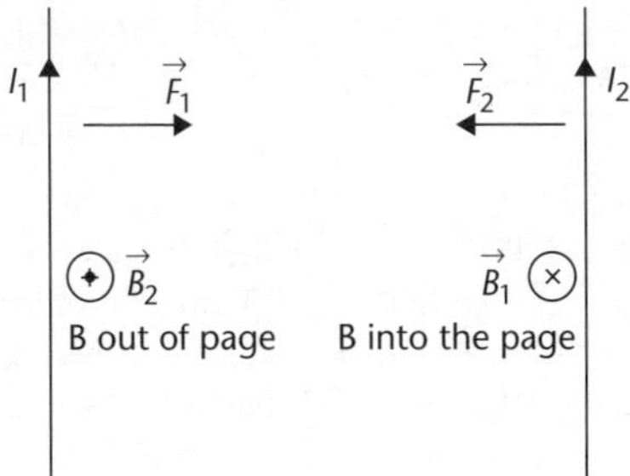

Force between two long straight parallel wires with currents flowing in the same direction

Wire number 2 will experience a magnetic field due to wire number 1, given by

$$B_1 = \frac{\mu_o I_1}{2\pi(r)}$$

in the direction indicated in the figure above.

Force experience by a 1-metre length of the wire is given by the equation

$$\frac{F_2}{l} = \frac{\mu_0 I_1 I_2}{2\pi(d)}.$$

So the direction of $\vec{F}_2$ on wire 2 will be attracted towards wire 1. In a similar manner, we can show that wire 1 will experience a force due to the magnetic field $\vec{B}_2$ of wire 2, and that this force $\vec{F}_1$ will have a magnitude equal to that of $\vec{F}_2$ given by the first equation above but opposite in direction. Thus, wire 1 will be attracted towards wire 2 and there exists an attractive force between the two parallel wires.

Case 2: Currents flowing in the opposite direction. Applying the right hand rule, we can determine the direction of each force experienced by each wire as in case number 1. The end result is that wire 1 will repel wire 2 and there exists a repulsive force between the two parallel wires.

If two long parallel wires 1 m apart each carry a current of 1 A, then the force per unit length on each wire is 2×10^{-7} N/m, with constant μ_0 given exactly as $4\pi \times 10^{-7}$ T • m/A.

Example F

Two straight, parallel, superconducting wires 4.5 mm apart carry equal currents of 15,000 A in opposite directions. Should we worry about the mechanical strength of these wires? This diagram illustrates the problem:

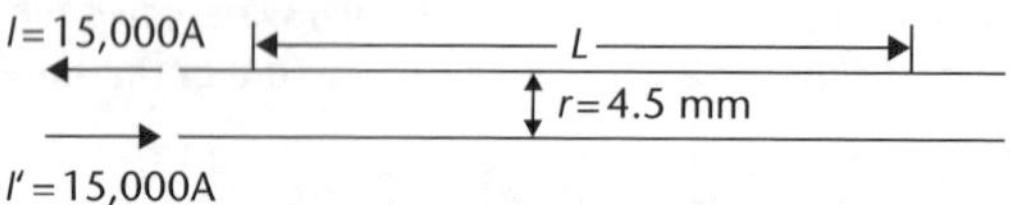

Unit 12.4 Activity 2B: Magnetic forces

1. A current of 5.0 A flows through horizontal wire in a magnetic field directed vertically upwards. The length of the wire in the magnetic field is 0.75 m and the wire experiences a magnetic force of 0.25 N.

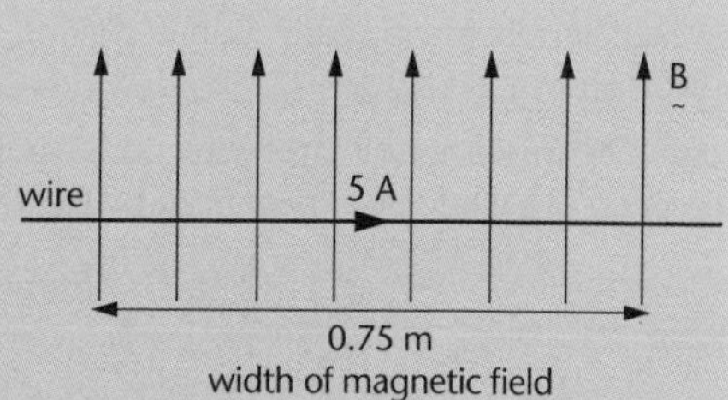

   a. Determine the size of the magnetic field, B.
   b. Determine in which direction the magnetic force acts on the wire.
2. A power cable slung between two pylons 0.50 km apart carries a current of 5.0 A. If the component of the Earth's magnetic field at right angles to the cable is 60 μT (6.0×10^{-5} T), calculate the total magnetic force on that length of cable.
3. A large powerful U-shaped magnet is moved into positions shown around two fixed horizontal brass rods.

A

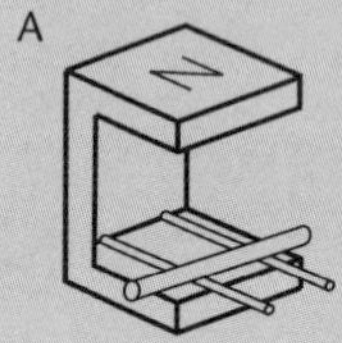

B

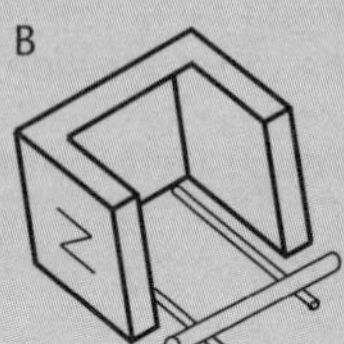

C

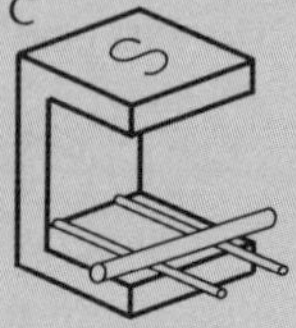

D

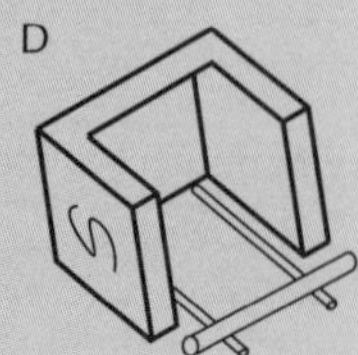

In all of the positions shown, a moveable metal roller is placed on the rods and the rods are connected to a special change-over switch (double pole double throw) as shown:

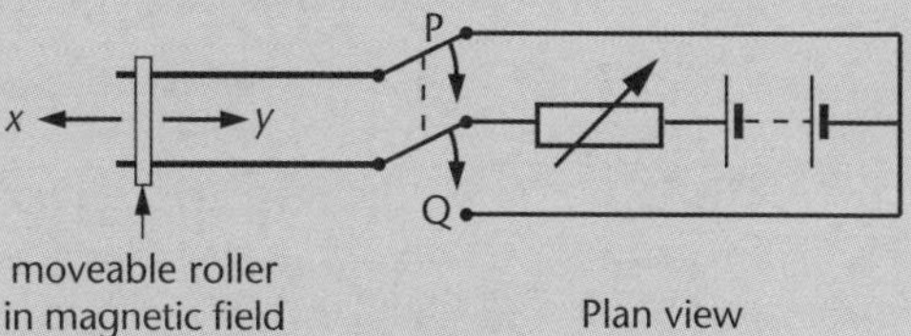

Copy and complete the following table using the words and symbols:

up, down, left, right, none, x, y.

Magnet position	Switch position	Magnetic field direction	Movement of roller.
A	P		
B	P		
C	P		
D	P		
A	Q		
B	Q		
C	Q		
D	Q		

4. A vertical uniform magnetic field acts in a direction upwards (ie 'out of the page') and a conductor PQ which is free to move rests on horizontal conducting runners X and Y, as shown.

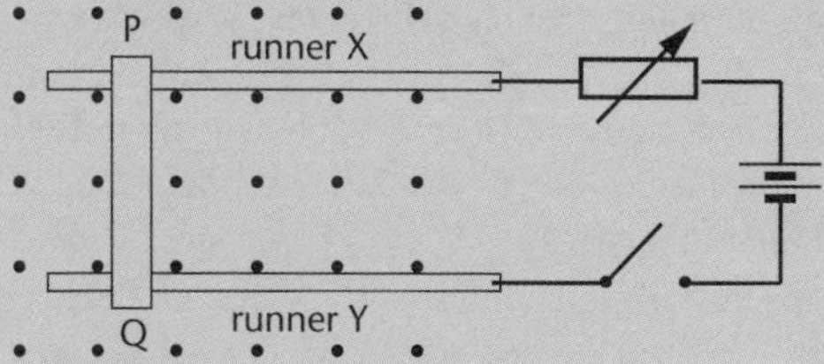

   a. When the switch is closed, conductor PQ experiences a force. In which direction is this force?
   b. State two ways to reverse the direction of movement of the conductor, PQ.
   c. State two ways to increase the force on the conductor PQ.

5. Several students performed the following current balance experiment using a solenoid as a source of magnetic field instead of a U-shaped magnet.

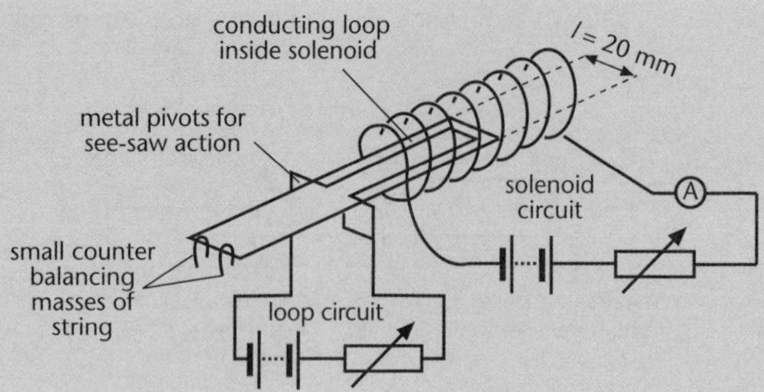

The students measured out 20.0 m of string and found that it had a mass of 200 g (2.0×10^{-1} kg). They held the current in the solenoid constant at 5.0 A and altered the current in the conducting loop, balancing it with small lengths of string to record the following measurements:

Length of string to balance (mm)	Balancing force (N)	Current in circuit loop (A)
100		0.98
200		1.97
300		3.05
400		3.98
500		5.05

a. In which direction must the end of the current balance inside the solenoid move in this experiment?

b. Suppose the current balance pivoted the wrong way. What could a student do to correct this?

c. **i**. Complete the table and draw a graph of balancing force (vertical axis) against current (horizontal axis). Draw in a straight line of best fit.

ii. Find the **slope** of the graph.

d. Show that the magnetic field strength of the coil (at 5.0 A) is given by B = 50 × slope.

e. Calculate the magnetic field strength inside the solenoid.

f. If the students keep the loop circuit current constant at 5.0 A and then alter the solenoid circuit current to balance, sketch a graph (force against solenoid current) of what they would find.

6. A thick current-carrying wire is suspended by a sensitive force meter in a magnetic field. Light flexible wires complete a circuit, as shown.

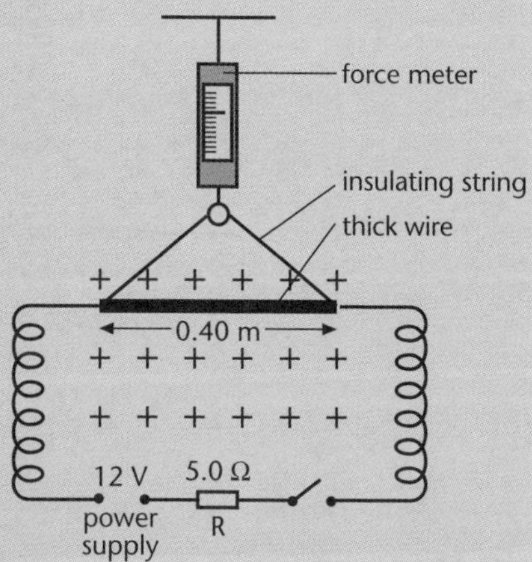

a. The force meter reads 0.80 N when the switch is open and 0.20 N when it is closed. Find

i. Which way round the current flows.

ii. The size of the current

iii. The size of the magnetic field.

b. What extra resistor would need to be connected in parallel to R for the force meter to read 0.80 N when the switch is open and 0 N when closed?

c. If the force meter reads 0.80 N when the switch is open and 2.0 N when it is closed, describe in detail what changes have been made to the magnetic field.

7. A spring with force constant $k = 40\ \text{N m}^{-1}$, is extended 5.0 cm (the force produced = kx, where x = extension) when a 1.2 m thick copper wire, mass 0.2 kg, is connected as shown:

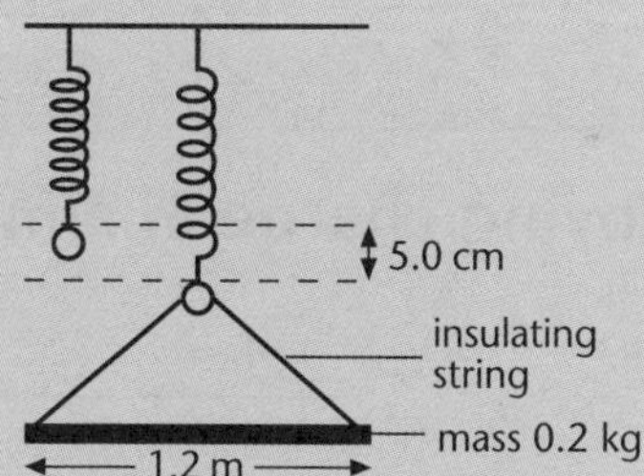

The extended spring and copper wire are placed in a magnetic field, and light flexible connecting wires complete an **electric circuit** as shown on the next page.

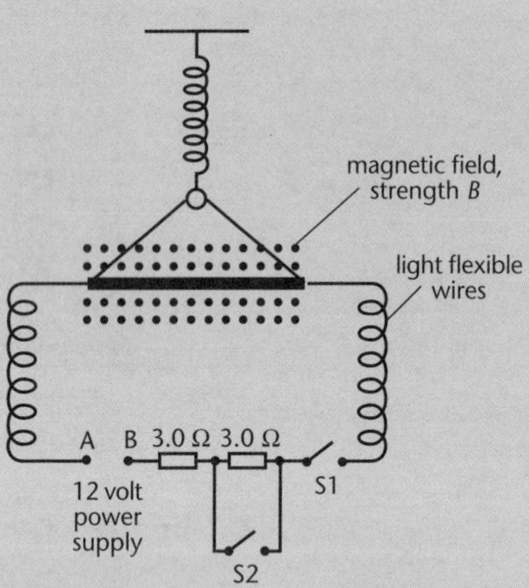

a. If switch S1 is closed, and S2 is open, and the extension of the spring becomes 4.0 cm:

i. Which is the positive terminal of the power supply, A or B?

ii. What is the size of the magnetic field B?

b. If both switches S1 and S2 are closed, what does the extension of the spring balance become?

The force on a moving charge in a magnetic field

A small charged object, such as a **proton** or electron, moving in a uniform magnetic field experiences a force.

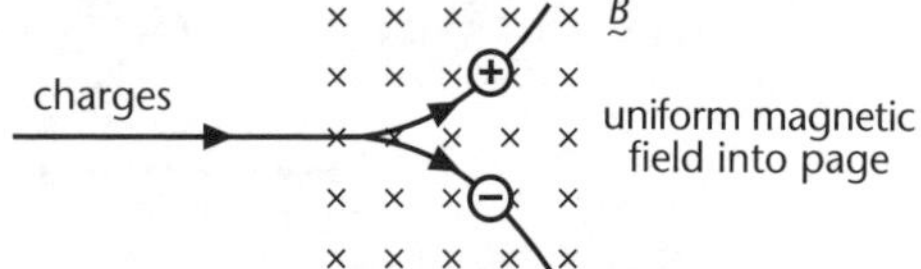

How charges are deflected in a magnetic field

Positive charges are deflected one way and negative charges in the opposite direction.

The force on a charge is constant and is at right angles to the direction of the charge's motion. This causes the charge to change direction but its speed remains the same. The charge moves in a circular arc.

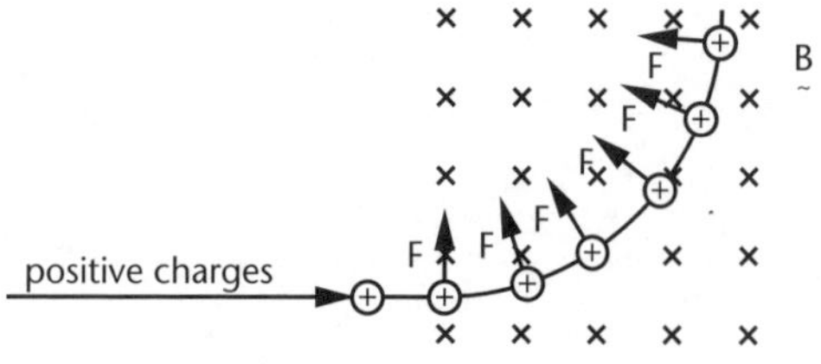

Explaining the deflection

To describe directions of current or magnetic field into and out of a page, a number of usual conventions are used:

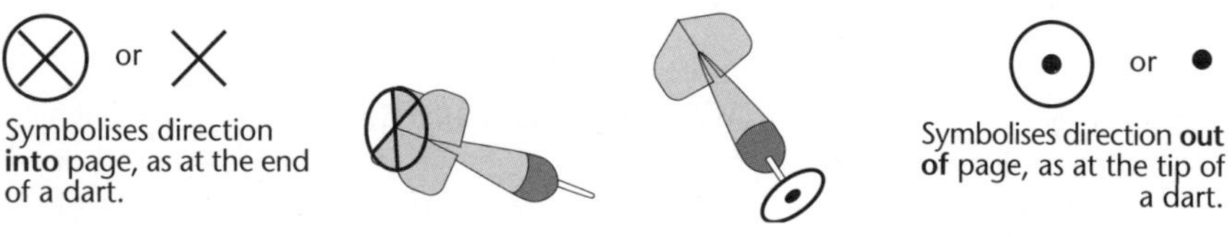

A beam of positive charges behaves like a small conventional current. The right-hand slap rule can be use to find the direction of the force on the positive charges.

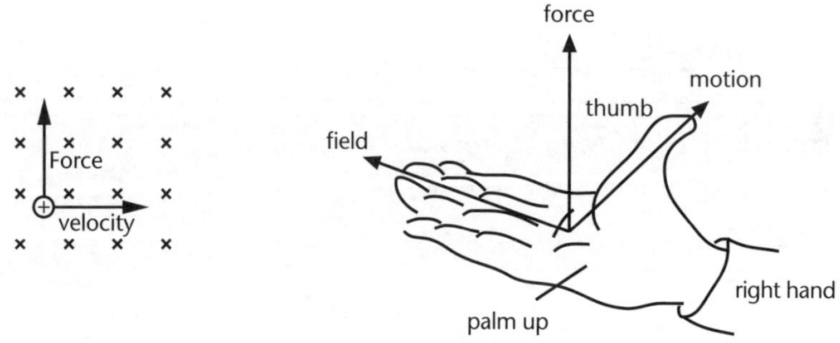

Using the right-hand slap rule to find direction of a force on positive charges

It can be shown by experiment that:

$$\boxed{F = Bqv}$$

where B is the size of the magnetic field in tesla, q is the amount of each charge in coulombs, v is the speed of the charges in m s^{-1}, and F is the force in **newtons**. In the special case where the charges are electrons or protons, q can be replaced by e, the elementary charge, giving $F = Bev$.

A TV tube uses deflecting coils to magnetically deflect the electron beam produced by the electron gun. The beam is deflected across the screen as a series of lines.

The TV adjusts the brightness of the spot as it moves according to the TV signal received, forming a picture 50 times a second.

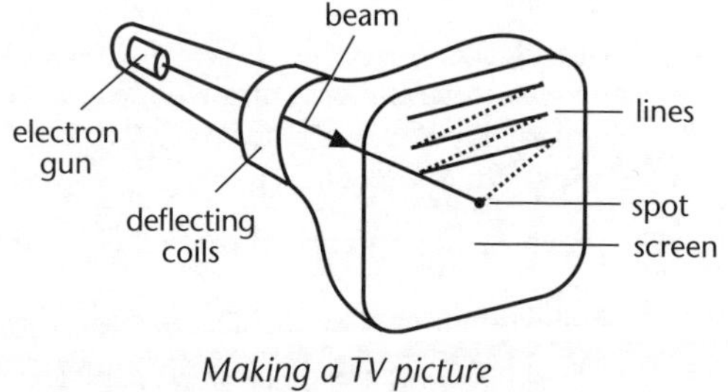

Making a TV picture

Example G

An electron in a TV tube is travelling at 10% of the speed of light and experiences a deflecting force of 9.6×10^{-14} N. Find the size of the deflecting magnetic field.

Solution:

Speed of light = 3×10^{8} m s^{-1}, electronic charge = 1.6×10^{-19} C.

10% speed of light = $0.1 \times 3 \times 10^{8}$ m s^{-1}

$= 3 \times 10^{-7}$ m s^{-1}

Since $F = Bev$, $B = \dfrac{F}{eV}$ [rearranging]

$= \dfrac{9.6 \times 10^{-14}}{1.6 \times 10^{-19} \times 3 \times 10^{7}}$ [substituting]

$= 2 \times 10^{-2}$ T

$= 20$ mT

Unit 12.4 Activity 2C: Moving charges and magnetism

1. Beams of charges enter magnetic fields, as shown.

a.

charges

b.

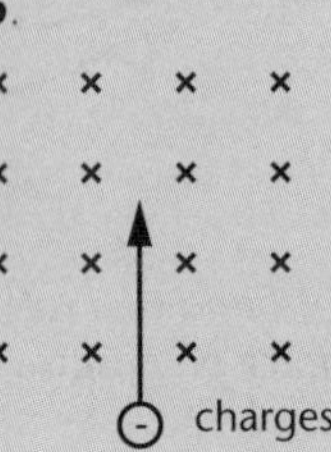

charges

Determine the direction of deflection of the charged particles in each case.

2. A narrow beam of protons (charge 1.6×10^{-19} C) moving with speed 2.0×10^{6} m s^{-1}, enters a uniform magnetic field of strength 0.20 T.
   a. Calculate the magnetic force on each proton.
   b. What is the size of a magnetic field that exerts a force of 1.0×10^{-14} N on helium **ions** ($q = 3.2 \times 10^{-19}$ C) travelling with a speed of 1.0×10^{7} m s^{-1} at right angles to the field?

3. A beam of electrons, each of mass m, charge e, and speed v, enters a strong uniform magnetic field B, and moves in a circular path of radius r, as shown.

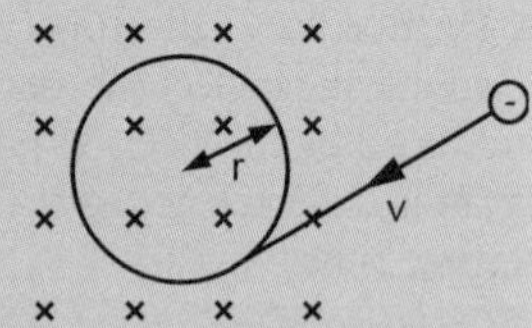

   a. The magnetic force on the electrons is $F = Bev$ and it provides the **centripetal force** $F = \dfrac{mv^2}{r}$ for their circular motion. Use this fact to show that the radius of the electrons' circular path is $r = \dfrac{mv}{Be}$.

b. Calculate the radius of the electrons' path if they enter a field of size 9.0×10^{-3} T at a speed of 3.2×10^{7} m s^{-1}. (Electronic charge $e = 1.6 \times 10^{-19}$ C, mass $m = 9.0 \times 10^{-31}$ kg.)

4. A metal strip is connected to voltage supply as shown; electrons can travel through the strip. Charge on the electron = 1.6×10^{-19} C. Mass of electron = 9.1×10^{-31} kg. Speed of the electron through the metal strip = 4.0×10^{-5} m s^{-1}. Size of the magnetic field = 0.20 T.

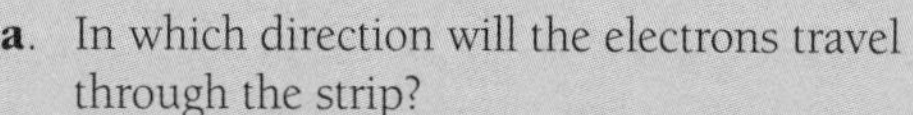

a. In which direction will the electrons travel through the strip?

b. Calculate the size of the force on the electron due to the magnetic field.

c. State which side of the strip the electrons move towards. Explain your answer.

d. State which side of the metal strip becomes positive. Explain your answer.

Unit 12.4 Activity 2D: Multiple choice questions

1. Which of the following fields can be created by a charge moving through the air?

A. **electric field**
B. gravitational field
C. magnetic field
D. electromagnetic field

2. Electromagnetic waves are created by variations of which two fields?

A. magnetic and gravitational
B. gravitational and electric
C. electromagnetic and magnetic fields
D. magnetic and electric fields

3. When an electric current is flowing through a wire a magnetic field will be created around it. If the wire were to move through an external magnetic field B, the wire will experience a magnetic force as a result of two opposing magnetic fields. The magnitude of the force is given by $F = IBl \sin\theta$, where θ is the angle between the current in the wire and the direction of external magnetic field. Which of the following positions between the current and external magnetic fields will give rise to maximum force?

A. parallel
B. perpendicular
C. antiparallel
D. at an angle of 45°

4. A wire carries a current of 10A in a direction of 90° with respect to the direction of an external magnetic field strength of 0.3 T. What is the magnitude of the magnetic force on a 5 m length of wire?

A. 0.15 N
B. 1.5 N
C. 15 N
D. 150 N

5. The diagram below shows a mass spectrometer. The electric field between the plates of the velocity selector is E = 1500 V/m, and the magnetic field B in both the velocity selector and the deflection chamber has a magnitude of 0.9 T. If the charge is an electron and the radius of orbit r for an electron of mass $m = 9.11 \times 10^{-31}$ kg in the deflection chamber, what is the velocity(v) of the electron?

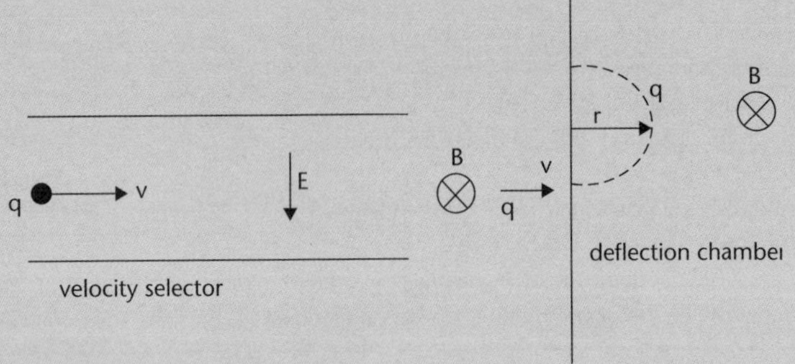

A. 1.05 ms^{-1}
B. 10.55 ms^{-1}
C. 105.55 ms^{-1}
D. 1,055.55 ms^{-1}

6. In the diagram of question 5 above, what is the radius (r) of the deflection inside the deflection chamber?
A. 6.68×10^{-7} m
B. 6.68×10^{-8} m
C. 6.68×10^{-9} m
D. 6.68×10^{-10} m

7. The two long straight wires in the diagram below each carry a current of I = 5 A in opposite directions and are separated by a distance d = 30 cm. What is the magnitude of the magnetic field at a distance L = 20 cm to the right of the wire on the right?
A. 3.00×10^{-5} T
B. 3.00×10^{-6} T
C. 3.00×10^{-7} T
D. 3.00×10^{-8} T

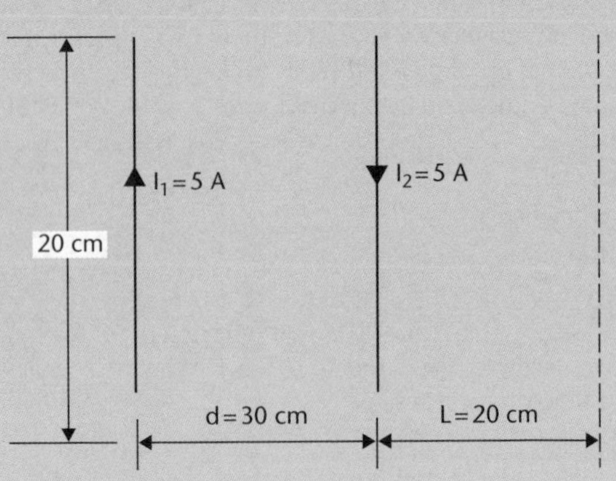

8. For the arrangement shown in the diagram below, the long straight wire carries a current of $I_1 = 5$ A. This wire is a distance $d = 0.1$ m away from a rectangular loop of dimensions $a = 0.3$ m and $b = 0.4$ m which carries a current $I_2 = 10$ A. What is the magnitude of the net force exerted on the rectangular loop by the long straight wire?
   A. 100×10^{-7} N
   B. 200×10^{-7}N
   C. 300×10^{-7} N
   D. 400×10^{-7} N

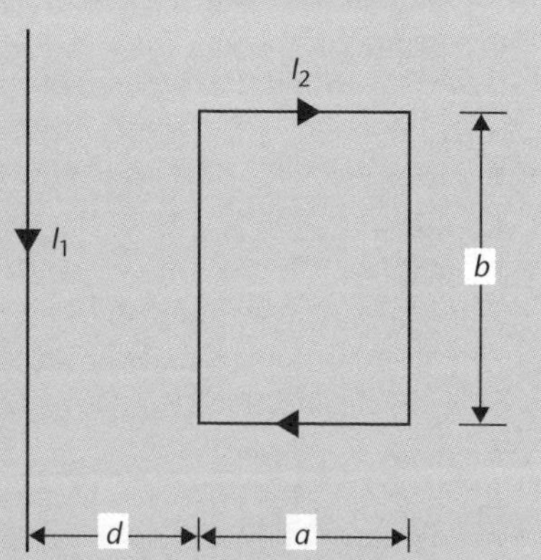

9. A proton is moving in a circular orbit of radius 14 cm in a uniform magnetic field of 0.35 T directed perpendicular to the velocity of the proton. What is the orbital speed of the proton?
   A. 4.7×10^5 ms^{-1}
   B. 4.7×10^6 ms^{-1}
   C. 4.7×10^7 ms^{-1}
   D. 4.7×10^8 ms^{-1}
10. A long straight wire carries a current of 5.0 A .At what distance from the wire does the field have a magnitude equal to 0.5×10^{-5} T?
   A. 0.02 m
   B. 0.2 m
   C. 2 m
   D. 20 m
11. A solenoid consists of 100 turns of wire and has a length of 10 cm. What would be the magnitude of the magnetic field inside the solenoid when it carries a current of 0.5 A?
   A. 6.3×10^{-2} Tesla
   B. 6.3×10^{-3} Tesla
   C. 6.3×10^{-4} Tesla
   D. 6.3×10^{-5} Tesla
12. Two wires, each having a weight per unit length of 1×10^{-4} N.m^{-1}, are strung parallel to one another above the surface of the Earth, with one being placed directly above the other. The wires are aligned in a north-south direction so that the Earth's magnetic field will not affect them. When their distance of separation is 0.1 m, what must be the current be in each in order for the lower wire to levitate the upper wire? Assume that each wire carries the same current travelling in opposite directions.
   A. 0.5 A
   B. 5.0 A
   C. 50 A
   D. 500 A

13. A proton moving with a speed of 5×10^7 ms^{-1} through a magnetic field of 2.0 T experiences a magnetic force of 3×10^{-12} N. What is the angle between the proton's velocity path and the direction of applied magnetic field?

A. 1.08°

B. 10.80°

C. 20.80°

D. 30.80°

14. A copper rod 10 m is placed in a north-east direction. It carries a current of 20 A in that direction. What would be the magnetic field *B* at a point S about 0.85 m from the wire and its direction?

A. 4.7×10^{-6} Tesla NE

B. 4.7×10^{-6} Tesla SE

C. 4.7×10^{-6} Tesla SW

D. 4.7×10^{-6} Tesla NW

15. An electric charge q = 25 μC and mass of 2.8×10^{-26} kg is moving with a velocity $v = 5 \times 10^7$ ms^{-1} in a north-east direction. What would be the magnitude of the magnetic field at a point 1.5 m from the path of the charge motion if **B** is situated at an angle $\theta = 65°$ to the direction of v?

A. 4.12×10^{-11} Tesla

B. 4.12×10^{-12} Tesla

C. 4.12×10^{-13} Tesla

D. 4.12×10^{-14} Tesla

Unit 12.4 Electromagnetism

Topic 3: The motor effect

Topic 3 deals with the motor effect. It covers:

- The motor effect.
- Symbols to indicate current direction.
- The principle of operation in electric motors.

A current-carrying wire which is at right angles to a magnetic field has a **magnetic force** exerted on it. This is called the **motor effect**. The size of the magnetic force can be calculated using the formula:

$$F = BIL$$

F is the magnetic force (N)
B is the magnetic field strength (N A^{-1} m^{-1}); 1 N A^{-1} m^{-1} is also called a **tesla** (T)
I is the current (A)
L is the length of wire in the field (in m)

The direction of the magnetic force may be determined by the 'Right Hand Slap Rule':

'Place your right hand so that your fingers point in the direction of the magnetic field (from N to S). Point your thumb in the direction of the conventional current (from +ve to –ve). The direction of the magnetic force is the direction in which your open palm will slap'.

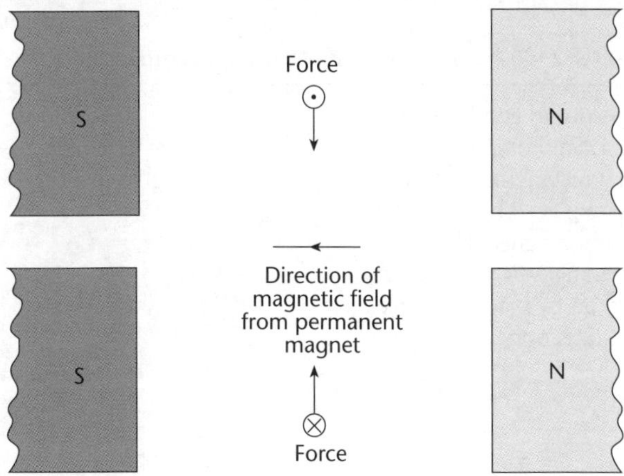

Force acting on a current-carrying conductor placed in an external magnetic field.

Symbols used to indicate direction of current and fields

Various symbols are used to indicate the direction of currents and magnetic fields in diagrams:

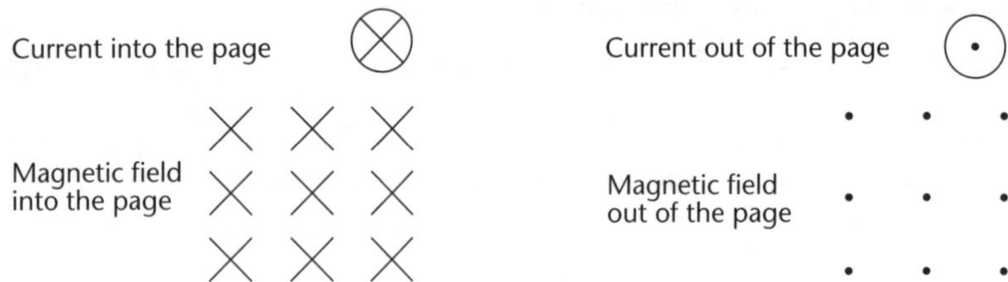

Symbols to indicate the directions of magnetic fields and currents.

Example A

Calculate the size and direction of the magnetic force on a 10 cm long wire that carries a current of 2 A (from left to right) at right angles to a magnetic field of strength 6×10^{-2} T (into the page), as shown in diagram.

$B = 6 \times 10^{-2}$ T

x x x
x x x
← 2 A
x x x
x x x

Magnetic force acting on a current-carrying wire.

Answer:

$F = BIL = 6 \times 10^{-2} \times 2 \times 0.1 = 0.012$ N = 12 mN acting upwards.

The motor effect is put to good use in the design of electric motors. An electric motor consists of a rectangular current-carrying coil of wire placed between the poles of a permanent magnet. The sides of the coil are at right angles to the magnetic field and are acted on by magnetic forces acting in opposite directions.

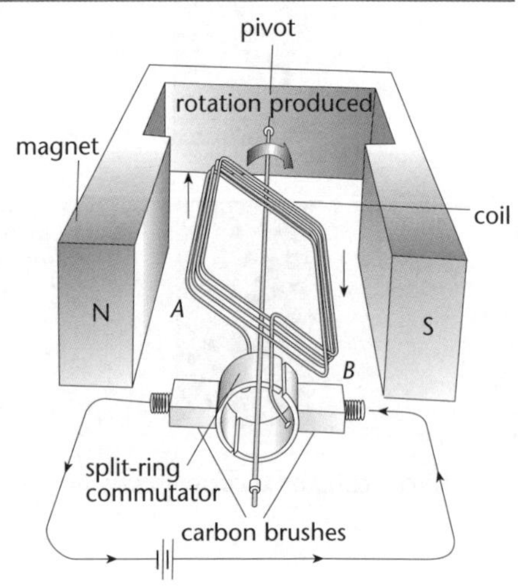

Design for a simple electric motor.

The forces acting on the sides of the coil constitute a force **couple** or **torque** that rotates the coil. The direction of **rotation** may be determined by applying the Right Hand Slap rule to the sides of the coil.

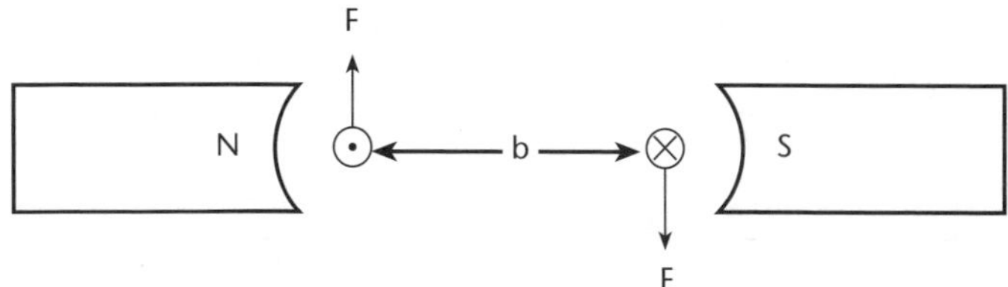

Forces acting on a rectangular current-carrying coil.

At the instant shown for the motor in the figure above, the torque of the motor may be calculated using the formula:

$$T = BIL\ \frac{b}{2} + BIL\ \frac{b}{2} = BILb$$

The **carbon brushes** press against the **commutator** and allow current to pass from the power source to the coil without anything becoming twisted or tangled. The brushes are made of carbon because carbon is an electrical conductor and is self-lubricating. The commutator reverses the direction through the current every half turn – this maintains the direction of rotation.

The strength of a motor can be increased by:

- Increasing the number of turns on the coil.
- Increasing the strength of the magnetic field.
- Increasing the current in the coil.

Unit 12.4 Activity 3A: Magnetic forces and motors

1. A 1.5 m long wire carries a current of 5 A at right angles to a magnetic field of strength 7 T. Calculate the size of the force exerted on the wire.
2. A 20 cm long wire carries a current of 3 A at right angles to a magnetic field. The wire has a force of 3 N exerted on it. Calculate the strength of the magnetic field.
3. A wire carries a current of 6 A at right angles to a magnetic field of strength 10 T. The wire has a force of 6 N exerted upon it. What is the length of wire in the field?
4. A 30 cm long wire carries a current of 2 A at right angles to the magnetic field of the Earth. The magnetic force acting on the wire is 4.8×10^{-3} N. What is the strength of the Earth's magnetic field?
5. Copy the following diagrams. Use the Right Hand Slap Rule to determine the direction of the magnetic force on the wires and mark the force on the diagram.

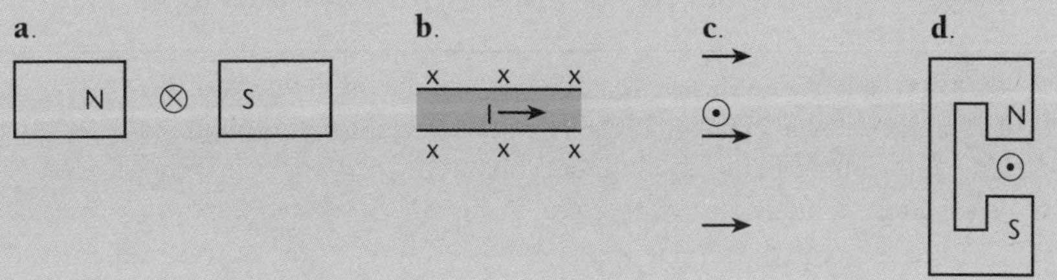

6. A power cable, slung between two pylons 400 m apart, carries a current of 10 A. The magnetic field strength of the Earth's magnetic field that acts at right angles to the cable is 6×10^{-5} T. Calculate the size of the force acting on the cable.
7. Describe the energy transformations that occur in an electric motor.
8. The diagram shows a cross-section of a simple electric motor:

   a. Use the Right Hand Slap Rule to determine whether the coil will rotate in a clockwise or anticlockwise direction.
   b. Describe the function of carbon brushes, spring contacts and the split ring commutator in a motor.
   c. What modifications can be made to an electric motor to make it more powerful?
9. The diagram shows a rectangular coil between the poles of a permanent magnet of strength 0.05 T. The current in the coil is 10 A. The coil is 10 cm long and 5 cm wide.

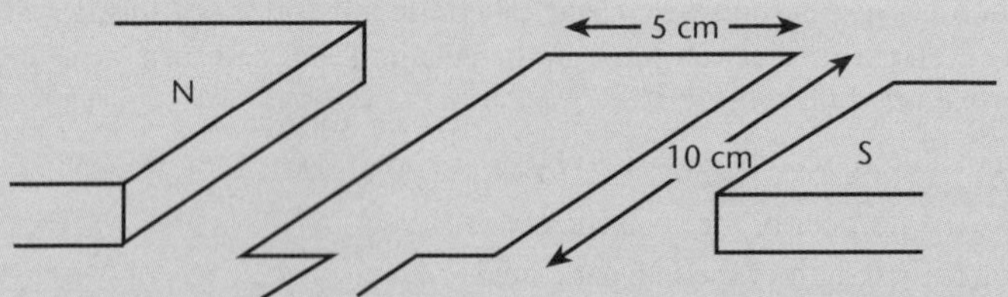

   Calculate the torque acting on the coil if the coil contains:
   a. 1 single turn. b. 200 turns.
10. What two properties of carbon make it suitable for the construction of brushes in electric motors?

The following information relates to Questions 11 and 12.

A vertical uniform magnetic field acts in a direction out of the paper and a conductor PQ which is free to move rests on horizontal conducting runners X and Y, as shown in the diagram.

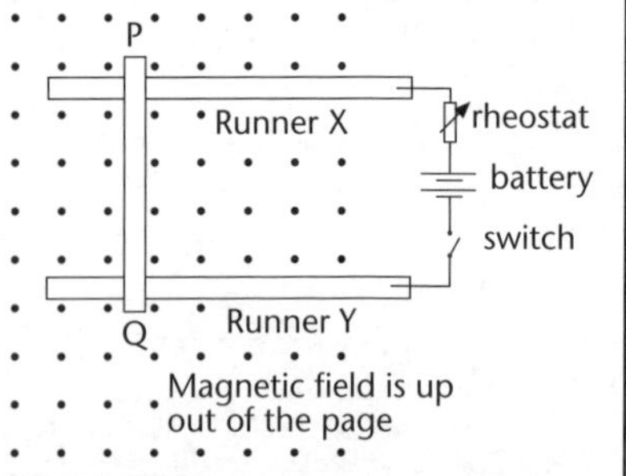

11. When the switch is closed, state the direction conductor PQ moves in.
12. Which of the following changes, when made separately, can result in the conductor moving in the opposite direction to that chosen in Question 11?
   i. The battery connections are reversed.
   ii. The conductor PQ is turned so that P rests on runner Y and Q rests on X.
   iii. The resistance of the **rheostat** is decreased.
   iv. The direction of the magnetic field is reversed.

13. Part of an electric motor is shown in the diagram. It consists of a rectangular coil which rotates between the poles of a magnet of strength 1.5 T. The current in the coil is in the direction shown.

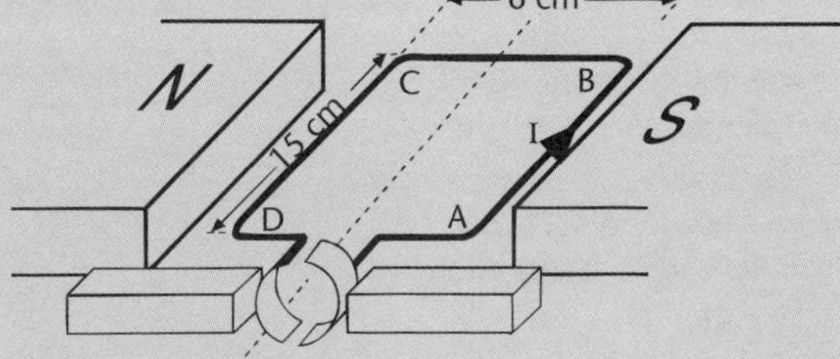

a. What is the function of:

i. The brushes?

ii. The commutator?

b. i. Copy the cross-sectional diagram below, then draw vectors to show the direction of the magnetic force on sides AB and CD.

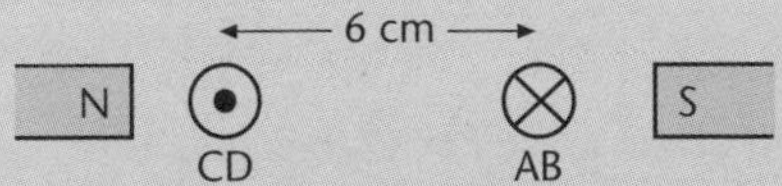

ii. Is the direction of the rotation of the coil clockwise or anticlockwise?

c. If the length of the sides AB and CD of the rectangular coil is 15 cm and the current through the coil is 8 A, calculate:

i. The force on side AB.

ii. The total torque (**moment**) on the coil about the pivot.

d. List *three* ways of improving the power output of the electric motor.

e. A DC motor in a crane on a building site is rated at 12 kW and 600 V DC.

i. What current does the motor draw?

ii. What is the resistance of the motor?

iii. How far can the crane raise a 150 kg concrete panel in 10 seconds?

14. The electric motor in the diagram contains a coil in a magnetic field. The motor spins clockwise when a **direct current** passes through it.

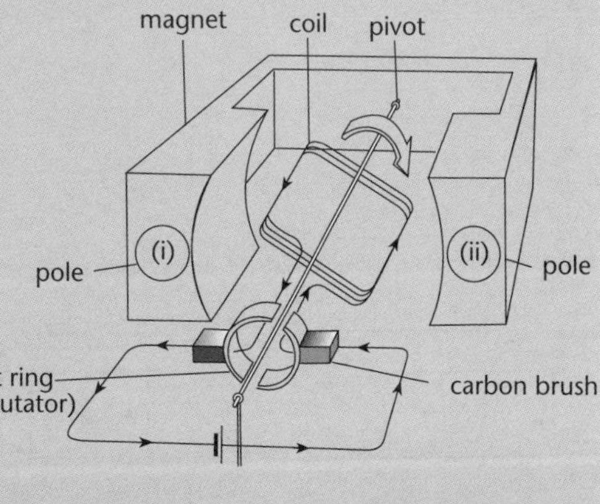

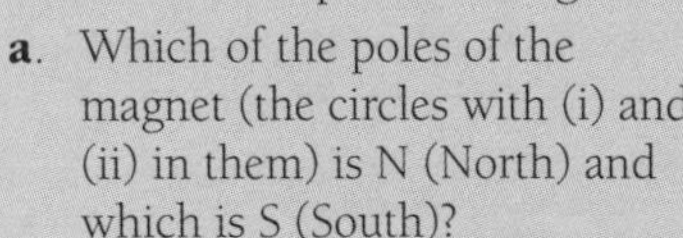

a. Which of the poles of the magnet (the circles with (i) and (ii) in them) is N (North) and which is S (South)?

b. What is the function of the split ring commutator?

c. The brushes allow current to pass in and out of the coil. Give two reasons why they are made of carbon.

d. Suggest two modifications which would make the motor more powerful.

15. An electric motor is used to lift a mass. Use the information in the diagram to answer the questions.

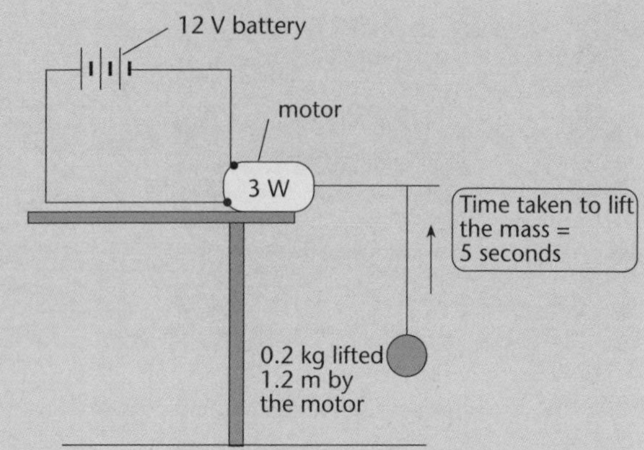

a. How much current is passing through the motor?

b. How much gravitational potential energy is gained by the mass?

c. How much electrical energy is used by the motor in the 5 second lift?

d. Discuss the energy changes which occur as the mass is being lifted.

16. On the following diagrams draw the direction of the force which acts on the current carrying wires in a magnetic field.

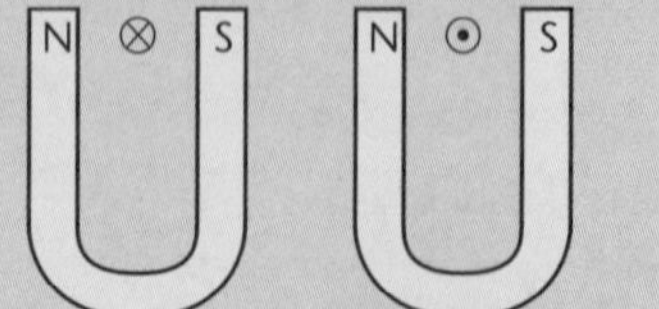

Key: ⊙ = current out of page.

Key: ⊗ = current into page.

17. A wire is suspended between the poles of a horseshoe magnet, as shown. Draw an arrow to show the direction the wire will move when the switch is closed.

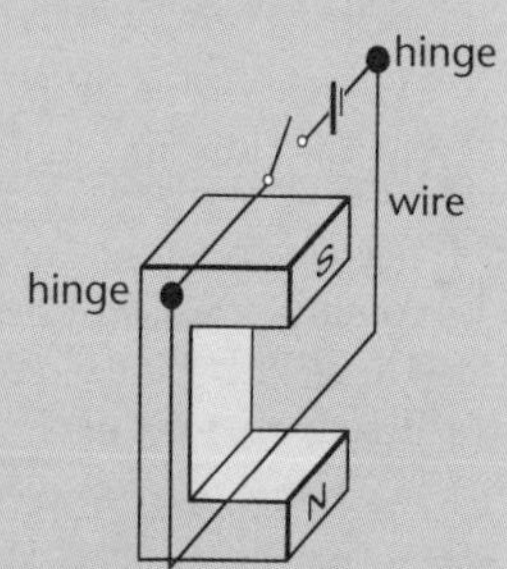

18. A coil carrying a current between the poles of a magnet can be simply drawn, as shown.

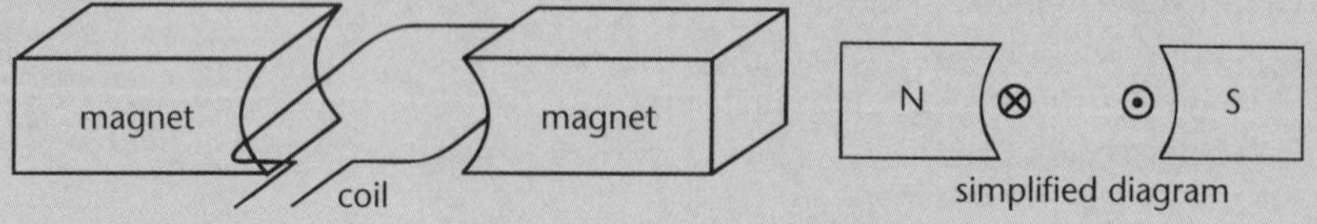

On the diagram draw arrows to correctly show the direction of the force on the coil.

19. **a**. The diagram shows a cross-section of a loudspeaker. What do the parts labelled A, B, and C represent?

b. The loudspeaker emits a high-pitched sound at low volume. Choose one of the options in each of **i–iii** to best describe the current which enters the speaker.

i. Type of current – AC *or* DC?

ii. Size of current – large *or* small?

iii. Frequency of current – high *or* low?

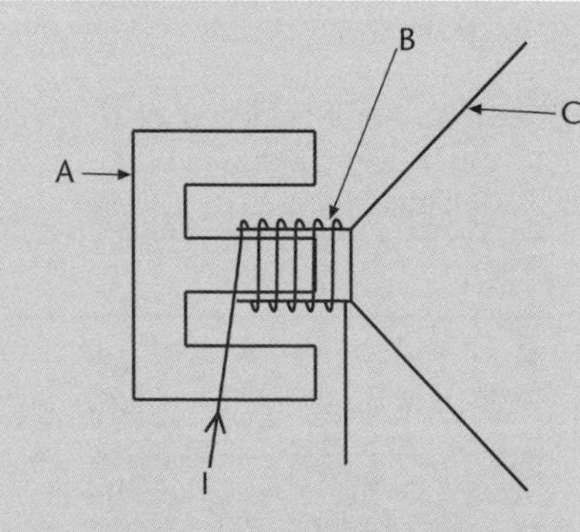

20. The diagram shows the working parts of a model railway signal.

a. Copy the diagram of the solenoid and mark on it the North and South poles when the switch is closed.

b. Does the right-hand end of the signal arm move up or down when the switch is closed?

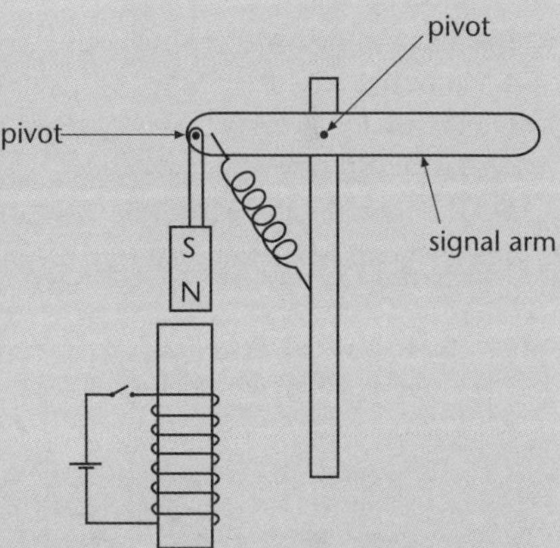

21. Sally uses a computer in her shop. A DC motor makes the computer fan rotate to keep the computer cool. A simplified diagram of a DC motor is shown below.

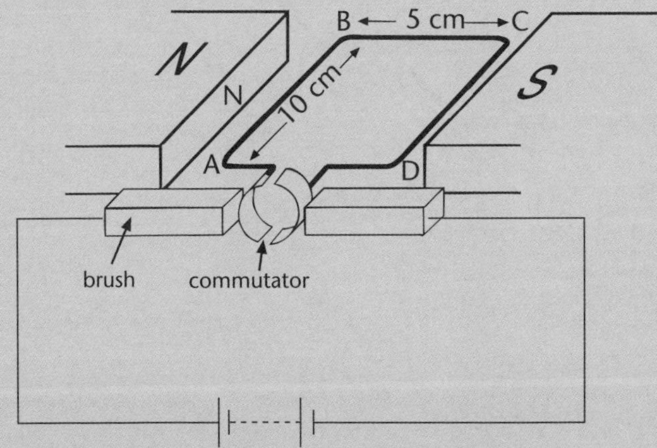

a. With the commutator in the position shown, the current in the coil flows in the direction ABCD. State the direction of the force on wire AB.

b. If the magnetic force on the wire is 0.050 N and the current through the wire is 2.0 A, calculate the magnetic field due to the magnets.
c. The motor can be made to spin faster. Describe ONE change that will make the motor spin faster and fully explain how the speed of the motor is altered by this change.

Unit 12.4 Activity 3B: Multiple choice questions

1. Which of the following statements briefly describes the difference between the DC motor and an AC motor?
   A. DC or AC refers to how the electrical current is transferred through and from the motor.
   B. AC motors work for situations where speed needs to be controlled and have a continuous current while DC has an alternating current.
   C. DC motors work well for systems that are hard to start because they need a lot of power upfront as compared to AC.
   D. DC motors can be two phases or three phases while AC is only a single phase.
2. The diagram below shows a DC motor. The rotation is in the clockwise direction.

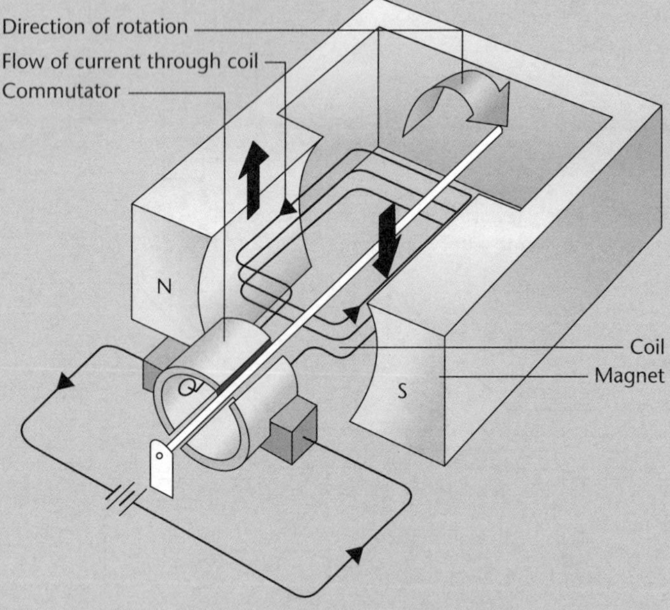

How many slip rings does it have?
   A. 1
   B. 2
   C. 3
   D. 4

3. If the DC motor in question 2 is replaced by an AC motor, how many slip rings does an AC motor have?
   A. 1
   B. 2
   C. 3
   D. 4
4. What would be the net torque on the electric motor in question 2 if the current through the rectangular coil is 5 A. The magnetic field strength is 0.5 T, the length of the coil is 8 cm and the width is 4 cm.
   A. 0.013 N.m clockwise
   B. 0.014 N.m clockwise
   C. 0.015 N.m clockwise
   D. 0.016 N.m clockwise
5. Which of the following types of motors are best suited for a system that is hard to start and requires a lot of power upfront?
   A. DC motor
   B. AC motor two phase
   C. AC motor three phase
   D. All of the above.

Unit 12.4 Electromagnetism
Topic 4: Electromagnetic induction

Topic 4 deals with electromagnetic induction. It covers:
- Generation of current.
- Factors determining magnitude of current.
- AC and DC generators.

Introduction

This chapter deals with the discovery made by Michael Faraday (1791–1867), an English physicist and chemist, that if a force moves a wire through a magnetic field, a voltage is *induced* across the ends of the wire.

This effect is called **electromagnetic induction** because electricity is generated from a magnetic field. Another name for this is the **generator effect**. In school situations, the induced current is often very small (a few microamps; 1 μA = 10^{-6} A), and to measure the induced current, a sensitive current measuring meter (such as a **galvanometer**, symbol Ⓖ) is used.

Induced voltage and current

Electromagnetic induction can be demonstrated using a bar magnet moving near a solenoid.

Example A

A bar magnet is moved near a solenoid (or coil) which is connected to a galvanometer.

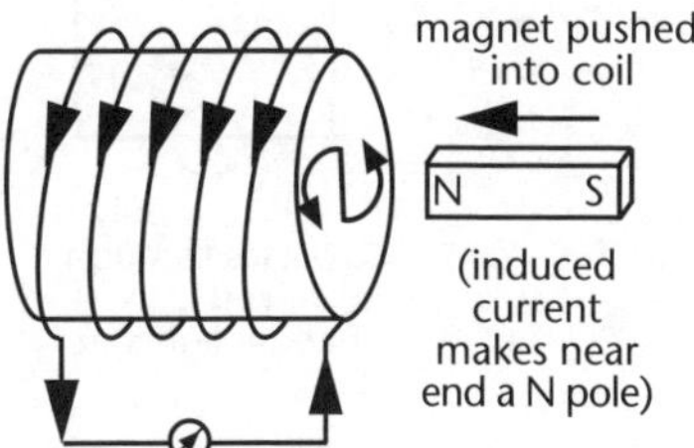

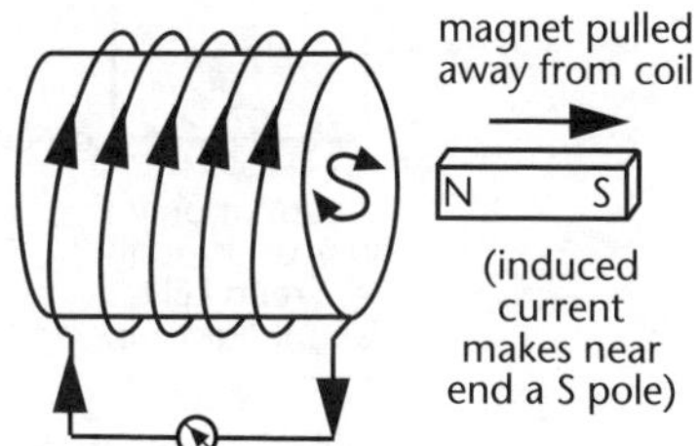

a. The north pole of a magnet is pushed toward the coil. A small current is induced in a direction around the solenoid which makes a north pole nearest the magnet.

b. A north pole is pulled away from the coil. A current is induced which makes the end of the solenoid closest to the magnet a south pole.

An electromagnetic induction demonstration

A small *opposing* force is caused by the magnet's movement. The magnet must be pushed with a small force to get it closer to the solenoid in **a** since *like poles repel,* and pulled with a small force from the solenoid in **b** since *unlike poles attract.*

If the magnet is pushed all the way through the coil and out the other side, the current will change direction as the magnet leaves the solenoid. The current direction is now the same as in **b**, when, instead, the magnet is pulled out of the solenoid.

The effect noted above is called **Lenz's Law** – the direction of the induced current always *opposes the change* producing it. This law is a result of *conservation of energy*, ie work has to be done to generate electrical energy.

> Students often get confused with this idea, thinking that the coil always sets up an opposing magnetic pole. This is clearly the case when the magnet is pushed into the coil, but not when it is pulled out. The key thought to keep in mind is that the coil always reacts to oppose the change.

A **light-emitting diode** (**LED**, symbol) produces light when a current flows in the forward direction (shown by the arrow) through it. No current is able to pass in the reverse direction through it. Light-emitting diodes can be used in place of a galvanometer in detecting the current induced in a solenoid.

Example B

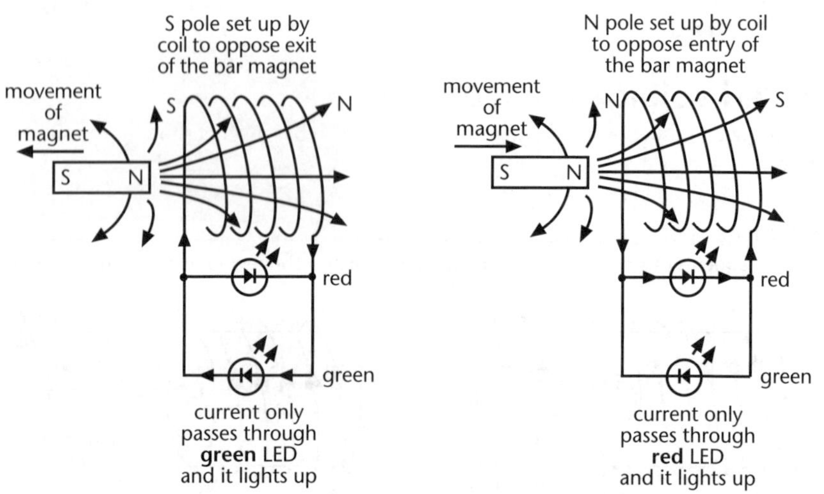

An LED electromagnetic induction demonstration

There are a number of induction fundamentals to be learned.

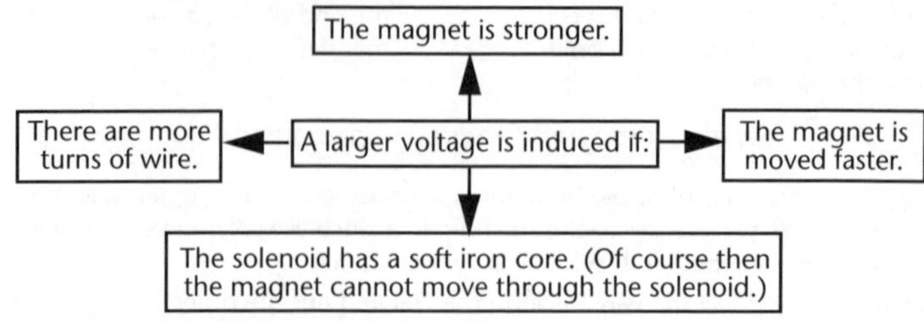

More turns of wire means more **resistance**, which will lower the current by **Ohm's Law**, so there is a 'play off' effect whether more current is induced in an external circuit. Likewise, thicker wire will mean less resistance, enabling a bigger induced current for the same induced voltage.

Unit 12.4 Activity 4A: An introduction to electromagnetic induction

1. A magnet is dropped vertically down through a solenoid and a graph of the induced current through the galvanometer is produced:

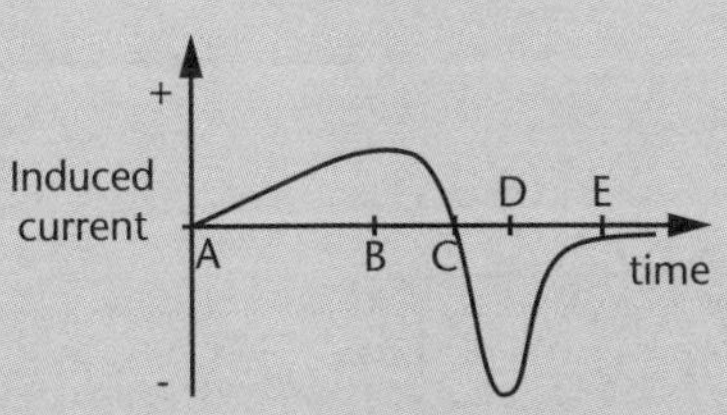

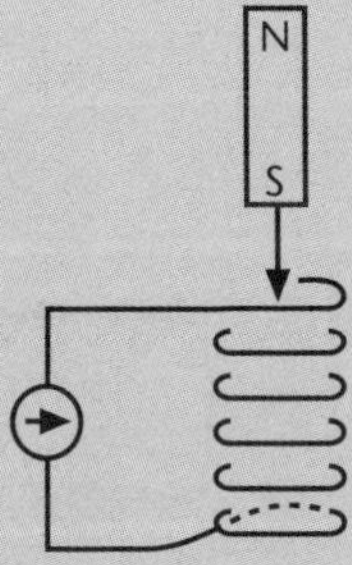

   a. At which point in time, A, B, C, D or E, is the magnet at the centre of the solenoid?
   b. Explain why the second (negative) pulse of induced current is shorter in duration.
   c. Explain why the second pulse of current has a greater magnitude than the first (positive) pulse of induced current.
2. A magnet moves towards a suspended metal ring, as shown.
   a. In which way does the induced current flow in the ring as the magnet approaches?
   b. Describe the motion of the ring as the magnet approaches.
   c. The magnet is held inside the metal ring and then withdrawn to the left.
      i. In which way does the induced current flow in the ring as the magnet is withdrawn?
      ii. Describe the motion of the ring as the magnet is withdrawn.
   d. The magnet is held inside the metal ring and then withdrawn to the right.
      i. In which way does the induced current flow in the ring as the magnet is withdrawn?
      ii. Describe the motion of the ring as the magnet is withdrawn.

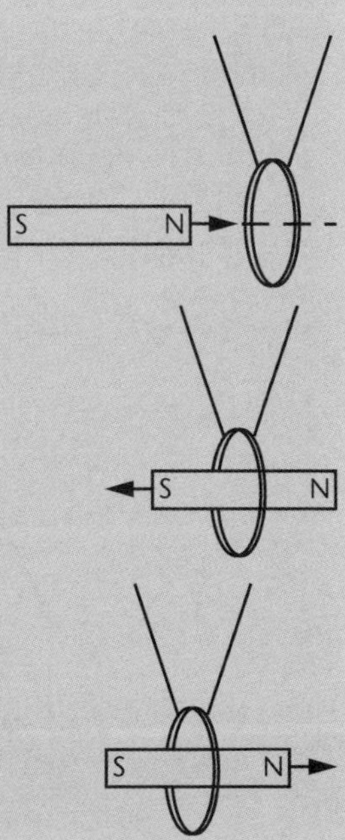

3. a. A bar magnet is pushed into a coil, stopped inside it, and pulled out as shown. Explain what will happen when:
   i. The bar magnet is pushed into the coil.
   ii. The bar magnet stops inside the coil.
   iii. The bar magnet is pulled out of the coil.

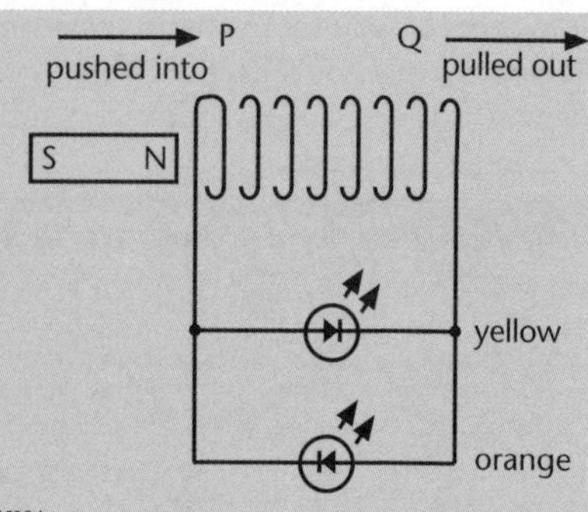

   b. A circuit is designed as shown in the diagram:

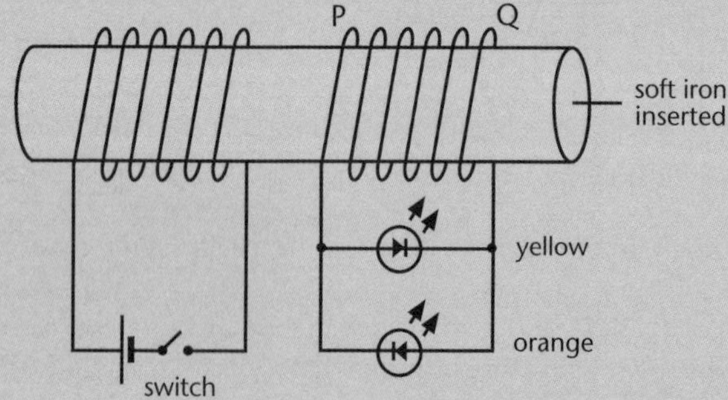

   Explain what will happen when:
   i. The switch is closed rapidly.
   ii. The switch remains closed.
   iii. The switch is opened rapidly.

4. A bar magnet is arranged as a pendulum to just swing into two coils as shown.

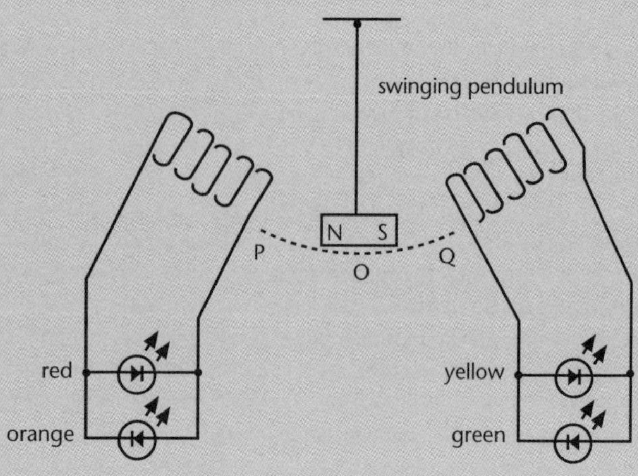

Use the words *red*, *orange*, *yellow*, *green*, *none*:

a. To describe the sequence that the LEDs light as the pendulum oscillates
O ⟶ P ⟶ O ⟶ Q ⟶ O.

b. Repeat part (a) for the situation where the bar magnet is reversed.

c. Carefully explain why in either parts **a** or **b** the two LEDs in each pair can never be on together.

How voltage and current are induced

Consider a wire of length l being moved through a magnetic field of size $\underset{\sim}{B}$ by an external force $\underset{\sim}{F}$. According to the right-hand slap rule (generator version), a voltage is induced, and in this closed circuit the induced current is clockwise. (An alternative is to use the left-hand slap rule, where your thumb points in the direction of induced current, your fingers point in the direction of the magnetic field, and your palm moves in the direction of motion of the wire.)

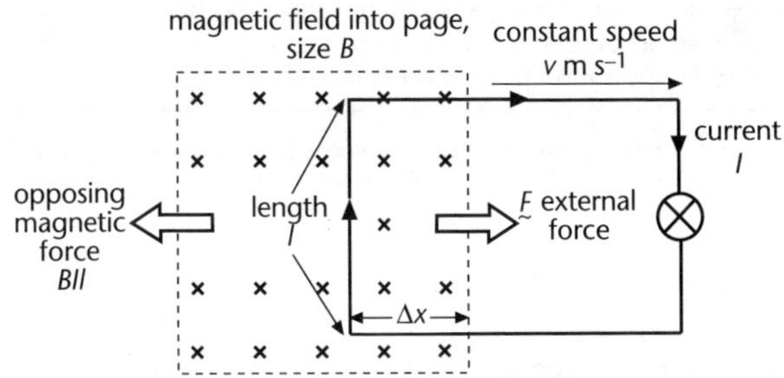

Further use of the right-hand slap rule

When a current flows in a magnetic field, the right hand-slap rule (motor version) has to be considered. This predicts a magnetic force ($F = BIl$) to the left opposing the external force to the right. An equilibrium situation is reached when the steady speed of the wire is v m s^{-1} to the right, so the external force is now equal to BIl. If the distance moved is Δx, then the mechanical work done is $BIl\Delta x$ in time Δt. The rate of work done is $BIl\dfrac{\Delta x}{\Delta t}$, and this must equal the rate of generation of electrical energy, IV.

$BIl\dfrac{\Delta x}{\Delta t} = IV$ and substituting $v = \dfrac{\Delta x}{\Delta t}$ gives:

$$\boxed{V = Bvl}$$

The moving wire acts like a **battery** with a voltage of Bvl volts.

Another commonly used term for induced voltage is **electromotive force (emf)**.

Example C

A small jet of wingspan 30 m is flying horizontally at a speed of 200 m s^{-1}. Find the induced voltage between the wing tips if the vertical component of the earth's magnetic field is 52 μT.

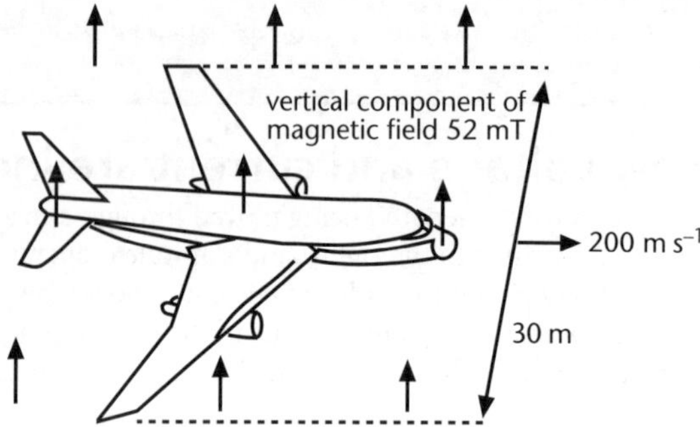

The induced voltage is:

$V = Bvl$

$= 52 \times 10^{-6} \times 200 \times 30$

$= 0.31$ volts

A **voltmeter** placed across the wingtips would not measure this voltage, since there would also be 0.31 V induced across the connecting wires and this would act like two identical batteries connected in opposition to each other.

An induced voltage will cause current to flow if a circuit is connected, outside the magnetic field.

Example D

A metal rod is pushed in a uniform magnetic field along contacts which are connected to a 2 Ω **resistor**. The length of rod in the magnetic field is 0.5 m. The rod moves at 10 m s^{-1}. The size of the magnetic field is 0.2 T. The induced voltage across the rod is:

$B = 0.2$ T into page

$l = 0.5$ m, $v = 10$ m s^{-1}, $R = 2\ \Omega$, I

region of constant magnetic field inside the dotted lines

$V = Bvl$

$= 0.2 \times 10 \times 0.5$

$= 1.0$ volt

The current, I, that flows is: $I = \dfrac{V}{R}$ [rearranging $V = IR$]

$= \dfrac{1.0}{2}$

$= 0.5$ A

The electrical power delivered to the circuit is: $P = VI$

$= 1.0 \times 0.5$

$= 0.5$ W

0.5 joules of electrical energy is produced each second. This energy comes from the work done in pushing the rod at a constant speed of 10 m s^{-1} against an opposing force.

As a current flows, a force, F, acts *to the left* to oppose the motion (Lenz's Law).

Example E

To keep the rod shown moving, a force equal in size and opposite in direction must be exerted. The size of the force is:

$F = BIl$

$= 0.2 \times 0.5 \times 0.5$

$= 0.05$ N

In one second, the rod would move 10 m and the work done in pushing would be:

$W = F \times d$

$= 0.05 \times 10$

$= 0.5$ J

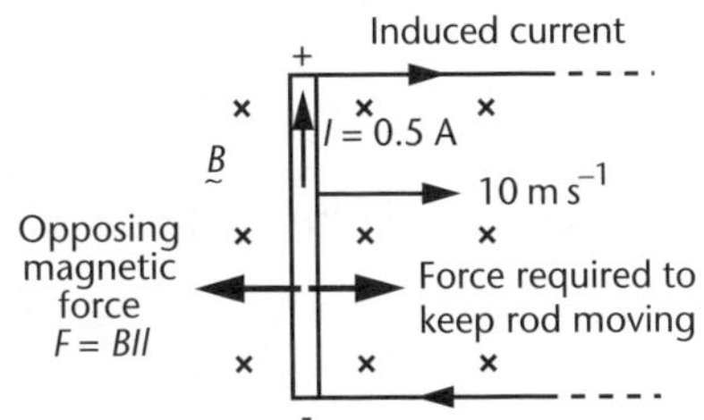

The work done per second to push the wire equals the electrical energy generated per second.

Unit 12.4 Activity 4B: Electromagnetic induction calculations

1. A particular place in the northern hemisphere has a very strong magnetic field described by the components shown.

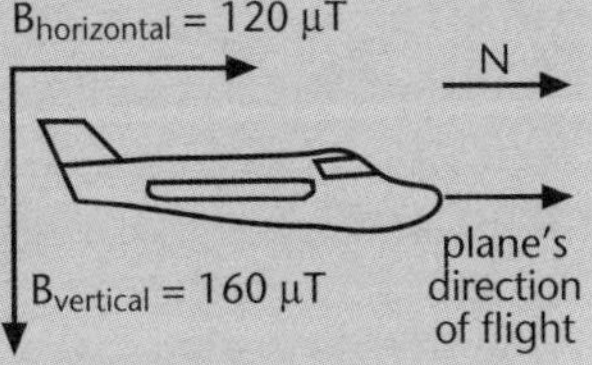

A military jet with a wingspan of 30 m is flying due N over the place. If the jet is flying at 500 m s^{-1} horizontally:

a. What is the voltage induced across the wing tips?

b. Which wing of the plane will have the higher voltage?

c. The pilot knows that this voltage is big enough to light green and red LEDs, so he has the two LEDs wired across the wing tips. Explain carefully which coloured LED will light.

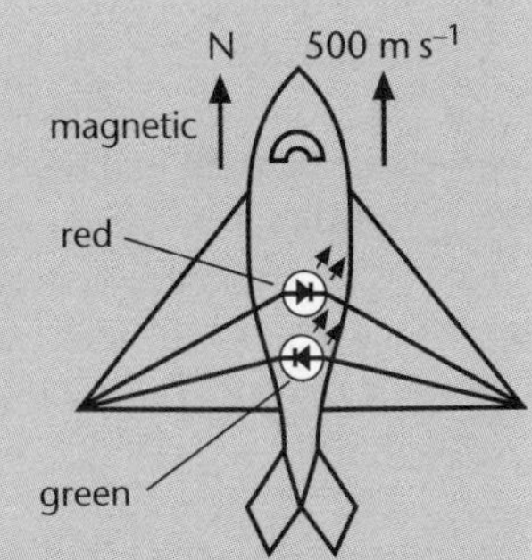

2. A wire of length 0.50 m is moving at a speed of 3.0 m s^{-1} at right angles to a uniform magnetic field of size 2.0 T. The ends of the wire are connected to a LED arrangement, as shown.

a. Which LED lights up?

b. What is the magnitude of the voltage generated while the wire is being moved?

c. What current flows while the wire is moving?

d. What is the magnitude of the applied force needed to move the wire at the steady speed of 3.0 m s^{-1}?

e. What happens to the energy that is needed to move the wire?

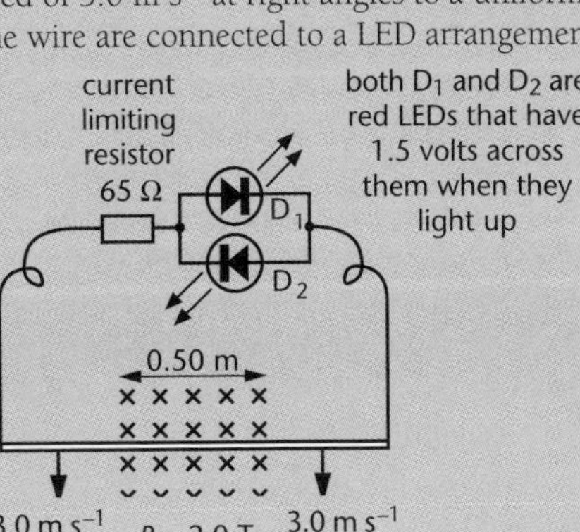

f. What will happen to D_1 and D_2:
   i. If the speed of movement is doubled?
   ii. If the wire moves at 3.0 m s^{-1} in the opposite direction?

3. A conducting rod RS makes contact with two metal rails, CD and EF, in a uniform magnetic field perpendicular to, and into, the plane of the paper.

   a. If the rod is moved to the right with a constant velocity, v, determine the direction of the force acting on an electron in the moving rod RS.
   b. If the size of the magnetic field is 0.8 T and the velocity of the 2.0 metre rod is 50 cm s^{-1} in the direction shown, determine the emf induced across the rod.

4. A wire of length 0.50 m is moved at a speed of 3.0 m s^{-1} at right angles to a uniform magnetic field of size 2.0 T. The ends of the wire are connected to a resistor of resistance 5.0 **ohms**.

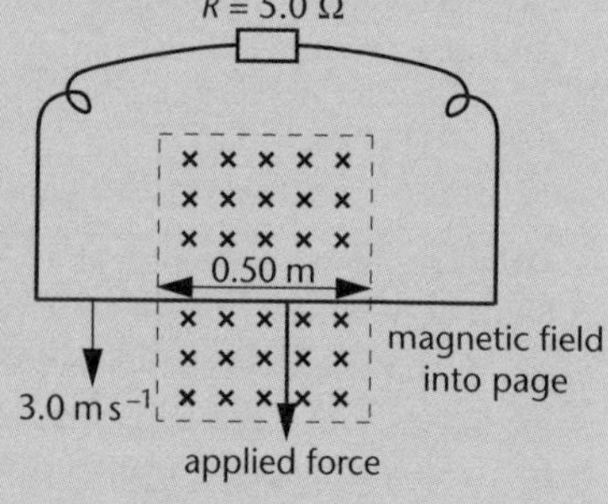

   a. What is the magnitude of the current flowing in the circuit while the wire is being moved?
   b. What is the magnitude of the applied force needed to move the wire at the steady speed of 3.0 m s^{-1}?
   c. What happens to the energy that is needed to move the wire?

5. Magnetic flux is defined as the *amount of magnetic field through a cross-sectional area*. Mathematically, magnetic flux is the product of the magnetic field strength, B, and the cross-sectional area, A, through which the magnetic field passes; ie magnetic flux = BA.

   Consider the situation of a metal rod moving in a **uniform field**, as shown.

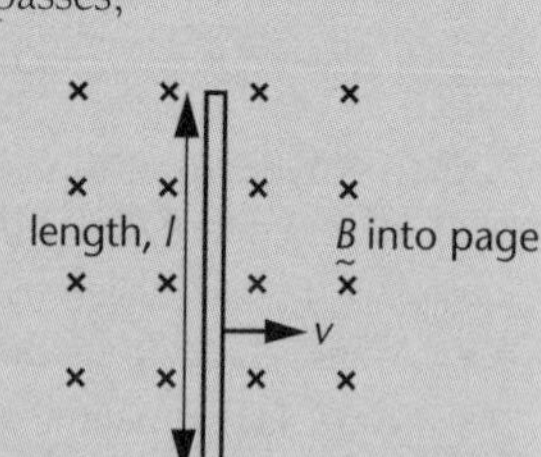

   a. Determine the area of magnetic field the rod crosses through in Δt seconds.
   b. Determine the amount of magnetic flux crossed through in Δt seconds.
   c. Show that the voltage, V, induced across the ends of the rod is found from:

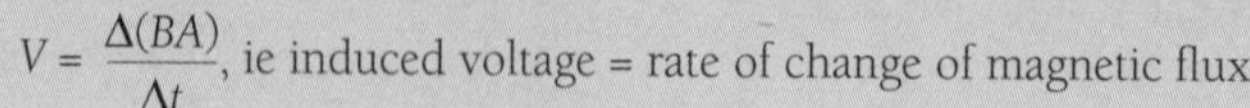

$V = \dfrac{\Delta(BA)}{\Delta t}$, ie induced voltage = rate of change of magnetic flux.

6. A boat travelling east at a speed of 2.5 m s^{-1} has a 5.0 m high aluminium mast. The magnetic field is 4.2×10^{-5} T northwards.

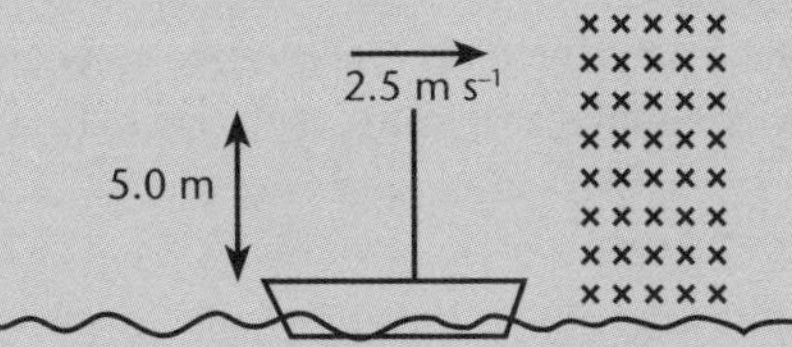

   a. Calculate the induced voltage across the mast.
   b. A piece of wire is connected to the top and bottom of the mast so that it is parallel to the mast. Explain why no current can flow in this circuit.
   c. Explain why, when the boat changes direction and sails northwards, the induced voltage is zero.

7. If a train moved across the magnetic field of the earth (as shown below), an induced voltage in the axles could be produced.

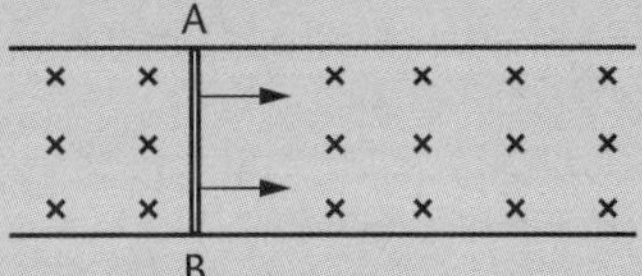

   The rails are 0.75 m apart and the Earth's magnetic field is 5.0×10^{-5} T.
   a. Calculate the speed of the train, if the induced voltage is 3.1 mV.
   b. If the train was moving east, state which end of the axle is positive.
   c. A lamp is placed into the circuit, as shown below.

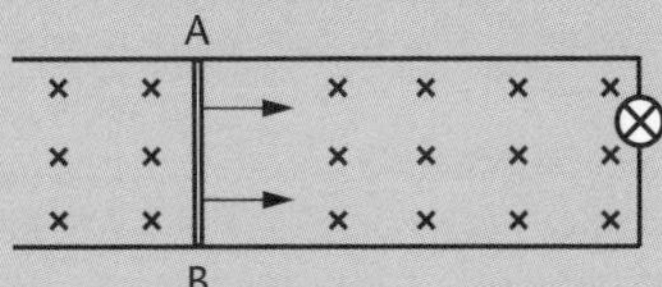

      i. Should the lamp have a high or a low resistance if the maximum power is to be produced? Explain.
      ii. If the resistance of the lamp is 1.1 Ω, calculate the power generated.

Unit 12.4 Activity 4C: Multiple choice questions

1. Which of the following statements best describes the term *electromagnetic induction*?
   A. Moving a length of conducting wire perpendicular to an external magnetic field, which disturbs the field, will cause the charged particles in the wire to drift. This is called current and voltage induction.
   B. Relative movement of a bar magnet close to a coil will cause opposing magnetic field reaction. This will cause the charged particles to flow through the wire. This is called current and voltage induction.

C. Alternation of current through a coil will alter its end polarities. This effect will cause the opposite reaction to another coil close to it. This will result in current flow in the nearby coil.

D. All of the above are various methods of electromagnetic induction and are correct.

2. Which of the following techniques will **not** result in a maximum induced voltage?

A. faster relative movement of a magnet to a coil

B. adding more turns to a coil

C. relative movement of magnet to a coil at an angle less than 90°

D. using a stronger magnet.

3. Which of the following definitions best briefly describes Faraday's Law?

A. The magnitude of the induced voltage is inversely proportional to the rate at which the external magnetic field lines are disturbed.

B. The magnitude of the induced voltage is directly proportional to the rate at which the external magnetic field lines are disturbed.

C. The magnitude of the induced voltage is directly proportional to the strength of the external magnetic field.

D. The magnitude of the induced voltage is inversely proportional to the strength of the external magnet field.

4. Which of the following definitions best briefly defines Lenz's Law?

A. An electromagnetic field interacting with a conductor will generate an electrical current that induces a counter magnetic field that opposes the magnetic field generating the current.

B. An electromagnetic field interacting with a conductor will generate an electrical current that induces a counter magnetic field that depends on the magnetic field generating the current.

C. An electromagnetic current interacting with a conductor will generate an electrical current that induces a counter magnetic field that opposes the magnetic field generating the current.

D. An electromagnetic current interacting with a conductor will generate a magnetic field that induces a counter magnetic field that opposes the magnetic field generating the current.

5. Which of the following factors will induce maximum current in a stationary coil?

A. the number of turns of a coil

B. the strength of the magnetic field

C. the speed of the magnet

D. all of the above.

6. A magnetic flow meter is measuring the induced voltage when a conductive liquid is moving with a velocity of 500 ms^{-1}. The magnetic field B due to a conductive liquid moving at that speed is 0.2 Tesla and the distance between electrodes in the magnetic flow meter is 5 cm: How much voltage is induced every second?

A. 0.05 V

B. 0.50 V

C. 5.00 V

D. 50.00 V

7. The diagram below shows a can of cola is wrapped around a solenoid. When the switch is closed a capacitor in the solenoid is discharged through the solenoid. What will happen to the can of cola?

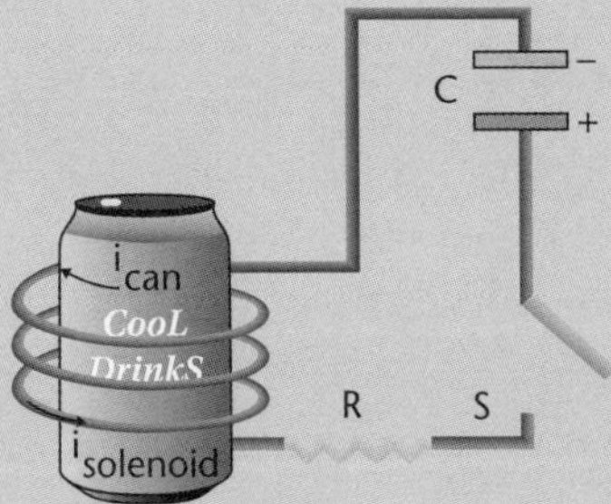

   A. The can will explode
   B. The can will be crushed
   C. The can will rotate inside the solenoid
   D. Nothing happens.

8. What would be the best explanation why the cola can in the diagram of question 7 would be crushed when the switch is closed?
   A. A variable flux is produced and passes through the can inside the solenoid, inducing in it a current in the opposite direction of the current through the solenoid (Lenz's Law). The two antiparallel currents repel each other, and since the solenoid is fixed the can will be crushed.
   B. A variable flux is produced and passes through the can inside the solenoid, inducing in it a current in the same direction of the current through the solenoid (Lenz's Law). The two parallel currents repel each other, and since the solenoid is fixed the can will be crushed.
   C. A variable flux is produced and passes through the can outside the solenoid, inducing in it a magnetic flux in the opposite direction of the current through the solenoid (Lenz's Law). The two antiparallel currents repel each other, and since the solenoid is fixed the can will be crushed.
   D. A variable flux is produced and passes through the can outside the solenoid, inducing in it a current in the opposite direction of the flux through the solenoid (Lenz's Law). The two parallel currents repel each other, and since the solenoid is fixed the can will be crushed.

Unit 12.4 Electromagnetism

Topic 5: More about induction

Topic 5 provides further information on induction. It covers:

- Lenz's Law.
- Magnetic flux and induction.
- Generating alternating current.
- AC and DC generators.

When a wire carrying a current is placed in a magnetic field it experiences a force. The effect involves the conversion of the electrical energy to mechanical energy and is the principle of the electric motor.

Electromagnetic induction is the opposite – the conversion of mechanical energy to electrical energy.

Induction is used to generate electricity in many different devices including a bicycle dynamo, a car generator and the turbines of a power station.

Induction always involves a *changing* magnetic field in or near a circuit or coil. An induced voltage is produced which can cause a current if the circuit is complete.

- A current is induced in the solenoid when the magnet is moved either in or out (but there is *no* current when the magnet is not moving).

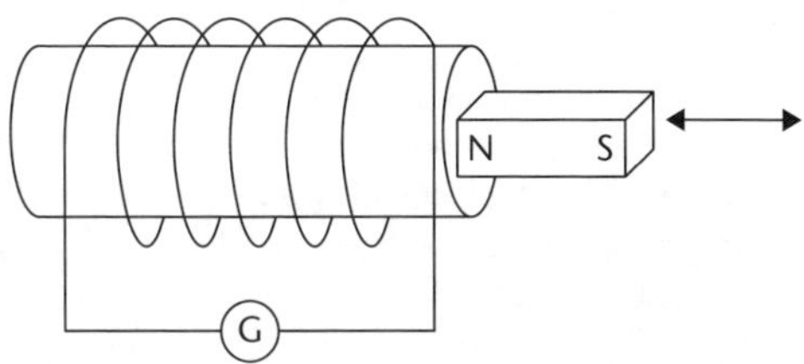

- When loop ABCD is pushed into the magnetic field a current is induced in the anticlockwise direction since BC is crossing the magnetic field lines. The size of the induced voltage V, across BC, is given by:

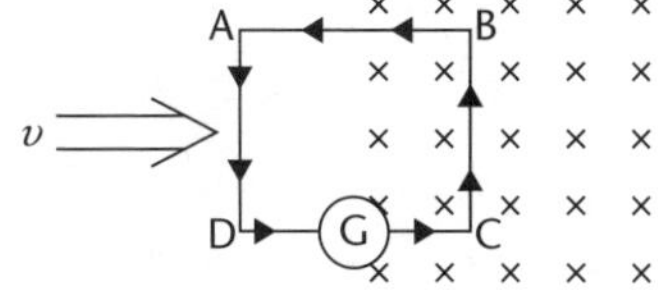

$$V = BvL$$

where B is the magnetic field strength, v is speed of movement across the field lines, and L is the length of wire in the field.

The direction of the induced current is given by the modified right hand slap rule.

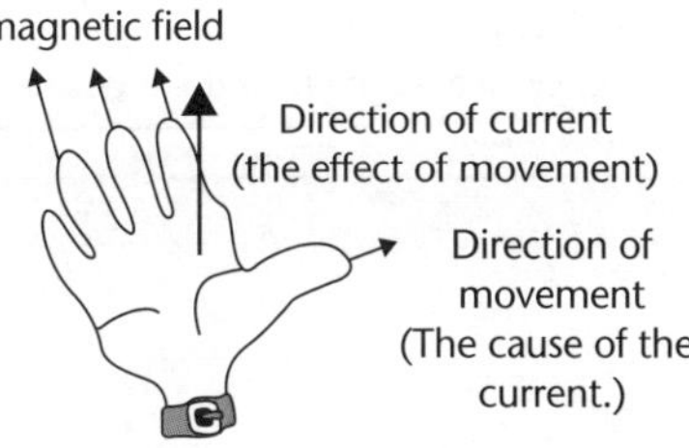

While the loop is completely in the magnetic field there is no induced current. The voltage induced across BC is balanced by an equal and opposite voltage induced across AD (see diagram below left).

As the loop leaves the magnetic field the induced current is clockwise, opposite to the direction in entering because the current is due to the voltage induced across AD (see diagram below right).

Note: No voltage is induced across AB or CD because they are not moving *across* the magnetic field lines.

Example A

A square loop of wire 0.5 m by 0.5 m moves into a uniform magnetic field of 0.8 *T*, at a steady speed of 0.1 ms^{-1}. An induced voltage *V* develops between points A and B on the wire loop.

Calculate the size of *V* and direction of the induced current in the loop as the loop is:

a. entering the magnetic filed.
b. completely in the field.
c. leaving the field.

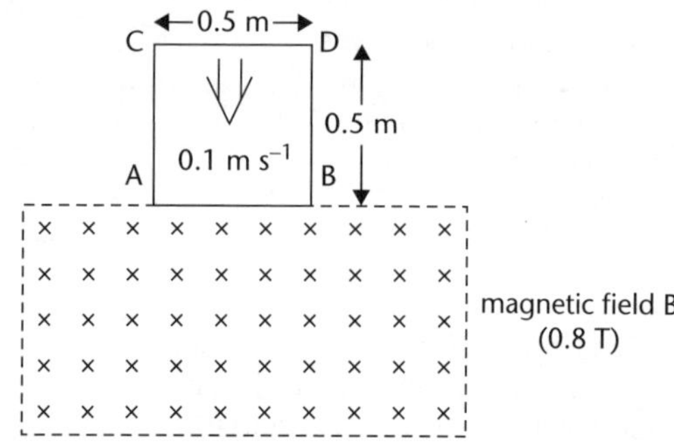

Solution:

a. As AB enters the field

$V = BvL$

$= 0.8 \times 0.1 \times 0.5$

$= 0.04$ *V*

Using the right hand slap rule the direction of the induced current will be anticlockwise.

b. When the coil is completely in the field there is no induced voltage or current.

c. As CD leaves the filed the induced voltage *V* is still equal to 0.04 *V* but the direction of induced current is clockwise.

Lenz's Law

The law of conservation of energy states that energy cannot be created but can only be changed from one form to another. So the energy produced by electromagnetic induction must have been supplied to the system in some other form. Lenz's Law is a consequence of this and states that an induced current causes a force to opposite the change which produced it. This means that work has to be done in order to produce electrical energy.

Lenz's law enables the direction of the induced current to be predicted using the right hand rule.

Example B

- The induced current in the coil creates a south pole to try to repel the magnet.

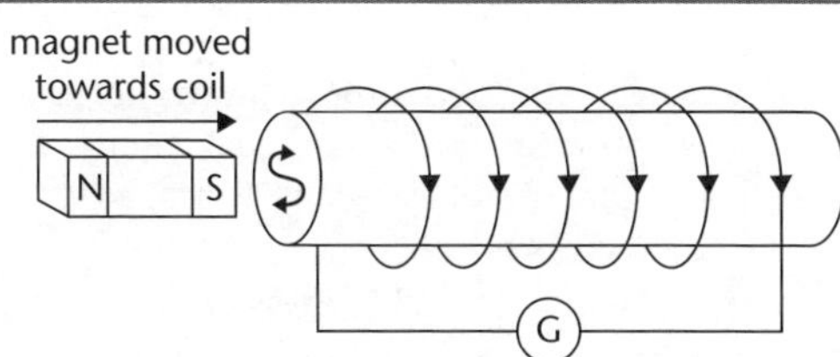

- The induced current in the coil creates a north pole to try to attract the magnet.

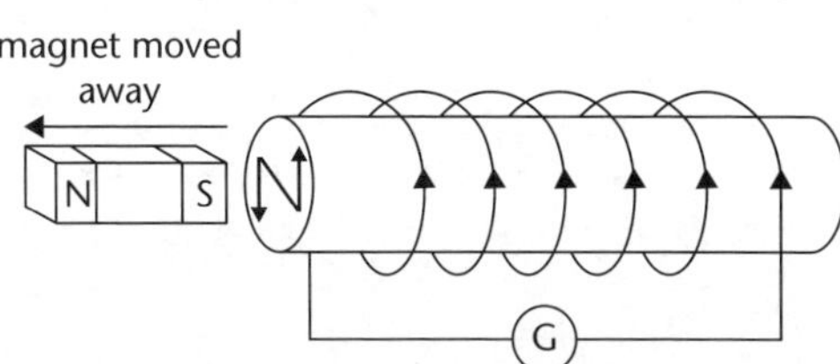

- When a wire is moved across the magnetic field *B*, it causes an induced current *I*. But the induced current causes a force *F* which opposes the movement. To keep the wire moving work must be done against this opposing force.

induced current, *I*

opposing force $F = BIL$

movement of wire

B

Note: 1. The opposing force *F* is given by the formula

$$F = BIL$$

2. When the light connected to a bicycle dynamo is turned on, an opposing force is created in the dynamo which makes it harder to pedal.

Unit 12.4 Activity 5A

1. XY is a wire, 0.40 m long, at right angles to a magnetic field of strength 6.0 T. XY is moving to the right at a steady speed of 5.0 m s^{-1}.
   a. Which end of the wire will become positively charged?
   b. Calculate the potential difference V, induced across the ends of the wire.
   c. If XY is connected to a circuit containing a resistance of 2.0 Ω, what is the induced current?

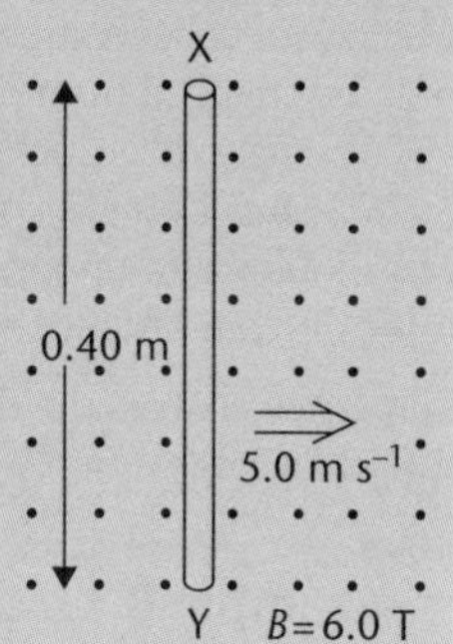

2. A magnet is dropped through a vertical coil as shown in the diagram. The graph shows a plot of voltage against time recorded during the fall.
   a. Explain in physical terms why the first voltage pulse during the time interval AB is produced.
   b. Explain in physical terms why the two voltage pulses shown are opposite in sign.
   c. Explain in physical terms why the second voltage pulse is larger in amplitude than the first.
   d. Explain why the voltage recorded during time interval BC is zero.
   e. Is the direction of current through the resistor R during the time interval AB from left to right or from right to left? Justify your answer.

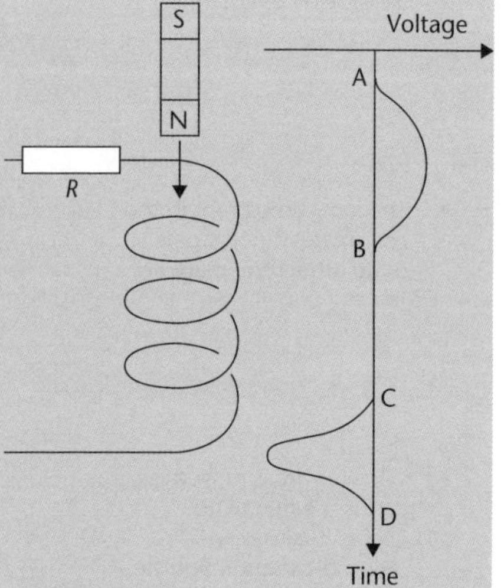

3.

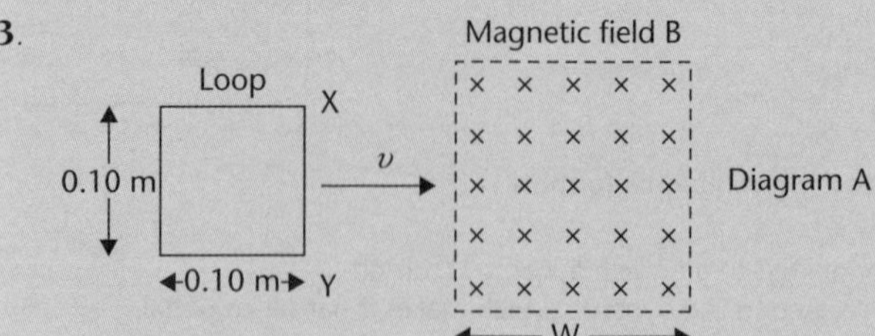

Diagram A shows a square loop of copper wire 0.10 m by 0.10 m, travelling to the right at a constant velocity v. The loop passes through a region of external magnetic field B, directed into the page. An induced voltage V is developed in the square loop. Diagram B shows a graph of the induced voltage V against the time taken t.

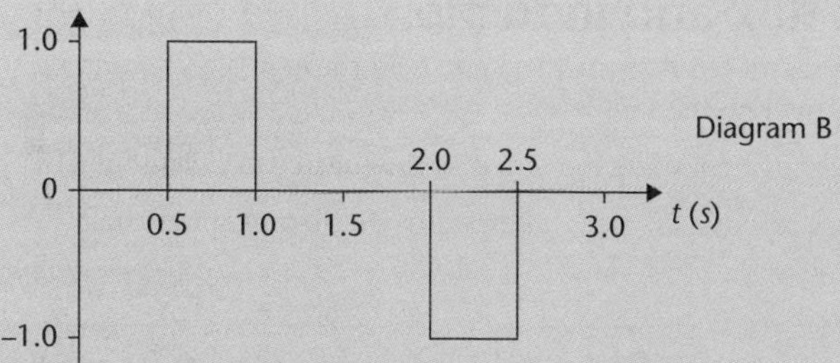

a. Explain as precisely as you can what the loop is doing:

i. during the time interval between $t = 0.5$ s and $t = 1.0$ s

ii. during the time interval between $t = 2.0$ s and $t = 2.5$ s

iii. during the time interval between $t = 1.0$ s and $t = 2.0$ s

b. What is the maximum value of the voltage V developed across XY?

c. By examining the graph:

i. determine the value of the velocity v, showing your working clearly.

ii. determine the width W of the region of the magnetic field.

d. Determine the value of the magnetic field B, showing your working clearly.

4. Use Lenz's Law to explain the following situations:

a. A magnet is not attracted to a piece of aluminium but when a magnet with cylindrical poles is dropped down a long aluminium pipe, it falls through slower than if it fell freely.

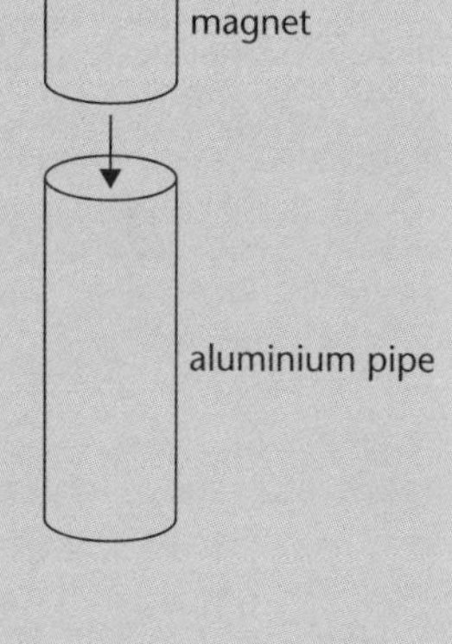

b.

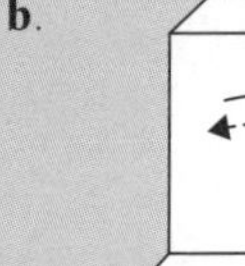

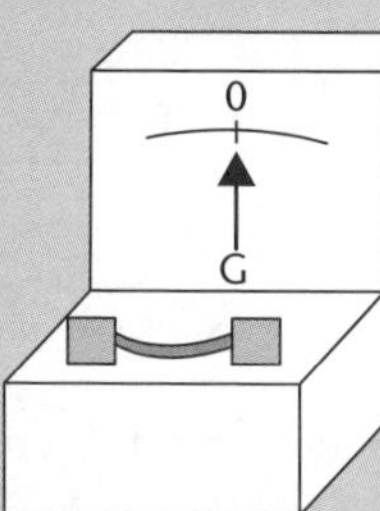

When a sensitive moving coil galvanometer is shaken, the needle moves rapidly from side to side. However, when the meter terminals are connected together by a wire, the needle hardly moves when the galvanometer is shaken.

c. A small aluminium wheel spins freely, but when it is placed between the poles of a magnet, it stops.

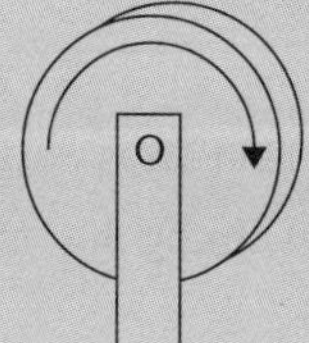

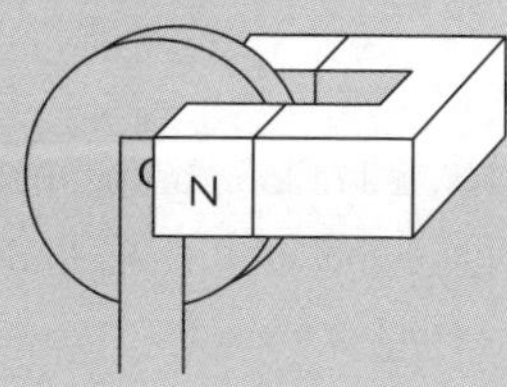

Magnetic flux and induction

On page 201 it was shown that the magnetic field through a coil or circuit must *change* to cause electromagnetic induction.

The magnetic field in a circuit is measured as **magnetic flux** (symbol ϕ).

The magnetic flux through an area A, perpendicular to a magnetic field B, is given by:

$$\phi = B \times A$$

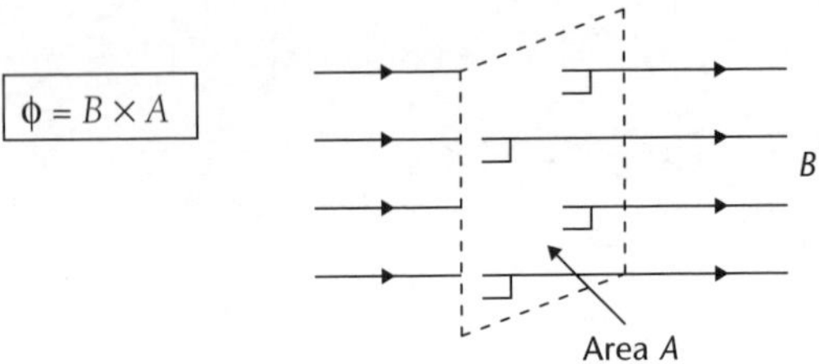

The unit for magnetic flux is the weber (symbol Wb).

Examples A and B have shown several different ways of demonstrating electromagnetic induction, but in all of them a change in the magnetic flux produces an induced voltage.

Michael Faraday, an English physicist and chemist, discovered a relationship between the induced voltage and the rate at which the magnetic flux changes.

In a situation such as Example A the induced voltage V in wire AB is given by $V = BvL$ (see diagram below).

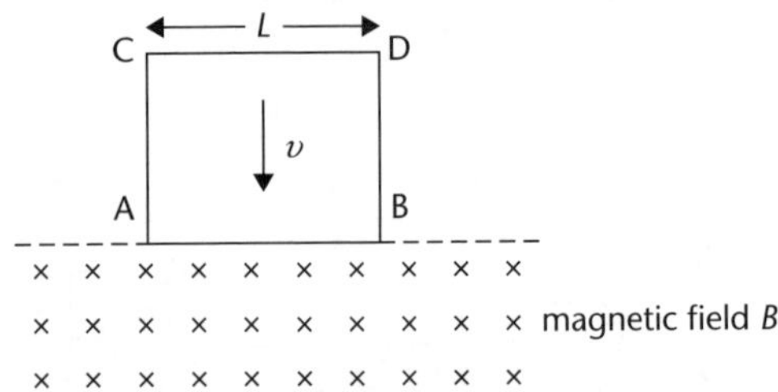

Since the loop is moving at v metres per second, in one second the loop moves distance v into the magnetic field (see diagram below).

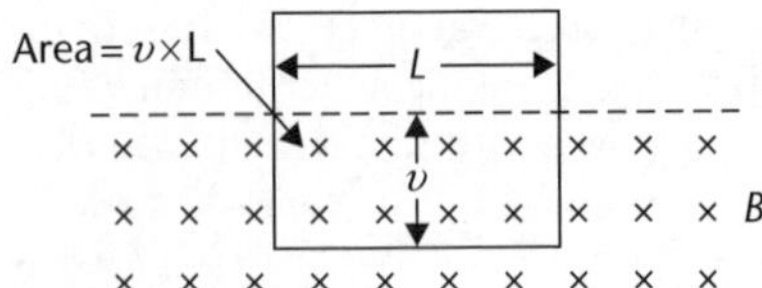

The area of loop in the magnetic field ΔA is $v \times L$.

The change in flux $\Delta\phi$ in the loop during this one second is

$$\begin{aligned}\Delta\phi &= B \times \Delta A \\ &= B \times v \times L\end{aligned}$$

We saw earlier that the induced voltage $V = BvL$, hence the change in flux in one second equals the induced voltage V.

This is an example of the law which Faraday discovered.

Faraday's Law states that the size of the induced voltage in a conductor equals the rate of change of magnetic flux.

This is written as:

$$V = -\frac{\Delta\phi}{\Delta t}$$

Where V is the induced voltage

$\Delta\phi$ is the change in flux

and Δt is the time taken for the flux to change.

The negative sign is a reminder of Lenz's law – that the induced current causes a force to oppose the change which produces it.

Example C

The diagram shows a circular wire of area 0.1 m^2, placed in a magnetic field of strength 0.8 T. The magnetic field is then reduced to zero, during a time of 0.4 seconds.

a. How much flux initially passes the coil?

b. What is the size of the induced EMF in the coil as the magnetic field strength is reduced?

c. What is the direction of the induced current?

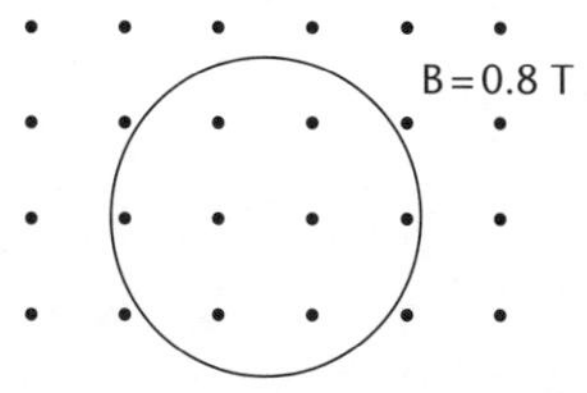

Solution:

a. $\phi = B \times A$
$= 0.8 \times 0.1$
$= 0.08$ Wb

b. $V = \frac{\Delta\phi}{\Delta t}$
$= \frac{0.08}{0.4}$
$= 0.2$ V

c. Since the magnetic field is being reduced, the direction of the induced current will oppose this change by maintaining a magnetic field out of the page. So, using the right hand rule the current direction is anticlockwise around the wire.

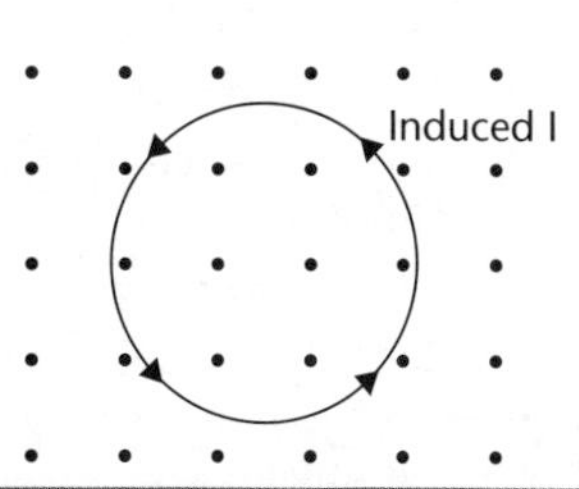

Note: For a coil of N turns, the induced voltage V is given by:

$$V = -N\frac{\Delta\phi}{\Delta t}$$

ie the induced voltage is N times greater than for a single turn, since there is an induced voltage in each turn, and each is in series.

Unit 12.4 Activity 5B

1. A square solenoid of 50 turns of wire and sides 5.0 cm long is placed perpendicular to a magnetic field of strength 0.50 T.

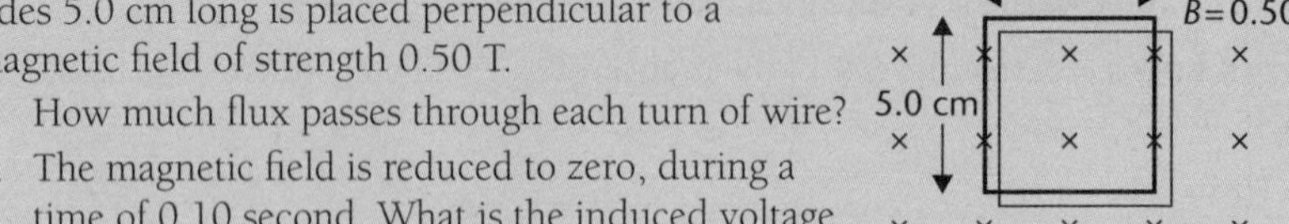

a. How much flux passes through each turn of wire?

b. The magnetic field is reduced to zero, during a time of 0.10 second. What is the induced voltage created in the solenoid?

c. In what direction does the induced current flow?

The following information is referred to in questions 2 to 5.

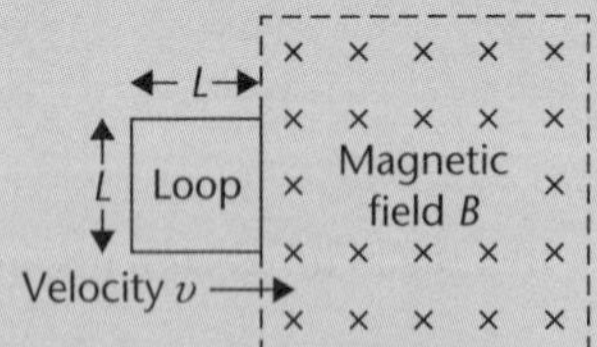

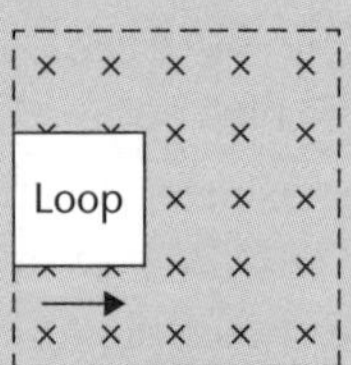

The diagrams show a square loop of copper wire, of side length L, moving at a constant velocity v, into a uniform magnetic field B directed into the page.

2. The time taken t for the loop to move between the two positions shown in the diagrams is given by:

A. $t = \frac{L}{v}$ **B.** $t = \frac{v}{L}$ **C.** $t = vL$ **D.** $t = \frac{v^2}{L}$ **E.** $\frac{L}{v^2}$

3. The change in magnetic flux $\Delta\phi$ through the loop during the time t that it moves between the two positions shown in the diagrams is given by:

A. $\Delta\phi = \frac{B}{L}$ **B.** $\Delta\phi = \frac{B}{L^2}$ **C.** $\Delta\phi = BL^2$ **D.** $\Delta\phi = \frac{L^2}{B}$ **E.** $\Delta\phi = \frac{L}{B}$

4. The voltage ε induced in the loop is given by:

A. $\varepsilon = Bv$ **B.** $\varepsilon = BLv$ **C.** $\varepsilon = \frac{B}{Lv}$ **D.** $\varepsilon = \frac{BL}{v}$ **E.** $\varepsilon = \frac{Bv}{L}$

5. Which of the following statements is true?

As shown in the diagrams, when the loop is entering the magnetic field, the direction of the induced current in the loop:

A. is clockwise around the loop, producing an induced magnetic field in a direction into the page.

B. is clockwise around the loop, producing an induced magnetic field in a direction out of the page.

C. is anticlockwise around the loop, producing an induced magnetic field in a direction into the page.

D. is anticlockwise around the loop, producing an induced magnetic field in a direction out of the page.

E. is both ways around the loop, producing a net induced magnetic field of zero.

Generating alternating current

The main parts of a generator are a magnet and a coil. When the coil rotates in the magnetic field, an alternating **potential difference** is induced across the terminals of the coil.

With the coil at the horizontal position (see diagram below) the output voltage is at its maximum value since the sides of the coil are moving at right angles to the magnetic field. There is no induced voltage caused by the ends of the coil since they don't cross the magnetic field lines.

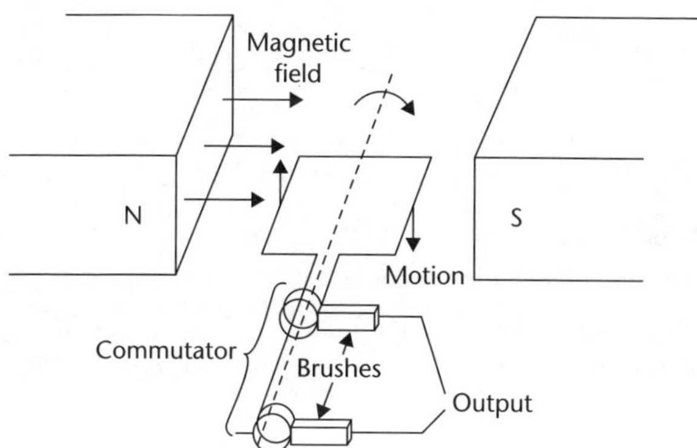

A quarter of a turn later the sides of the coil are moving parallel to the field lines (see diagram below), and the output voltage is zero.

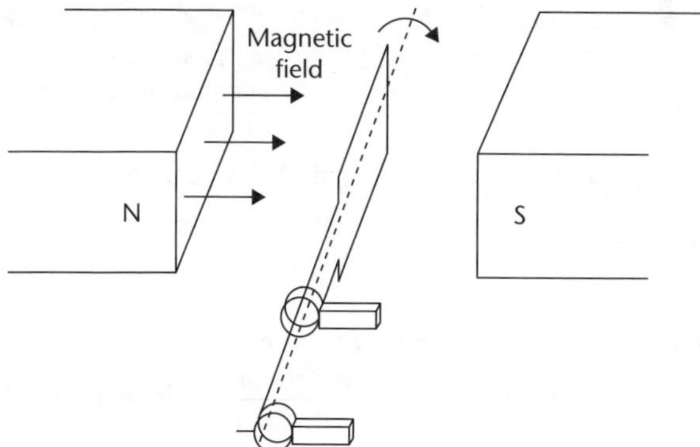

A graph of the output voltage during one cycle is shown with the positions of the coil (see diagram on next page).

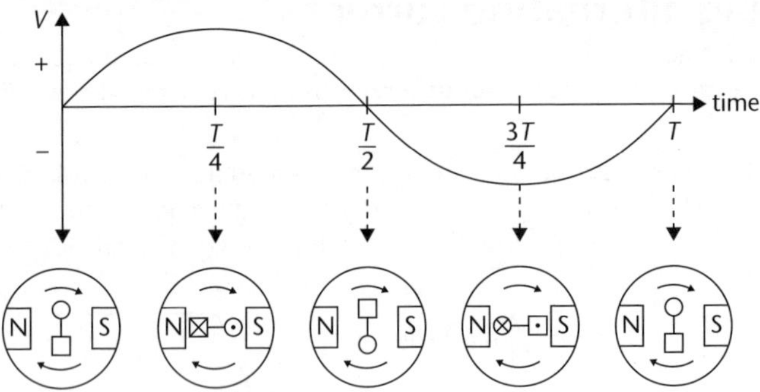

The induced voltage is produced because the magnetic flux through the coil is constantly changing.

The magnetic flux ϕ through the coil is given by:

ϕ = component of B perpendicular to the coil $\times A$

If timing starts with the coil in the vertical position ($t = 0$), then at that time $\phi = B \times A$

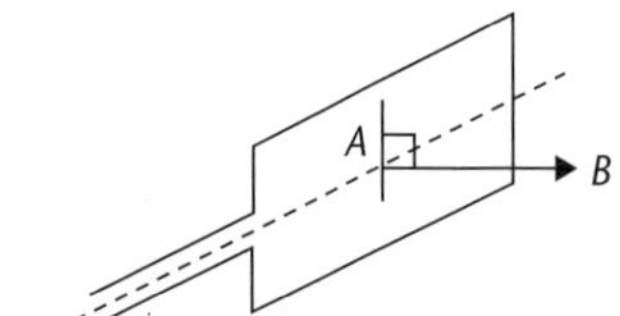

The coil rotates with an angular velocity ω. After time t, the coil has turned through angle ωt.

[since $\omega = \frac{\theta}{t}$ then $\theta = \omega t$]

The flux through the coil after time t is given by:

$$\phi = B \cos \omega t \times A$$

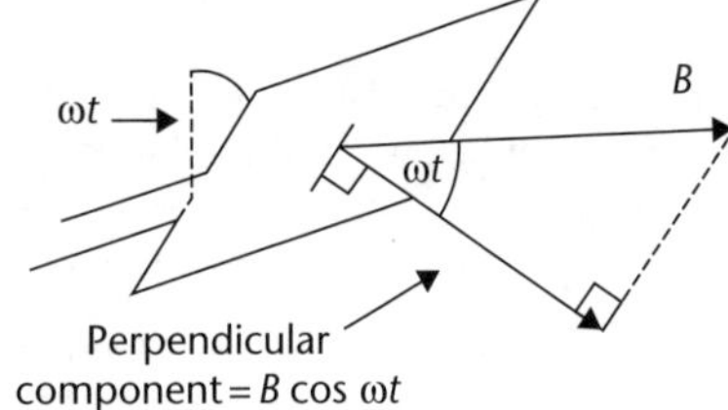

To calculate the induced voltage of the coil, Faraday's Law is used in the calculus form.

$$v = \frac{-d\phi}{dt}$$

$$v = \frac{-d(BA\cos\omega t)}{dt} \qquad \text{[substituting } \phi = BA\cos\omega t\text{]}$$

$$V = BA\,\omega \sin \omega t \qquad \text{[by differentiation } \frac{d(\cos\omega t)}{dt} = -\omega \sin \omega t\text{]}$$

For a coil with N turns the induced voltage at time t is given by:

$$V = BAN\,\omega \sin \omega t$$

This alternating voltage formula is often written as:

$$V = V_{max} \sin \omega t$$
$$\text{where } V_{max} = BAN\omega$$

This gives a voltage-time graph with the general shape of a sine curve.

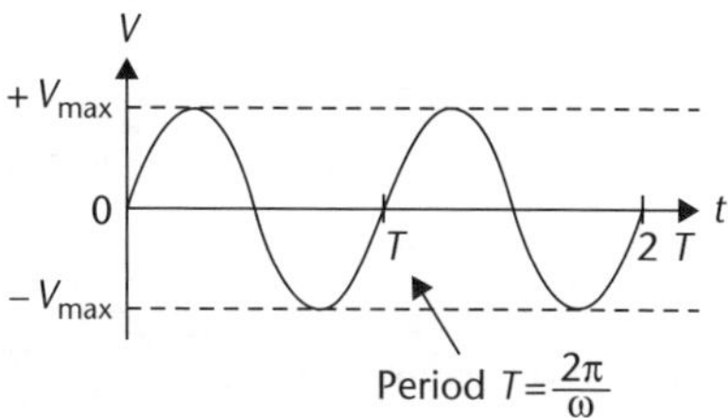

Example D

A simple generator is made from a 50 turn coil placed in a magnetic field of 0.50 T. The area of the coil is 0.0010 m^2. What is the maximum voltage of the coil when it rotates with a frequency of 100 Hz?

Solution:

$\omega = 2\pi f$
$= 2\pi \times 100$
$= 200\pi$ rad s^{-1}

$V_{max} = BAN\omega$
$= 0.50 \times 0.0010 \times 50 \times 200\pi$
$= 15.7$ V
$= 16$ V

AC and DC generators

The basic structure of AC and DC generators is shown in the diagram below:

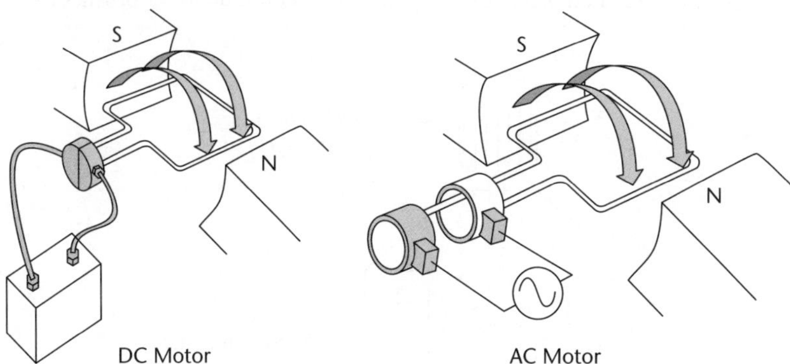

DC Motor AC Motor

The difference is that the DC motor has only one slip ring and the AC motor has two slip rings, as the diagram above illustrates.

All parts of the motor are labelled in the diagram below:

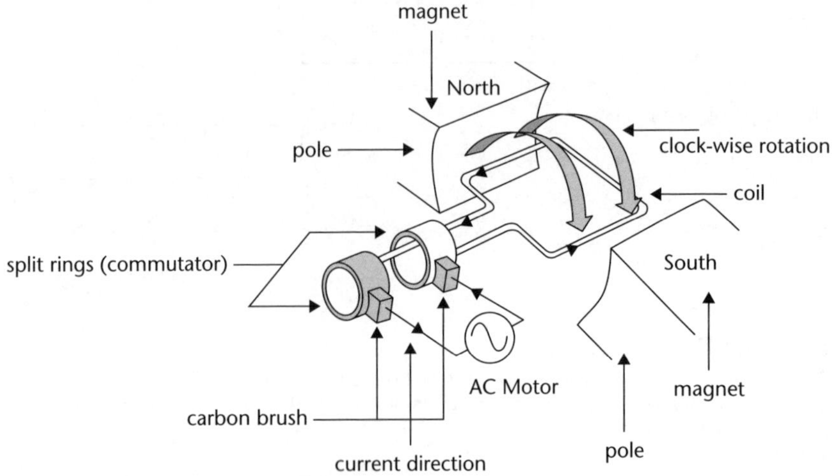

The bicycle dynamo

A bicycle dynamo works by changing the magnetic flux through a coil, but it moves the magnet while leaving the coil still.

As the magnet rotates the magnetic field in the coil changes. This induces a current in the coil.

An advantage of this method is that the dynamo has no commutator or brushes to wear out.

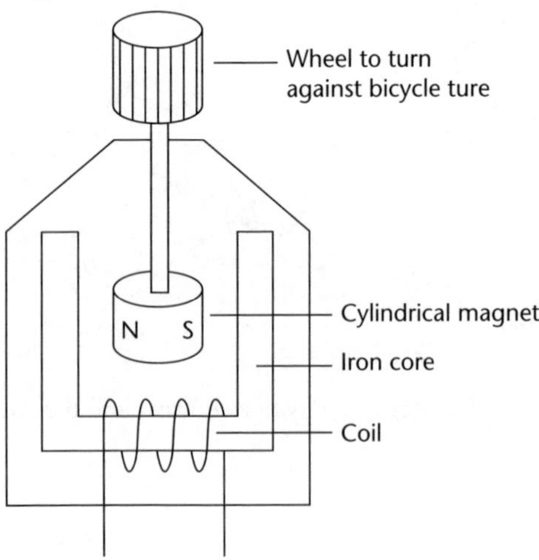

The microphone

To change sound into an electrical signal requires a microphone. The basic parts of a microphone are the same as for a loudspeaker (see Page 269), in fact most speakers can act as a primitive microphone.

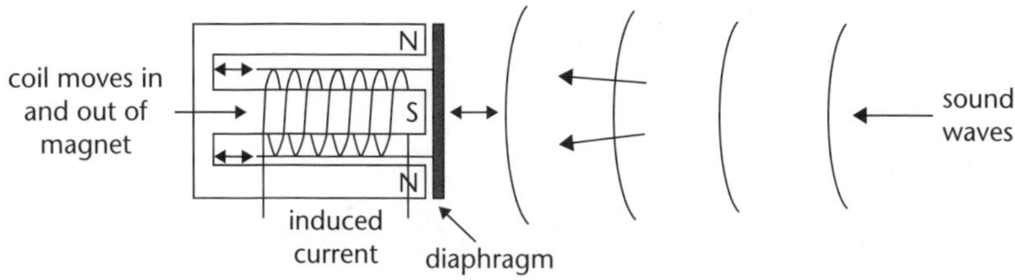

The moving coil is inside a cylindrical magnet, and attached to a flat diaphragm. Sound waves make the diaphragm vibrate and this moves the coil in and out of the magnetic field causing a tiny induced current in the coil.

Unit 12.4 Activity 5C

1. A generator is made from a 100 turn coil, 5.0 cm by 4.0 cm, placed in a magnetic field of strength 0.20 T. The coil rotates at 50 **revolutions** per second.

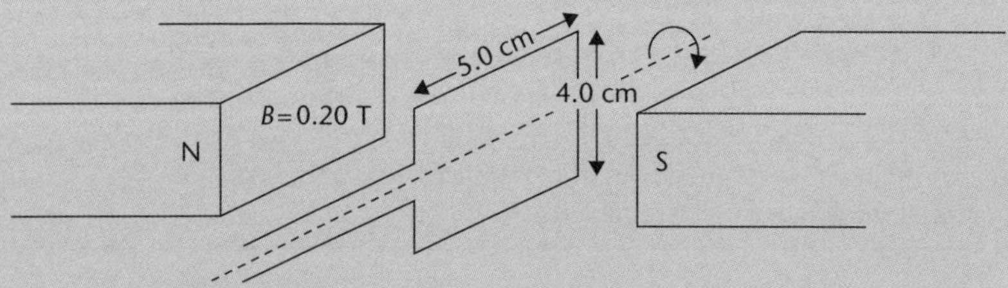

   a. At the position shown in the diagram the magnetic field is perpendicular to the coil. How much magnetic flux passes through each turn?
   b. What is the period of the rotation?
   c. What is the angular velocity of the coil?
   d. What is the peak voltage generated by the coil as it rotates?
   e. Sketch a graph of the output voltage of the coil for one cycle if timing begins at the position shown in the diagram. Indicate the appropriate values on the voltage and time scale.

The following information refers to questions 2 and 3.

This graph shows the voltage from a small generator against time.

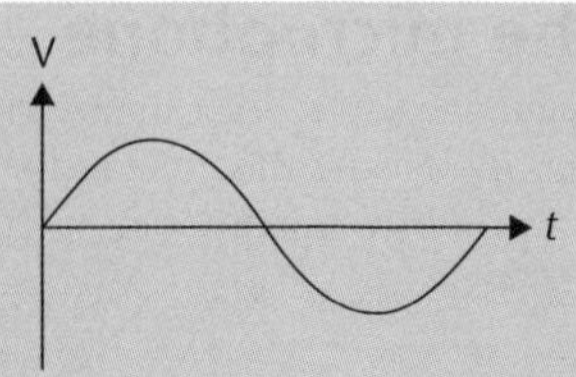

2. Which of the graphs below shows the output voltage if the number of turns on the coil is doubled?

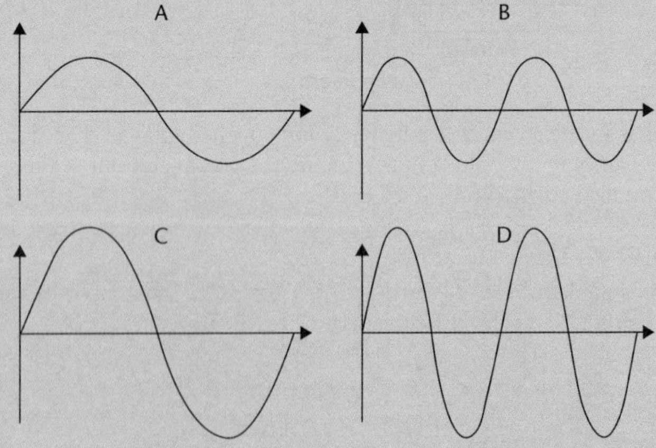

3. Which graph in Question 2 shows the output voltage if the speed of rotation of the original coil is doubled?

Unit 12.4 Electromagnetism
Topic 6: Transformers and inductance

Introduction

The previous chapter showed how a changing magnetic field through a coil will cause an induced voltage. This effect is used in a **transformer**.

The transformer

A transformer consists of two coils wound onto an iron core.

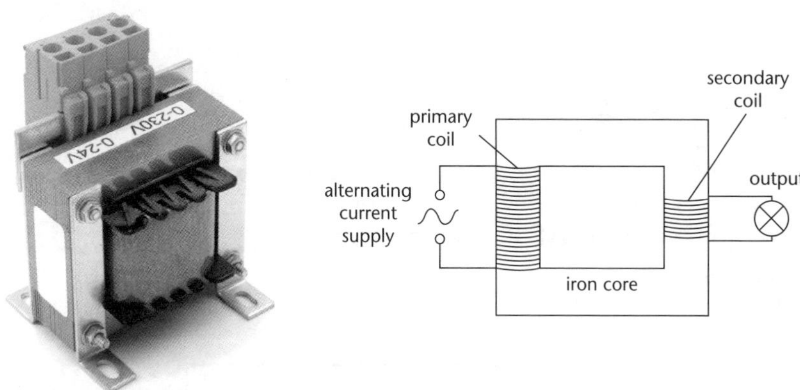

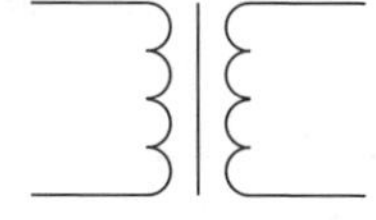

The circuit symbol for a transformer is shown at right.

When an AC supply is connected to the primary coil, it creates a changing magnetic flux in the iron core. This changing flux passes through the secondary coil where it induces an alternating voltage.

The main purpose of a transformer is to change (transform) the voltage in an AC circuit.

The voltage in the primary V_p and the voltage in the secondary V_s are related to the number of turns on the primary N_p and the number of turns on the secondary N_s.

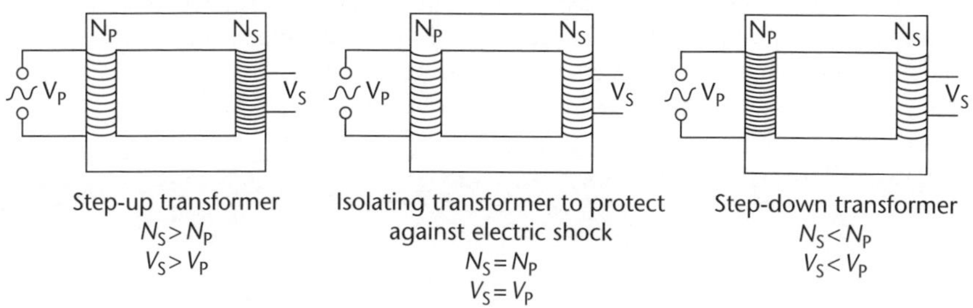

Step-up transformer
$N_S > N_P$
$V_S > V_P$

Isolating transformer to protect against electric shock
$N_S = N_P$
$V_S = V_P$

Step-down transformer
$N_S < N_P$
$V_S < V_P$

In an ideal transformer the voltage and number of turns are related by:

$$\frac{V_s}{V_p} = \frac{N_s}{N_p}$$

Example A

A battery charger contains a transformer to convert 240 V to 12 V. If the primary coil consists of 1 200 turns. How many are on the secondary?

Solution:

$$\frac{V_s}{V_p} = \frac{N_s}{N_p}$$

$$N_s = \frac{N_p V_s}{V_p} \quad \text{[rearranging]}$$

$$= \frac{1200 \times 12}{240}$$

$$= 60 \text{ turns}$$

The efficiency of a transformer

In any situation where energy is changed from one form to another, efficiency is important. Efficiency is the percentage of the input energy which is available as output energy.

$$\text{Efficiency} = \frac{\text{output energy}}{\text{input energy}} \times \frac{100}{1}$$

A transformer, like any other machine, is never completely efficient. The output energy will always be less than the input energy due to dissipation of energy, mainly in the form of heat.

However, careful design can make transformers up to 99% efficient.

Energy losses are reduced by:

- using an iron core to ensure a strong magnetic field between the primary and secondary coils.
- Making the core from many flat electrically insulated sheets or laminations. Rather than solid iron. Laminations reduce 'eddy' currents (induced currents in the core) which would waste energy by causing heating.
- Using low resistance copper wire for the primary and secondary coils.

In an ideal transformer

Power input = Power output

$$V_p I_p = V_s I_s$$

Since $P = V \times I$ this gives

where I_p and I_s are the current in the primary and secondary coils.

Note: This formula shows that when a transformer *increases* the voltage there is a corresponding *decrease* in current. Similarly a *decrease* in voltage has a matching *increase* in current.

Example B

The transformer in Example A draws 0.20 A of current from the 240 V supply.

a. If the transformer were ideal, how much current would be supplied by the secondary?

b. In practice, the current supplied by the secondary is 3.8 A. How efficient is the transformer?

Solution:

a.
$$V_p I_p = V_s I_s \quad \text{[for an ideal transformer]}$$
$$240 \times 0.20 = 12 \times I_s$$
$$I_s = \frac{240 \times 0.20}{12} \quad \text{[rearranging]}$$
$$= 4.0 \text{ A}$$

b.
$$\text{Efficiency} = \frac{\text{power output}}{\text{power input}} \times \frac{100}{1}$$
$$= \frac{12 \times 3.8}{240 \times 0.2} \times \frac{100}{1} \quad \text{[since } P = V \times I\text{]}$$
$$= 95\%$$

The induction coil

An induction coil is a step-up transformer which uses a switched DC supply to the primary.

In a simple petrol engine the induction coil is used to change 12 V from the battery to several thousand volts for the spark plug.

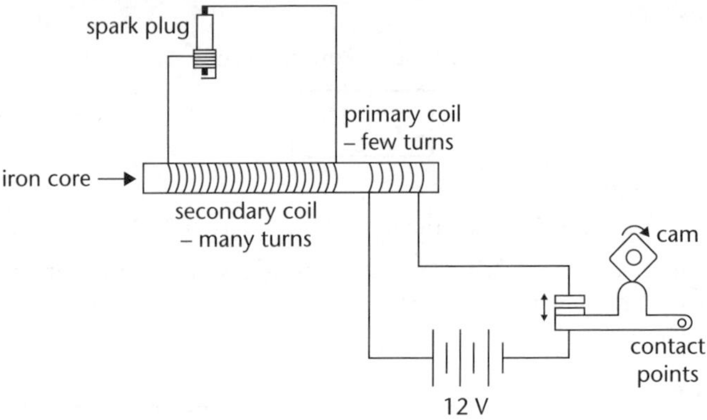

Since a changing magnetic field is needed to induce a voltage in the secondary coil, the primary current must be switched on and off. This is done by the contact points in the distributor.

Unit 12.4 Activity 6A

1. A transformer is used to convert 240 V AC to 6.0 V AC to power a cassette recorder. If the primary coil has 600 turns, how many are in the secondary?
2. A step-up transformer has 200 turns in the primary and 1 000 in the secondary. What is the output voltage if the primary is connected to 240 V AC?
3. A bathroom razor plug contains a small transformer with two secondary windings, one to supply 110 V and the other 240 V. The primary coil has 600 turns and operates at 240 V AC.

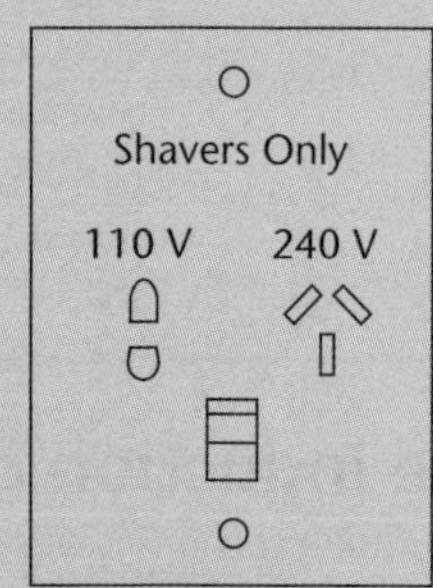

   a. How many turns are on the coil supplying
      i. 110 V? **ii**. 240 V?
   b. What is the purpose of the transformer which supplies 240 V when the voltage is unchanged?
   c. A razor has a power rating of 50 W and operates on 110 V.
      i. How much current does it draw?
      ii. How much current will flow in the primary coil when the razor is being used?

4. An isolating transformer is used with an electric floor polisher.
   The mains plug supplies 10 A of current at 240 V.

   10 A
   240 V mains

   a. How much power is supplied by the mains?
   b. What voltage does the transformer supply to the floor polisher?
   c. If the transformer is 97% efficient, how much power and current are supplied to the floor polisher?

Transmission of electricity

One of the major uses of transformers is in the national distribution system (or National Grid) which carries electricity from power stations to the users.

Electricity in New Zealand is generated from three main energy sources:

- **hydroelectric** uses **gravitational potential energy** from lakes and rivers (about 80%) of New Zealand's power is hydroelectric).
- **geothermal** uses heat energy from underground.
- **thermal** uses heat energy from burning fuels (oil, coal or natural gas).

In each case, the energy is used to drive generators which produce alternating current (as described in the previous chapter).

Many power stations are located far from the cities and factories which use the power, so it must be carried over long distances, using **transmission lines**.

Since the transmission lines have some resistance, energy will be lost in the wires. However, this loss of energy will be made as small as possible if the power is transmitted using a high voltage and low current.

Example C

A transmission line is required to carry 10 kilowatts of power to a factory. The transmission line has a resistance of 5.0 Ω. The power can be transmitted at two different voltages, 240 V or 220 000 V.

Which voltage will cause the smallest power loss in the transmission line?

Solution:

Calculate the loss at each voltage.

At 240 V

Transmission current

$I = \frac{P}{V}$ [since $P = V \times I$]

$= \frac{10000}{240}$ [10 KW = 10 000 W]

$= 41.7$ A

$= 42$ A

Power loss on transmission

$P = I^2R$

$= (41.7)^2 \times 5$

$= 8694$ W

$= 8700$ W

At 220 000 V

Transmission current

$I = \frac{P}{V}$

$= \frac{10000}{220000}$

$= 0.45$ A

Power loss on transmission

$P = I^2R$

$= (0.45)^2 \times 5$

$= 1$ W

The power loss is reduced dramatically from almost 87% to only 0.01% by using a high voltage and low current in the transmission line.

At power stations, step-up transformers are used to raise the voltage to 220 kV for transmission. While high voltages are more efficient for transmission they are dangerous and require careful insulation, so a series of step-down transformers at nearby substations are used to reduce the voltage to 240 V for the domestic supply.

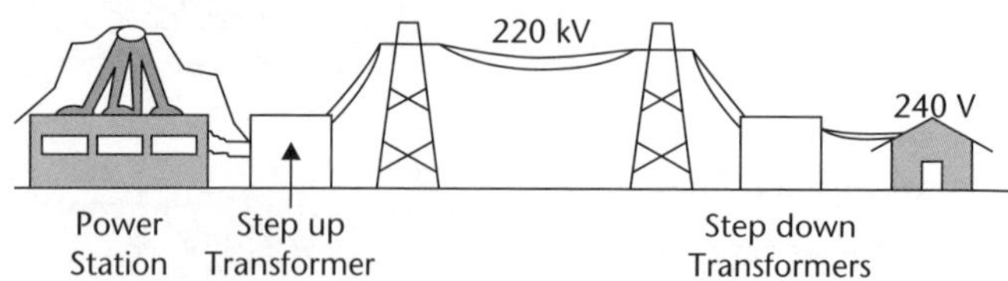

A transformer-switching station in the National Grid system

Unit 12.4 Activity 6B

1. A 10.0 Ω transmission cable is used to carry 5 000 W of electrical power. Calculate both the current in the cable, and the power loss during transmission when the voltage is:
   a. 240 V **b**. 11 000 V

Mutual inductance

In a transformer, changing the current in the primary coil causes an induced voltage in the secondary coil. This is called **mutual inductance** and happens because the magnetic flux from the primary coil passes into the secondary coil. Mutual inductance occurs when changing current in one coil produces an induced voltage in another coil.

Example D

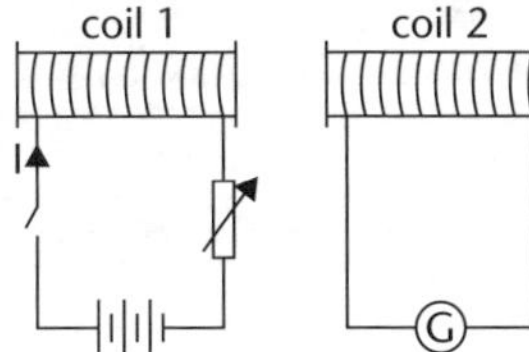

Two coils are positioned so that the magnetic field from coil 1 passes into coil 2. The current in coil 1 can be changed by either:

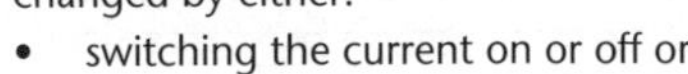

- switching the current on or off or
- changing the size of the current using the rheostat

Changing the current in coil 1 changes the magnetic flux in coil 2 and this causes and induced voltage. The effect can be increased by placing an iron core between the two coils. This makes the magnetic filed passing between the coils much stronger.

The magnetic flux in coil 2 is proportional to the current in coil 1.

$$\phi \propto I$$

$$\text{or } \phi = \text{constant} \times I$$

The proportionality constant is called the mutual inductance (symbol *M*), so

$$\phi = M \times I$$

If the current *I* changes, the flux ϕ changes, causing an induced voltage *V* in coil 2.

Faraday's Law states that:

$$V = \frac{-\Delta\phi}{\Delta t}$$

$$= \frac{-\Delta(M \times I)}{\Delta t} \qquad \text{[substituting } \phi = M \times I\text{]}$$

$$= \frac{-M\Delta I}{\Delta t} \qquad \text{[since } M \text{ is a constant]}$$

The induced voltage V in the second coil equals the mutual inductance *M* multiplied by the rate of change of the current in the first coil.

$$V = \frac{-M\Delta I}{\Delta t}$$

Note: 1. The unit of mutual inductance is the **henry** (Symbol H). Two coils have a mutual inductance of I H when a change in the current in one coil of 1 amp per second causes an induced voltage of 1 volt in the other coil.
2. The value of *M* depends on the size and position of two coils and the presence of any materials such as an iron core between them.
3. The negative sign in the formula shows that the induced voltage opposes the change of current (Lenz's Law).

Example E

Using two coils from Example D, when the current in coil 1 is increased from 0 to 5 A in 2 seconds, an induced voltage of 100 mV is induced in coil 2.

The mutual inductance between the coils is calculated from:

$$V = \frac{-M\Delta I}{\Delta t}$$

$$0.1 = \frac{M \times 5}{2} \qquad \text{[substituting 100 mV = 0.1 V]}$$

$$M = \frac{0.1 \times 2}{5} \qquad \text{[rearranging]}$$

$$M = 0.04 \text{ H}$$

Unit 12.4 Activity 6C

1. The diagram below shows how a search coil can be used to measure the magnetic field in a solenoid.

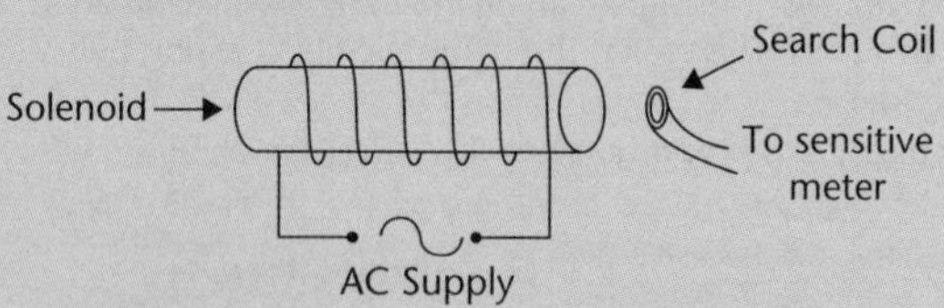

When the search coil is placed in the solenoid a small voltage is induced in it. The size of the induced voltage indicates the strength of the magnetic field in that position.

a. Why is it necessary to use an AC supply to the solenoid?

b. Is the voltage induced in the search coil AC or DC? Explain why.

2. When the current in one coil of a transformer is increasing by 10 amps each second, an induced voltage of 120 volts is measured in the other coil. Calculate the mutual inductance.

3. A pair of coils wound onto an empty cardboard tube have a mutual inductance of 0.05 H.

a. The current in one coil increases from 0 to 20 A in 0.5 seconds. What voltage is induced in the other coil?

b. How would the answer to (a) be affected if the cardboard tube was filled with iron powder?

Self inductance

When a coil is connected to a power supply and the current is switched on, a magnetic field is produced in the coil. The creation of this magnetic field will itself cause an induced voltage in the coil. This effect is called **self-inductance** since the changing current in the coil produces an induced voltage across the *same* coil.

Example F

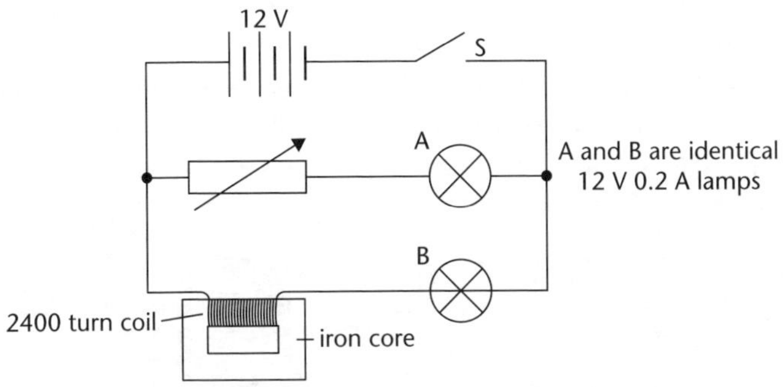

The variable resistor is adjusted until both lamps glow equally, then switch S is opened.

When switch S is now closed lamp A glows immediately, while lamp B takes a little time to reach full brightness.

The increasing current in the coil produces an opposing voltage which slows the buildup of current in the coil.

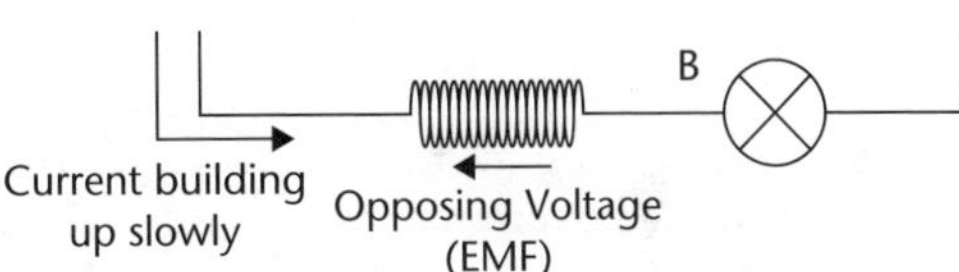

Example G

When switch S is closed, there is current in the coil, but the neon lamp doesn't glow. Neons require a voltage of approximately 90 V to glow and this is much more than the voltage of the supply.

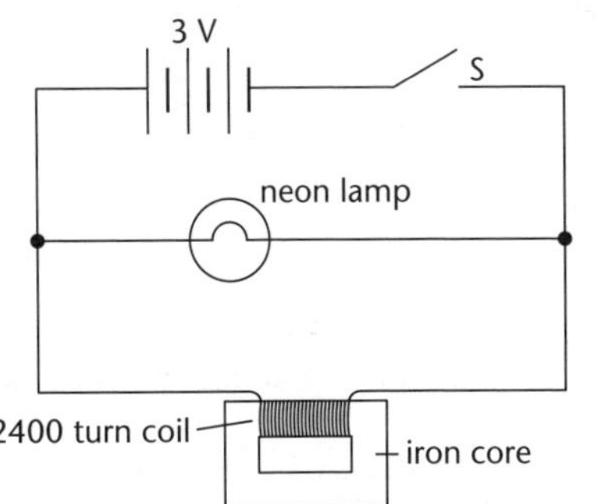

However, when switch S is opened, the neon will briefly flash.

When the current is switched off in the coil, the magnetic field starts to collapse. This induces an EMF of at least 90 V, to oppose the change by continuing the current in the coil.

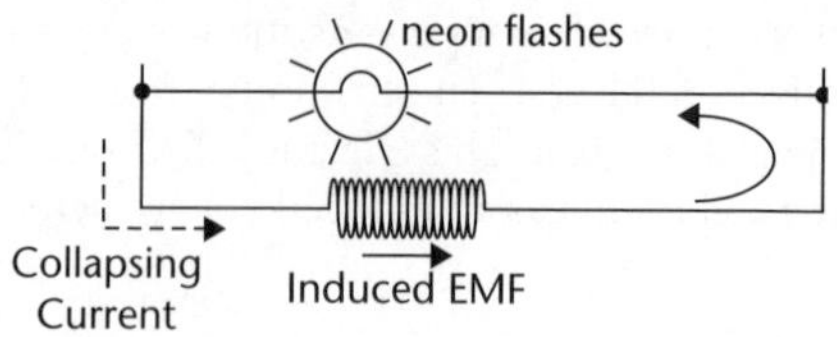

An **inductor** is a device which produces an opposing EMF (voltage) when the current (and so also the magnetic field) changes.

In Examples F and G, the coil acted as an inductor. When there is an inductor in a circuit the current takes longer to build up to maximum or to fall to Zero.

An inductor is made from a wire coil usually wound around an iron core.

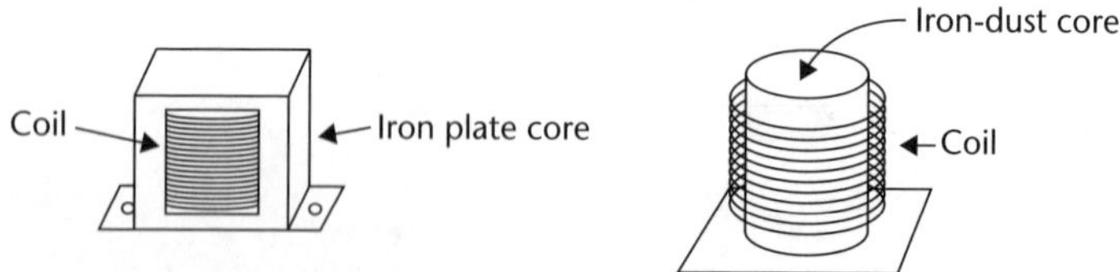

The circuit symbol for an inductor is

Inductors have resistance as well since they are made from many turns of wire. This can be shown as

Inductors are mainly used in circuits carrying alternating current, especially in tuning circuits for radio receivers.

Inductance

The effect of an inductor on a circuit is measured by a quantity called **self inductance**. Current I in a coil produces a magnetic flux ϕ.

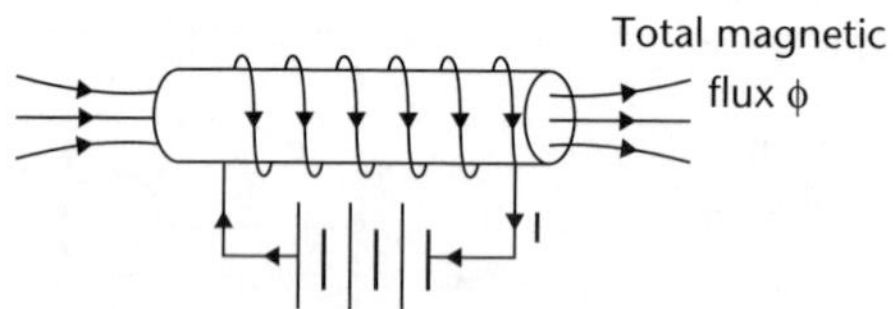

The total flux ϕ in the inductor is proportional to the current I.

$$\phi \propto I \qquad \text{[since } B \propto I \text{ and } \phi = B \times A\text{]}$$

$$\text{or } \phi = \text{constant} \times I$$

This proportionality constant is called the self inductance (symbol L),

$$\text{so } \phi = L \times I$$

If the current I changes, the flux ϕ also changes, causing an induced voltage V in the coil.

Faraday's Law states that

$$V = \frac{-\Delta\phi}{\Delta t}$$

$$= \frac{-\Delta(L \times I)}{\Delta t} \qquad \text{[substituting } \phi = L \times I\text{]}$$

$$= -L\frac{\Delta I}{\Delta t} \qquad \text{[since } L \text{ is constant]}$$

The induced voltage V in a coil equals the inductance L multiplied by the rate of change of current.

so

$$\boxed{V = -L\frac{\Delta I}{\Delta t}}$$

The unit of self inductance is the henry (symbol H), the same unit as for mutual inductance.

Note: 1. An inductor only affects a circuit while the current is changing, ie as in an AC circuit, or in a DC circuit being switched on or off. When the current is steady, only the resistance of the coil will have any effect on the circuit.

2. The negative sign in the formula shows that the induced voltage opposes the change of current (Lenz's Law).

Example H

In the circuit shown in Example G, the coil has an inductance of 0.2 H and the supply voltage is 3 V. When switch S is opened, the current falls from 5 A to zero in 0.01 s.

a. What is the induced voltage?

b. What is the resistance of the coil?

Solution:

a.
$$V = -L\frac{\Delta I}{\Delta t} = \frac{0.2 \times 5}{0.01} = 100\ \text{V}$$

b.
$$R = \frac{V}{I} = \frac{3}{5} = 0.6\ \Omega$$

Energy stored in an inductor

Current in an inductor causes a magnetic field to form. Energy is stored in the magnetic field of an inductor.

In Example F, lamp B took some time to reach full brightness because energy from the power supply was being stored in the magnetic field of the inductor, rather than causing the lamp to glow.

Example G showed how energy stored in the magnetic field of the inductor was released as light when the current in the circuit was switched off.

The energy stored in an inductor L, carrying current I, is given by:

$$E = \frac{1}{2}LI^2$$

Example I

The circuit in Example H contained a 0.20 H inductor carrying a current of 5.0 A. The energy stored in the inductor is:

$$\begin{aligned} E &= \frac{1}{2}LI^2 \\ &= \frac{1}{2}\times 0.20\times 5.0^2 \\ &= 2.5\text{ J} \end{aligned}$$

Unit 12.4 Activity 6D

1. Care must be taken when disconnecting a circuit which contains a large inductance as an electric shock could occur. Explain how this could happen when the circuit is disconnected.
2. a. Find the induced voltage when the current in a 2.0 H inductor is reduced from 10 A to zero in 0.50 s.
   b. Find the energy stored in the inductor before the current was reduced to zero.
3. a. When the current in an inductor is switched on, it increases to 4.0 A in 0.20 s. If this causes an induced voltage of 5.0 V, calculate the inductance.
   b. How much energy is stored in the inductor?

Changing current through an inductor

The current in an inductor can be investigated using a large value inductor (approximately 35 000 turns), a milliammeter and a power supply.

The current and time are recorded after switch S is closed.

The maximum current occurs when the current is steady.

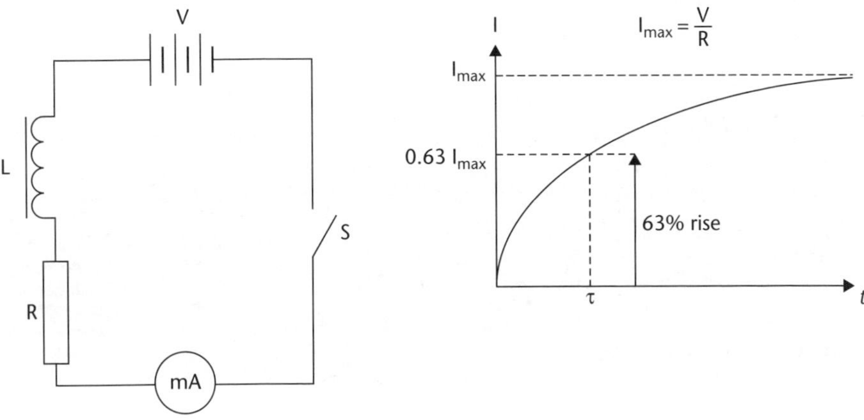

The results show that the growth in current is exponentially related.

The time taken for the current to increase is measured using a time constant τ (similar to the time constant for charging or discharging a capacitor).

The time constant is the time taken for the current to change by 63% of the maximum.

The time constant depends on the value of the inductor L and the resistance R by the formula:

$$\tau = \frac{L}{R}$$

Example J

An inductor of 0.50 H is connected to a 12 V supply. A steady current of 4.8 A is recorded.

Find:

a. The resistance of the inductor.
b. The time constant.
c. The current one time constant after switch S is opened.

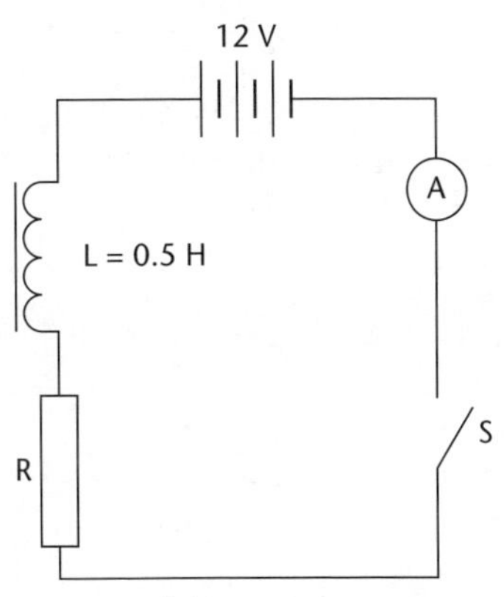

Solution:

a. $R = \frac{V}{I}$

$= \frac{12}{4.8}$

$= 2.5\ \Omega$

b. $\tau = \frac{L}{R}$

$= \frac{0.50}{2.5}$

$= 0.2\ \text{s}$

c. After 0.2 s the current will fall by 63% of 4.8 A.
63% of 4.8 A = 3.0 A
Current after 0.2 s = 4.8 – 3.0
= 1.8 A

Unit 12.4 Activity 6E

1. An inductor is placed in a circuit as shown. The switch is closed. The final current as shown by the ammeter depends on:

i. the voltage.

ii. the self-inductance *L* of the inductor.

iii. the resistance *R* of the inductor.

Which of the above statements is/are true?

A. I only **B**. II only **C**. III only

D. I and II only **E**. I and III only

2.

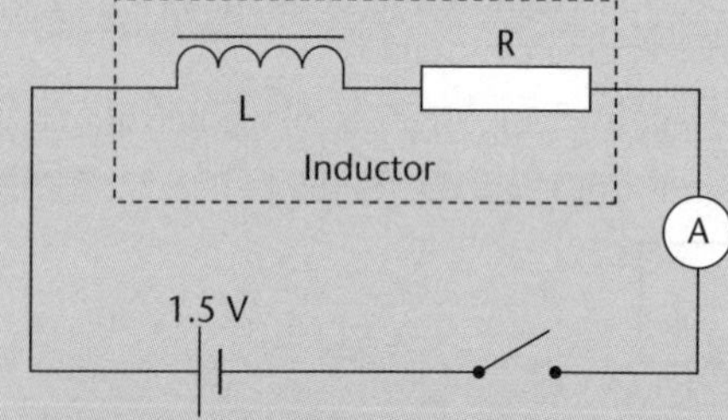

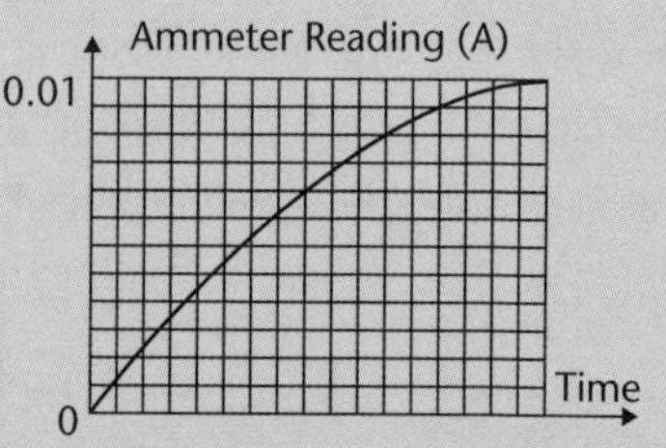

The circuit in the diagram shows an inductor of inductance *L* and resistance *R*. The inductor is in series with an **ammeter**, an open switch and a cell which may be assumed to have a voltage of 1.5 V at all times. As soon as the switch is closed, readings are taken of the current each second. A sketch graph of the results is shown in the diagram.

a. Give an explanation as to why the current slowly grows with time rather than reaching its maximum value of 0.01 A instantly.

b. Showing your working clearly, calculate the resistance R of the inductor.

3. An inductor of 2.0 H and resistance of 5.0 Ω is connected to a 6.0 V battery.
   a. Calculate the time constant of the circuit.
   b. What is the maximum current which will flow in the circuit?
   c. What will be the current one time constant after switch S is closed?

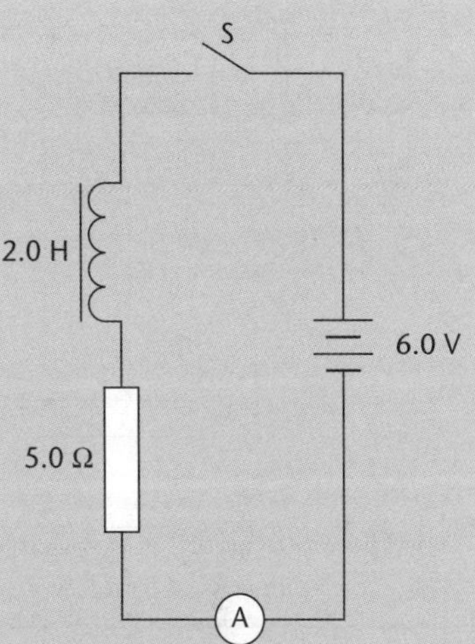

4. In an experiment, the current in an circuit containing an inductor *L* and a 12 V power supply is measured after the switch S is closed.
   Here are the results:

I (mA)	*t* (s)	*I* (mA)	*t* (s)
0.4	1.0	1.4	5.0
0.7	1.8	1.6	6.2
0.8	2.3	1.7	7.3
1.0	3.0	1.8	8.8
1.3	4.3	1.8	10.0

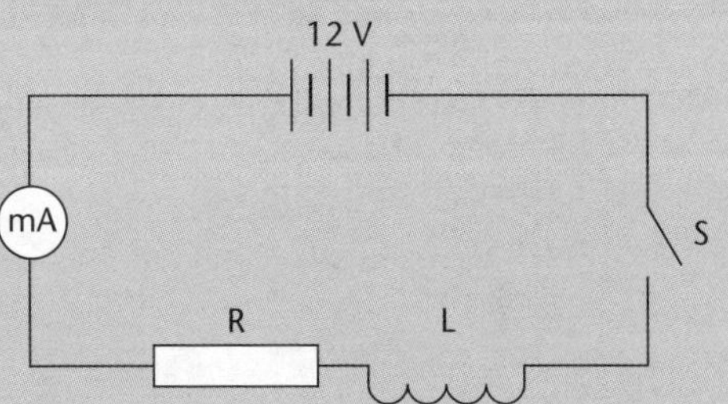

   a. Plot these results on a graph of Current against Time and draw a curve of best fit.
   b. What is the maximum current I_{max}?
   c. Calculate the resistance *R* of the inductor.
   d. Use the graph to estimate the time constant τ for this circuit.
   e. Use the values of *R* and τ to estimate the inductance *L*.

5. Here is the graph of Current against Time for a circuit containing an inductor *L*, with resistance *R* and a battery, voltage *V*. From the graphs below, choose the one which best shows the graph of Current against Time when these changes are made separately.

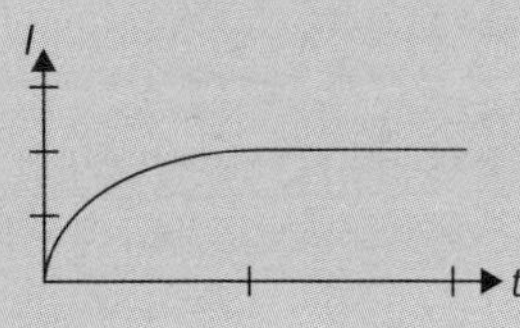

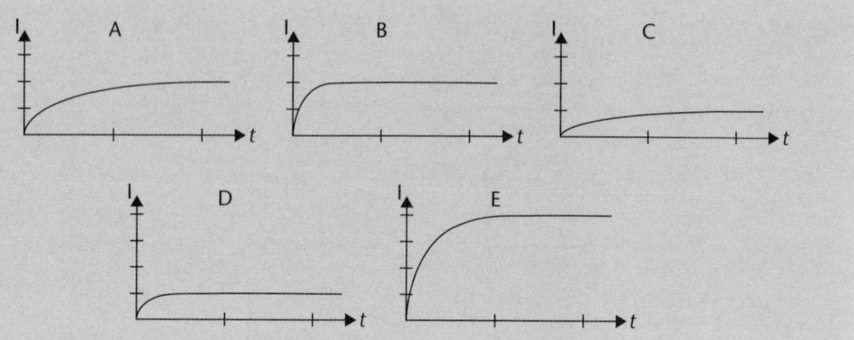

i. The resistance R is doubled.

ii. The inductance L is doubled (but R remains constant).

iii. The voltage V is doubled.

Unit 12.4 Activity 6F: Multiple choice questions

1. The diagram below shows Faraday's experiment on induction. The momentary deflection in the galvanometer of coil B is observed when the switch in coil A is switched on and off. This is caused by the instant flow of current in coil B, which is induced due to a change in magnetic field in coil A.

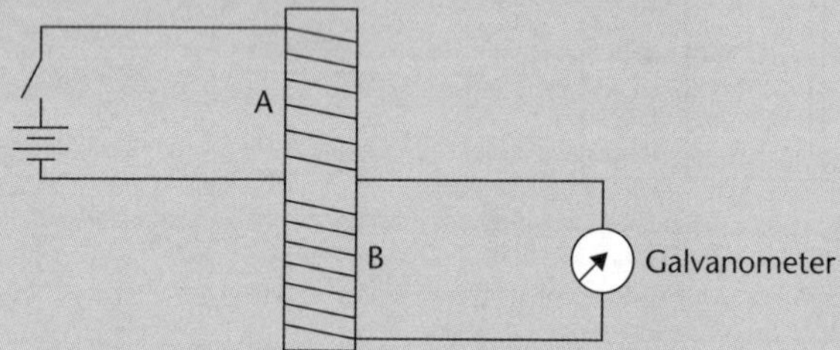

The phenomenon described above is called:

A. self-induction

B. mutual inductance

C. primary inductance

D. secondary inductance

2. The diagram below shows circuit 1 with coil 1 and circuit 2 with coil 2.

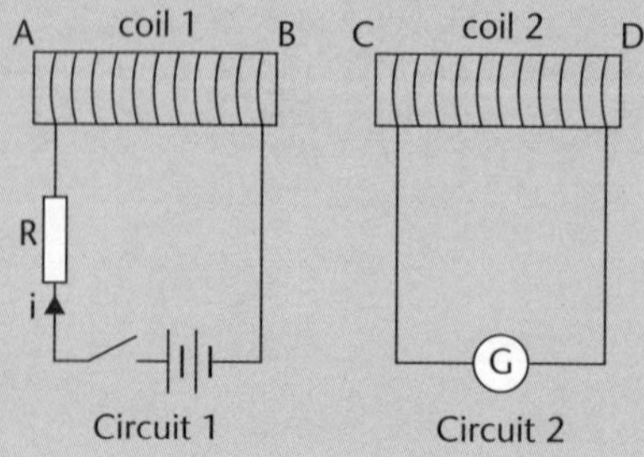

When the switch in coil 1 is closed, the direction of the current will be as shown. Which of this statements below correctly gives the polarities of A, B, C and D?

A. A= South, B= North, C = South and D= North.

B. A= North, B= South, C = North and D= South.

C. A= South, B= North, C = North and D= South.

D. A= North, B= South, C = South and D= North.

3. What will be the polarities when the switch in circuit 1 of question 2 above is off?

A. A= South, B= North, C = South and D= North.

B. A= North, B= South, C = North and D= South.

C. A= South, B= North, C = North and D= South.

D. A= North, B= South, C = South and D= North.

4. The diagram below shows that the induced electromagnetic force, E, is directly proportional to the number of magnetic flux lines passing through the area.

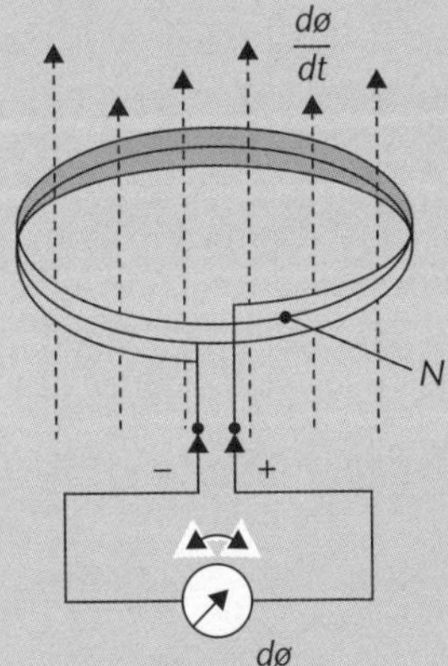

E: Magnitude of the induced e.m.f.
N: Number of turns of the coil
$\frac{dø}{dt}$: Rate of change in the number of Lines of Force

If there are 20 magnetic flux lines passing through the area disturbed in every second with the coil having 20 turns, what would be the magnitude of induced emf per second?

A. 0.040 kV

B. 0.4 k V

C. 40 k V

D. 400 kV

5. Which of the following statements best describes a power transformer?

A. A transformer is a device that converts electrical energy from one circuit to mechanical energy in another through the transformer's coils.

B. A transformer is a device that transfers electrical energy from one circuit to heat and light in another through the transformer's coils.

C. A transformer is a device that produces electrical energy from one circuit to another through the transformer's coils.

D. A transformer is a device that transfers electrical energy from one circuit to another through inductively coupled conductors called the transformer's coils.

6. The diagram below shows a step down transformer. The number of turns in the primary is 400 and in the secondary is 300. What would be the power in the primary and secondary if the primary voltage is 20 V and the current in the secondary is 5A?

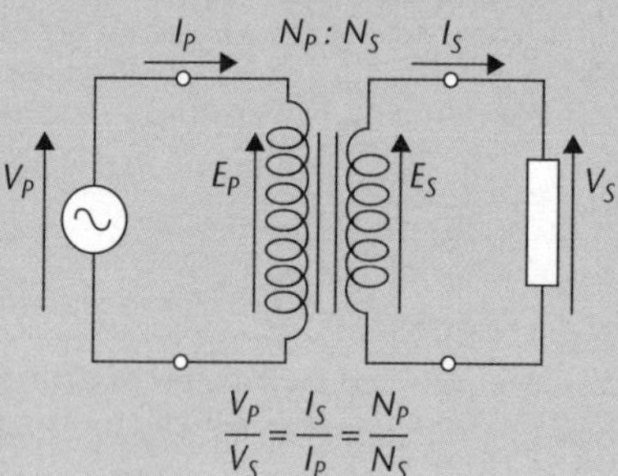

$$\frac{V_P}{V_S} = \frac{I_S}{I_P} = \frac{N_P}{N_S}$$

A. 15 watts
B. 150 watts
C. 1500 watts
D. 15,000 watts

7. Power is transmitted from Yonki Hydro Power Station to Taraka in Lae over hundred kilometres to be distributed to the house holds and industries in Lae. Assuming along the way the average resistance for the power lines was 50 Ω and the current through the line is 50 A. How much power would have been lost as it is transmitted along power lines from Yonki to Lae.

A. 10.125 kW
B. 1.25 kW
C. 12.5 kW
D. 125 kW

Unit 12.5 Radioactivity and Nuclear Energy

Topic 1: The atom and nuclear reactions

Topic 1 deals with the atom and nuclear reactions. It covers:

- Structure of the atom.
- Atomic number and mass number.
- Isotopes.
- Fission.
- Fusion.

An **atom** is the smallest part of an **element**. It consists of a positively charged **nucleus** which is surrounded by a cloud of negatively charged electrons. The nucleus contains positively charged **protons** and neutral **neutrons**.

The **atomic number**, Z, is the number of protons in the nucleus of an atom. In a **neutral** atom the atomic number is equal to the number of electrons. All atoms of a particular element have the same atomic number. The **mass number**, A, is the number of protons plus the number of neutrons in the nucleus.

An atom can be represented by the notation $^{A}_{Z}X$, where X is the symbol that denotes the element, Z is the atomic number and A is the mass number. The number of neutrons = $A - Z$.

Example A

An atom of uranium may be represented by the notation: $^{235}_{92}\text{U}$

The atom of uranium consists of a nucleus that contains 92 protons and 143 (235 – 92) neutrons. There are 92 electrons in orbit around the nucleus.

A helium nucleus (ie a helium atom with no electrons) is called an alpha (α) particle; it is represented by the notation: $^{4}_{2}\text{He}$

Isotopes are atoms of the same element. They have the same atomic number but a different mass number. This means that isotopes have identical numbers of protons but different numbers of neutrons.

Example B

There are three isotopes of hydrogen.

Hydrogen	Deuterium	Tritium
Protons = 1 Electron = 1 Neutrons = 0	Protons = 1 Electron = 1 Neutrons = 1	Protons = 1 Electron = 1 Neutrons = 2

The isotopes of hydrogen.

Unit 12.5 Activity 1A: The atom

1. Draw diagrams of the following atoms. Clearly label all the particles that make up each atom and state what charge, if any, each particle has.
   a. An oxygen atom. b. A nitrogen atom. c. A boron atom.
2. Copy and complete the following table:

Element	Mass number	Atomic number	Number of protons in the nucleus	Number of neutrons in the neucleus	Number of electrons
Po	209	a	b	c	84
Es	254	99	d	e	f
At	g	h	i	125	85
Cm	247	j	k	151	l
Md	m	101	n	157	o
U	235	p	92	q	r

3. What does the notation ${}^{1}_{1}X$ describe?
4. How many protons, electrons and neutrons are there in each of the following atoms?
   a. ${}^{238}_{92}\mathrm{U}$ b. ${}^{229}_{90}\mathrm{Th}$ c. ${}^{4}_{2}\mathrm{He}$ d. ${}^{225}_{89}\mathrm{Ac}$
5. In U-235 (uranium isotope of mass number 235), the fission reaction products consist of two smaller nuclei. What are they?
6. Uranium-235 and uranium-238 are two isotopes of uranium. The atomic number of uranium is 92. For each of the isotopes, describe how many protons, neutrons and electrons they contain.
7. Which of the following pairs of atoms are isotopes of each other?
   A. ${}^{178}_{68}X$ and ${}^{170}_{70}X$ B. ${}^{156}_{66}X$ and ${}^{160}_{66}X$ C. ${}^{172}_{68}X$ and ${}^{160}_{60}X$
   D. ${}^{156}_{66}X$ and ${}^{156}_{64}X$ E. ${}^{170}_{70}X$ and ${}^{168}_{68}X$
8. What is the number of neutrons in a neutral atom of Krypton gas (${}^{85}_{36}\mathrm{Kr}$)?

Nuclear reactions

There are two main rules that govern **nuclear reactions**:

- The sum of the mass numbers of the reactants is equal to the sum of the mass numbers of the products.
- The sum of the atomic numbers of the reactants is equal to the sum of the atomic numbers of the products.

Nuclear reactions may involve **fission** or **fusion**.

In a fission reaction, a large atom is split into smaller particles and energy is released.

Example C

The fission reaction in a **nuclear reactor** is:

$${}^{235}_{92}U + {}^{1}_{0}n \rightarrow {}^{141}_{56}\mathrm{Ba} + {}^{92}_{36}\mathrm{Kr} + 3{}^{1}_{0}n + \gamma$$

If there is sufficient uranium in a nuclear reactor, neutrons released collide with other uranium atoms and the reaction keeps repeating itself and a **chain reaction** occurs. The minimum amount of uranium for this to occur is called the **critical mass**.

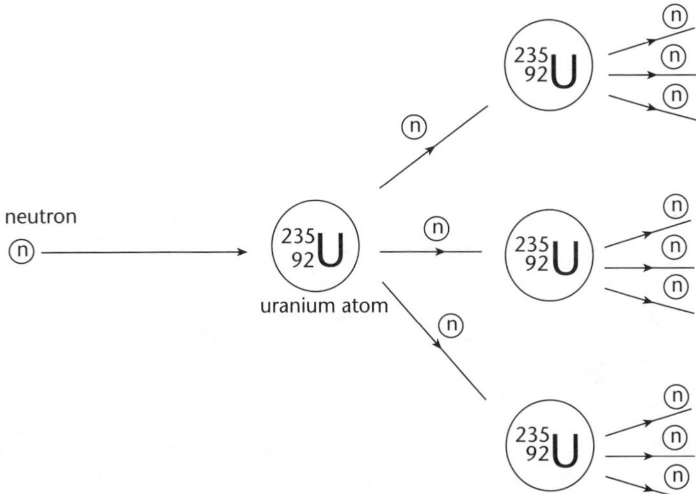

A nuclear chain reaction.

In an **atomic bomb,** an **implosion** triggered by conventional explosives forces two pieces of uranium together. After this happens, the total mass exceeds the critical mass and an uncontrolled chain reaction commences.

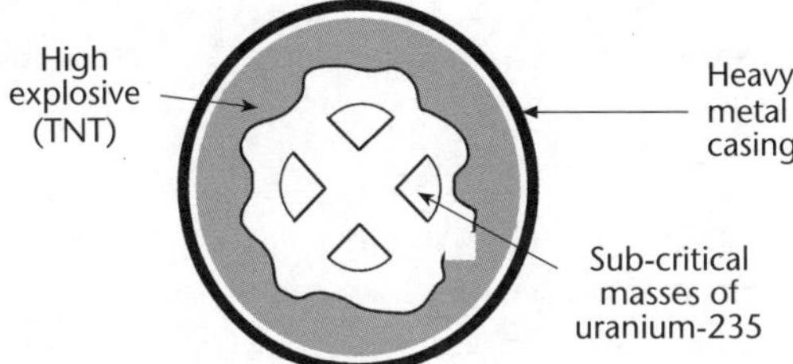

The atomic bomb.

In a fusion reaction, smaller atoms combine to form a larger atom and a large amount of energy is released.

Example D

Fusion reactions occur in the Sun.

$^{2}_{1}H + ^{3}_{1}H \rightarrow ^{4}_{2}He + ^{1}_{0}n$ + energy

Unit 12.5 Activity 1B: Nuclear reactions

1. Use the periodic table (on the following page) and the rules governing nuclear reactions to find the missing quantities labelled (a)–(e) in the following reactions.

a. ${}^{238}_{92}U \rightarrow {}^{234}_{\mathbf{a}}T + {}^{\mathbf{b}}_{2}He$ **b.** ${}^{\mathbf{c}}_{6}d \rightarrow {}^{14}_{7}N + {}^{0}_{-1}\beta$ **c.** ${}^{225}_{89}Ac \rightarrow {}^{\mathbf{e}}_{87} + {}^{4}_{2}He$

2. Complete the following equations:

a. ${}^{23}_{11}Na + {}^{4}_{2}He \rightarrow {}^{26}_{12}Mg + ?$ **b.** ${}^{106}_{47}Ag \rightarrow {}^{106}_{48}Cd + ?$

c. ${}^{10}_{5}B + {}^{4}_{2}He \rightarrow {}^{13}_{7}N + ?$ **d.** ${}^{238}_{92}U \rightarrow {}^{234}_{90}Th + ?$

3. Where does the process of nuclear fission occur?

4. Uranium-238 (${}^{238}_{92}U$) is a radioactive isotope of uranium that undergoes nuclear fission.

a. Explain what is meant by the term 'isotope'.

b. Explain what is meant by the term 'fission'.

One of uranium's other isotopes, uranium-235, is used in nuclear bombs.

c. Balance the reaction below by stating the values of X, Y, and Z.

${}^{235}_{X}U + {}^{1}_{0}n \rightarrow {}^{141}_{Y}Ba + {}^{Z}_{36}Kr + 3{}^{1}_{0}n$

d. State two peaceful uses for uranium-235.

5. Natural uranium is mainly ${}^{238}_{92}U$. About 0.7% of natural uranium is ${}^{235}_{92}U$. These two atoms are known as isotopes.

a. Explain what is meant by the term isotope.

b. Complete the following table for U-238.

Symbol	Number of protons	Number of neutrons
${}^{235}_{92}U$		

Plutonium is a fuel sometimes used in nuclear reactors. One possible nuclear fission reaction for plutonium is:

${}^{239}_{x}Pu + {}^{1}_{0}n \rightarrow {}^{147}_{56}Ba + {}^{90}_{38}Sr + y{}^{1}_{0}n + \text{energy}$

c. Calculate the value of x and state the conservation law used to calculate its value.

d. Calculate the value of y and state the conservation law used to calculate its value.

Stars produce their energy by nuclear fusion. In stars larger than our sun this is achieved by the Carbon–Nitrogen–Oxygen cycle. One possible reaction for this cycle is:

${}^{15}_{7}N + \text{proton} \rightarrow {}^{m}_{6}C + {}^{4}_{n}X + \text{energy}$

e. Complete the following equation by writing the correct symbols for proton and X; and the correct numbers for m and n in the given square brackets.

${}^{15}_{7}N + {}^{[\ ]}_{[\ ]}[\quad] \rightarrow {}^{[\ ]}_{6}C + {}^{4}_{[\ ]}[\quad] + \text{energy}$

f. Explain why the fusion process is difficult to reproduce on Earth.

(Source: NCEA Examination paper)

6. Geiger and Marsden, Rutherford's co-workers, were investigating the structure of the atom. They found that when **alpha particles** were fired at a thin gold foil most particles passed through and a few were deflected back towards the source. Alpha particles (α) are small positively charged particles.

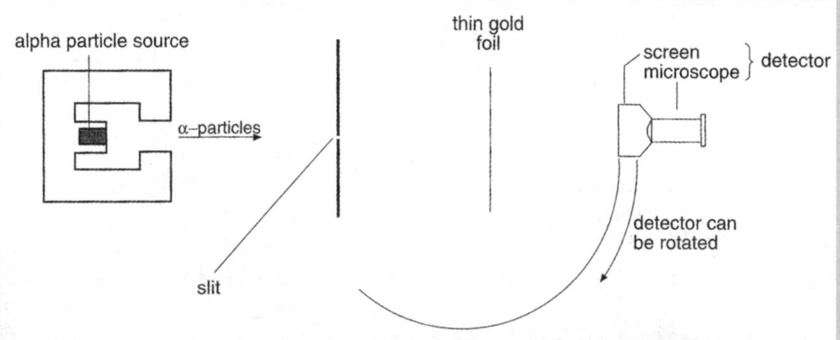

   a. What does the observation that most α-particles passed through indicate about the size of the nucleus of the gold atom?

   b. State which particles are found in the nucleus of the gold atom.

   The symbol for gold may be written $^{197}_{79}Au$.

   c. How many protons are there in the nucleus of a gold atom?

   d. How many neutrons are there in the nucleus of a gold atom?

   e. Apart from protons and neutrons, what other type of particle is found in a gold atom?

   f. How many of these other particles are found in a gold atom?

   Uranium 238 ($^{238}_{92}U$) is a radioactive isotope of uranium that undergoes nuclear fission.

   g. Explain what is meant by the term 'isotope'.

   h. Explain what is meant by the term 'fission'.

   One of uranium's other isotopes, uranium 235, is used in nuclear bombs.

   i. Balance the reaction below by stating the values of X, Y, and Z.

$$^{235}_{\mathbf{X}}U + ^{1}_{0}n \rightarrow ^{141}_{\mathbf{Y}}Ba + ^{\mathbf{Z}}_{36}Kr + 3^{1}_{0}n$$

   X = ____________ **Y** = ____________ **Z** = ____________

   j. State two peaceful uses for uranium 235.

(Source: NCEA Examination paper)

Periodic Table of the Elements

Key

Z — atomic mass
X — symbol
A — atomic number

1	2	3	4	5	6	7	8	9	10	11	12	13	14	15	16	17	18
1 H 1																	4 He 2
7 Li 3	9 Be 4											11 B 5	12 C 6	14 N 7	16 O 8	19 F 9	20 Ne 10
23 Na 11	24 Mg 12											27 Al 13	28 Si 14	31 P 15	32 S 16	35.5 Cl 17	40 Ar 18
39 K 19	40 Ca 20	45 Sc 21	48 Ti 22	51 V 23	52 Cr 24	55 Mn 25	56 Fe 26	59 Co 27	58.7 Ni 28	63.5 Cu 29	65.3 Zn 30	70 Ga 31	72.6 Ge 32	75 As 33	79 Se 34	80 Br 35	84 Kr 36
85.5 Rb 37	87.6 Sr 38	89 Y 39	91 Zr 40	93 Nb 41	96 Mo 42	99 Tc 43	101 Ru 44	103 Rh 45	106 Pd 46	108 Ag 47	112 Cd 48	115 In 49	119 Sn 50	122 Sb 51	128 Te 52	127 I 53	131 Xe 54
133 Cs 55	137 Ba 56	139 La* 57	178.5 Hf 72	181 Ta 73	184 W 74	186 Re 75	190 Os 76	192 Ir 77	195 Pt 78	197 Au 79	201 Hg 80	204 Tl 81	207 Pb 82	209 Bi 83	209 Po 84	210 At 85	222 Rn 86
223 Fr 87	226 Ra 88	227 Ac** 89	261 Rf 104	262 Db 105	263 Sg 106	264 Bh 107	265 Hs 108	268 Mt 109									

139 La* 57	140 Ce 58	141 Pr 59	144 Nd 60	145 Pm 61	150 Sm 62	152 Eu 63	157 Gd 64	159 Tb 65	162.5 Dy 66	165 Ho 67	167 Er 68	169 Tm 69	173 Yb 70	175 Lu 71
227 Ac** 89	232 Th 90	231 Pa 91	238 U 92	237 Np 93	244 Pu 94	243 Am 95	247 Cm 96	247 Bk 97	251 Cf 98	254 Es 99	253 Fm 100	256 Md 101	259 No 102	257 Lr 103

Unit 12.5 Radioactivity and Nuclear Energy
Topic 2: The nuclear reactor and safety

Topic 2 discusses the nuclear reactor and safety. It covers:

- The nuclear reactor.
- Safety precautions necessary to protect radiation workers and the environment.

As the world population continues to grow there will be an increased demand for electricity. Since there is a limited supply of non-renewable fossil fuels such as coal and oil, other sources of power need to be used. Some countries use hydroelectric and **geothermal power**. In many countries nuclear reactors are used to generate electricity.

The figure below shows the parts of a typical nuclear reactor.

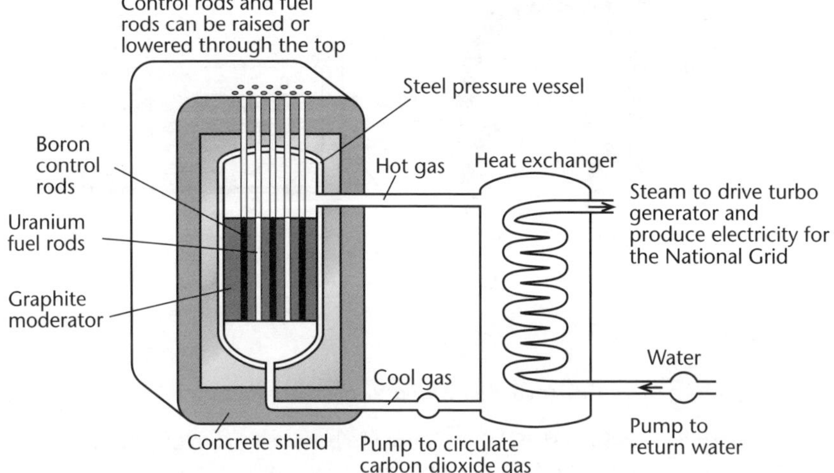

The nuclear reactor.

The **fuel rods** are made of uranium-238 enriched with about 3% of uranium-235. The fission of uranium-235 produces energy and releases neutrons that continue the chain reaction.

The **moderator** slows down the neutrons to the correct speed for fission to occur. If the neutrons move too fast they bounce off the atom instead of causing fission. The moderator is usually *graphite* or water.

The nuclear reaction is controlled by **control rods** made of cadmium or boron. These can be lowered into the **core** between the fuel rods to absorb neutrons to slow or stop the chain reaction.

The **coolant** circulates through tubes in the core. Its purpose is to absorb heat produced by the fission reaction and carry it to the **heat exchanger**. In the heat exchange unit the heat from the coolant is used to heat water to steam, which drives steam turbines that produce electricity.

For nuclear reactors to operate safely, all radioactive materials must be shielded to prevent harmful radiation from entering the environment. For this reason, the core is enclosed in a steel container surrounded by a thick concrete and lead casing. This prevents harmful radiation from entering the environment. **Geiger counters** are regularly used to check radiation levels.

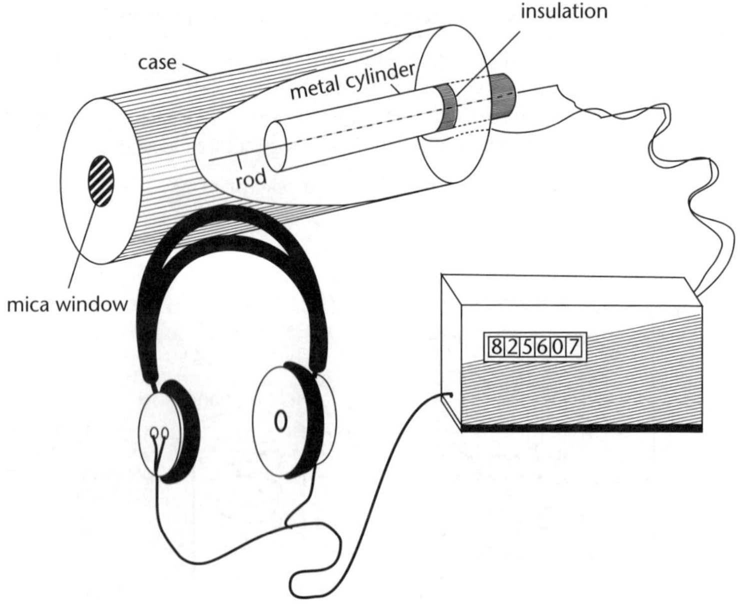

A Geiger counter with attached earphones.

People that work with **radioactive** materials need to be monitored regularly to ensure they are not subjected to unsafe levels of radiation. They wear radiation detection badges that are regularly monitored by the National Radiation Laboratory.

The radioactive byproducts of nuclear reactors need to be disposed of safely. In **breeder reactors** the radioactive waste is treated and used as fuel for other nuclear reactors. In other reactors, it is encased in lead and concrete and stored in remote underground locations, such as disused salt mines.

Nuclear reactors rely on controlling a nuclear chain reaction (see the figure on p. 242).

Unit 12.5 Activity 2A: The nuclear reactor

1. Copy and complete the following table by stating what material each part of the nuclear reactor is made of and describe its function.

Part of the nuclear reactor	Material	Function
Fuel rods	a	b
Control rods	c	d
Moderator	e	f
Coolant	g	h
Shielding	i	j

2. The following questions refer to the figure on p. 242 (the diagram of the nuclear reactor).
   a. Why is the core of the nuclear reactor surrounded by a thick concrete shield?
   b. Why is water pumped through the reactor?
   c. What radioactive isotope are the fuel rods made of?
   d. Will the fuel rods last forever?
   e. What material is the moderator made of?
   f. Why do the neutrons need to be slowed down by the moderator?
   g. Why is it important to monitor the core temperature carefully?
3. The core of a nuclear reactor contains 1 500 fuel rods. The mass of each fuel rod is 14 kg. Three per cent of each fuel rod contains uranium-235. The amount of heat produced by the fission of 1 kg of uranium-235 is 1 014 J of heat energy. The power produced by the reactor is 2 000 MW.
   a. What is the total mass of uranium in the core?
   b. What is the total amount of heat energy that can be produced by the fission of all the uranium in the core?
   c. If the reactor produces power continuously, how long will it take before the fuel rods need to be replaced?
4. What is the advantage of a breeder reactor over an ordinary nuclear reactor?
5. Describe:
   a. What is meant by the term 'chain reaction'.
   b. The main difference between the chain reactions in an atomic bomb and a nuclear reactor.
6. At present there is a ban on nuclear-powered ships coming into New Zealand harbours. Write a paragraph stating whether you agree with this ban or not. Support your opinion with a reasoned argument.

7. Radiographers wear badges to monitor their exposure to radiation. The badge contains a photographic film.

photographic film
case

   a. Why would it be wrong to keep the badge in a pocket?
   b. How does the photographic film show the amount of radiation the person has received?
   c. Why is the film in a light-proof packet?

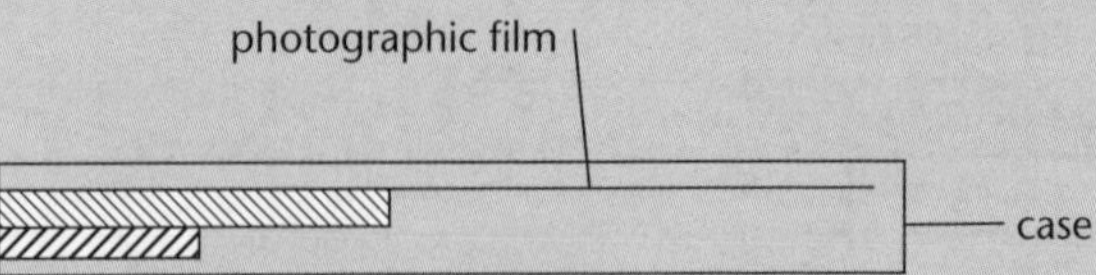

8. The following symbols represent hypothetical atoms:

   $^{58}_{29}A$ $^{54}_{27}B$ $^{59}_{29}C$ $^{58}_{31}D$ $^{59}_{30}E$

   a. Which atoms are isotopes of each other?
   b. Atom ______ i ______ could be produced from atom ______ ii ______ by the emission of an alpha particle ($^{4}_{2}He$).
   c. Which atom contains the largest number of neutrons?

9. Allocate the following labels in the correct position on the diagram of the nuclear reactor.

   Labels: control rods, moderator, coolant, shielding

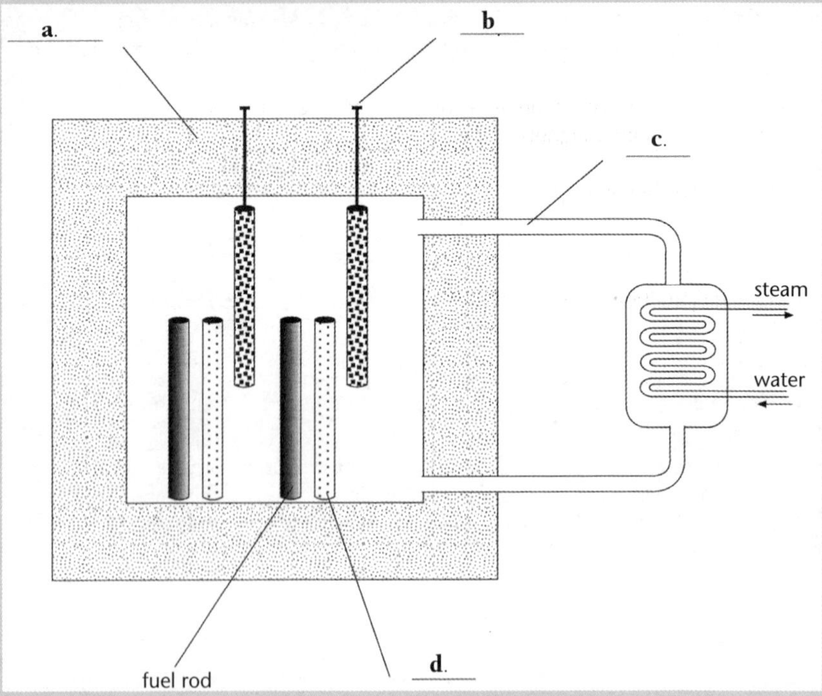

a.
b
c.
steam
water
fuel rod
d.

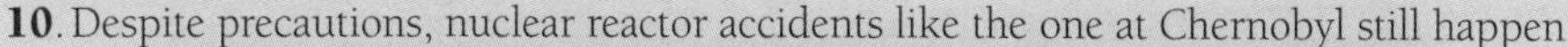

10. Despite precautions, nuclear reactor accidents like the one at Chernobyl still happen.

Describe what would happen in a nuclear reactor if the following were to occur, and explain why.

a. The control rods were **all** fully inserted into the reactor core.

b. The control rods were removed from the reactor core.

c. The reactor core had no moderator.

(Source: NCEA Examination paper)

Unit 12.5 Activity 2B: Multiple choice questions

1. Which of the following definitions best defines the term *radioactivity*?

A. emission of particles from nuclei as a result of nuclear stability

B. emission of particles from nuclei as a result of atomic **collisions**

C. emission of particles from nuclei as a result of carbon dating

D. emission of particles from nuclei as a result of nuclear instability

2. Which of the following particles is not a natural radioactive particle?

A. gamma

B. beta

C. alpha

D. photon

3. If a magnetic field is directed perpendicular to the path of radioactive particle travel, which of the radioactive particles will **not** suffer any deflection at all?
   A. gamma
   B. beta
   C. alpha
   D. all of the above
4. Which of the following statements best defines the term half-life of a radioisotope?
   A. time for a naturally radioactive nucleus to decay completely
   B. time for a naturally radioactive nucleus to emit radioactive particles
   C. time for a naturally radioactive nucleus to decay to half its original value
   D. time for a naturally radioactive nucleus to give off large amounts of energy
5. The graph below shows strontium-90 which has a half-life of approximately 28 years. How much of the activity remains after its third half-life has lapsed?

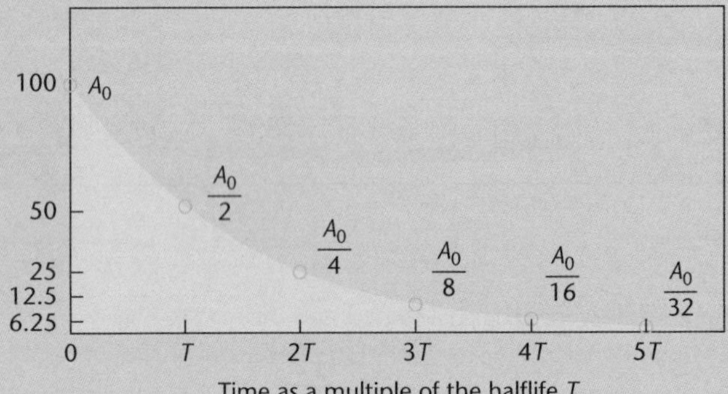

   A. 50%
   B. 25%
   C. 12.5%
   D. 6.25%
6. From the graph in question 5 above, what percentage of strontium-90 would remain after 280 years? You will need to do a calculation, as it is impossible to get the answer from the graph. Use the equation $N_t = N_o \times (0.5)^{\text{number of half-lives}}$
   A. 0%
   B. 0.098%
   C. 0.10%
   D. 1.00%
7. What is the main difference between the two processes of nuclear reaction fission and fusion?
   A. Fission is the fusing of two or more smaller atoms while fusion is the splitting of an atom into two or more smaller ones.
   B. Fission is the splitting of an atom into two or more smaller ones and requires a large amount of energy while fusion is the fusing of two or more smaller atoms into a larger one and requires less energy.

C. Fission is the splitting of an atom into two or more smaller ones and normally occurs in nature while fusion is the fusing of two or more smaller atoms into a larger one and does not occur in nature.

D. Fission is the splitting of an atom into two or more smaller ones while fusion is the fusing of two or more smaller atoms into a larger one.

8. Nuclear fission is the splitting of a massive nucleus into photons in the form of gamma rays, free neutrons, and other **subatomic** particles. In a typical nuclear reaction involving ^{235}U being split into two subatomic particles, free neutrons and photons:

$^{236}_{92}U = ^{144}_{56}Ba + ^{89}_{36}Kr + An + 177$ MeV

What is the value of A?

A. 1

B. 2

C. 3

D. 4

9. In question 8 above, the mass of $^{236}_{92}U$ is reduced in the form of energy released by how much?

A. 3.15×10^{-26} kg

B. 3.15×10^{-27} kg

C. 3.15×10^{-28} kg

D. 3.15×10^{-29} kg

10. The core of a reactor has three main parts. Which of the three main parts is made of boron or cadmium to absorb neutrons? It is inserted in the core to slow down the reaction or removed to speed it up.

A. moderator

B. fuel rods

C. radiator

D. control rods

11. The damage caused by an atomic explosion may come from which of the factors below?

A. an intense burst of radiation at the time of the explosion

B. the release of radioactive fallout

C. the intense heat and destructive force of the explosion itself

D. all of the above

12. Many nuclear power plants around the world are nearing the end of their operating lives. The disposal of nuclear waste is becoming a concern. Which of the following methods are currently under consideration for storing the wastes of the reactors when they are decommissioned?

A. short term storage

B. long term storage

C. transmutation

D. all of the above.

Unit 12.5 Radioactivity and Nuclear Energy

Topic 3: Nuclear reactions

Topic 3 deals with nuclear reactions. It covers:

- Main radiation types.
- Radioactivity units.
- Half-life of a radioactive substance.
- Equations for radioactive decay.
- Nuclear fission and fusion.
- Development of nuclear energy for peaceful and non-peaceful purposes.

Introduction

In 1896 Henri Becquerel, a French scientist, was studying minerals. He left a piece of uranium ore in his desk drawer on top of a photographic plate wrapped in light-proof paper. When he developed the plate he was surprised to discover that the uranium had given off some form of energy which passed through the light-proof paper and darkened the film. This form of energy given off by the uranium was called **radioactivity**.

It was soon found that a few other very heavy elements were also radioactive. In 1898, after two years of exhausting work, the Curies had separated radium and polonium. Radium is several million times more radioactive than uranium. Since radioactivity is not affected by physical or chemical changes it was realised that it must be a property of the nucleus of the atom.

The nucleus

Protons and neutrons are called **nucleons** since they are found in the nucleus.

Atomic number (Symbol Z) is the number of protons in a nucleus.

Mass number or **nucleon number** (Symbol A) is the number of protons *plus* neutrons in a nucleus.

Nuclei with the same atomic number can have different mass numbers. These are called **isotopes** of an element. A **nuclide** is the name used when referring to a specific isotope of an element.

If X is the symbol for an element, the information about the atom is written:

Example A

The atom $^{7}_{3}$Li is a lithium atom with 3 protons, since Z = 3. Since atoms are neutral, there must be 3 electrons as well. The '7' shows that there is a total of 7 nucleons (protons and neutrons). Thus:

number of protons + number of neutrons = 7
3 + number of neutrons = 7
number of neutrons = 7 – 3
= 4

There are 3 protons and 4 neutrons in $^{7}_{3}$Li.

Example B

Carbon has three isotopes.

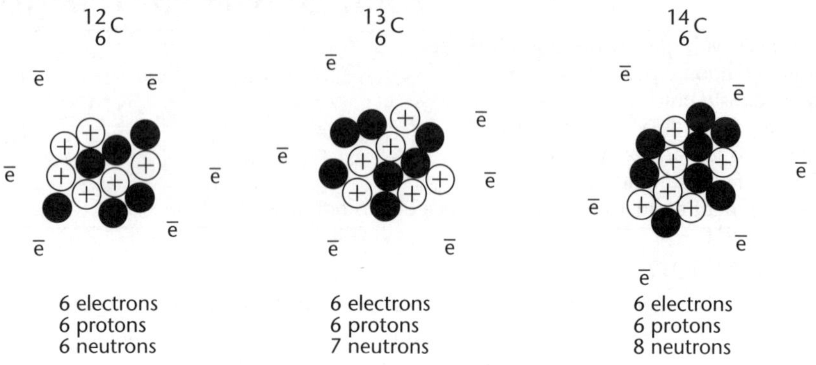

Note: Often isotopes are referred to by their name and mass number, ie $^{14}_{6}C$ is called carbon-14.

Radioactivity

In a **nuclear reaction**, the number of protons or neutrons in an atom changes so that the atom is changed from one element to another. One type of nuclear reaction is called **radioactive decay**. Other types include **fusion** and **fission** reactions.

Some elements, usually those with an atomic number greater than that of lead, *spontaneously* emit particles or radiation from their nuclei. Such elements are said to be **unstable** and **radioactive**.

Radioactivity was discovered by accident by Jacques Becquerel in 1896. Pierre and Marie Curie, French chemists, did much of the early work on radioactive substances and eventually showed that radium and uranium were among the most active elements.

Types of radioactivity

In 1900 Rutherford experimented with the penetrating power of radiation. He found that one type of radiation (called alpha α) could be absorbed by a single sheet of thin aluminium, while a second type (beta β) needed 50 times as much aluminium to be absorbed.

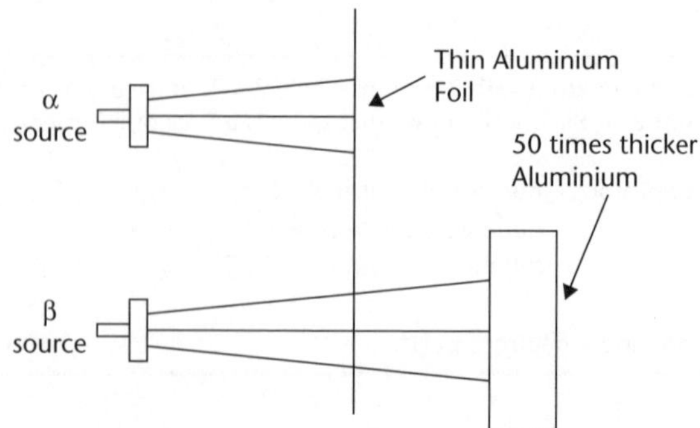

Paul Villard discovered a third type of radiation, much more penetrating than α or β. This was named **gamma** (or γ).

Passing the radiation through a magnetic field shows that α and β radiation is electrically charged, and gamma radiation is uncharged.

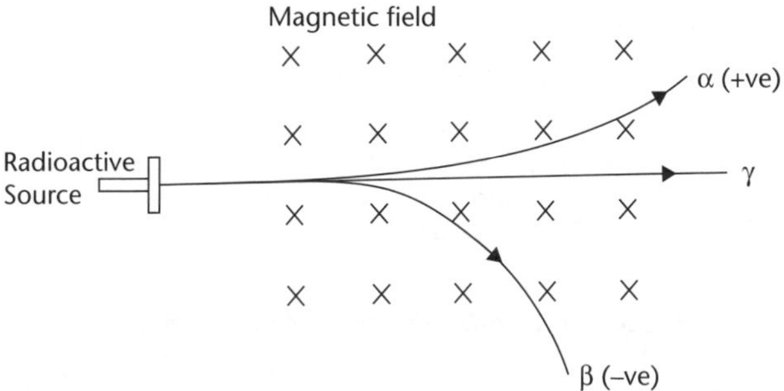

Alpha radiation, α

By 1909 Rutherford and Royds proved that alpha particles consist of Helium nuclei, ${}^{4}_{2}He$. **Alpha particles** can travel at up to 10% of light speed, but have a range of only a few centimetres in air.

Beta radiation, β

This is a stream of very high speed electrons. **Beta particles** (written ${}^{0}_{-1}e$) can travel close to light speed and penetrate up to 30 cm in air.

Gamma radiation, γ

This is very high frequency electromagnetic radiation. Gamma rays travel at the same speed as light and are very penetrating. They are emitted when a nucleus is left in an excited (high energy) state after some other type of radioactive decay. This excess energy is given out as a photon of gamma radiation.

Measuring radioactivity

There are two common units to measure radioactive activity:

- 1 becquerel (Bq) = number of radio active particles disintegration per second.
- 1 curie (Ci) = 3.7×10^{10} becquerel (Bq).

The samples with a higher rate of disintegration (decaying) have more particles emitted in a given period of time.

Radioactive half-life

The radioactive half-life is the time when a radioactive nucleus splits into half its original (parent) size, into two 'daughter' isotopes. The two 'daughter' radioactive nuclei then split into two half-lives, each of which is one quarter of the original sample. After three half-lives each half-life will be one eighth of the original sample, and so on until it disintegrates completely. The graphs below illustrate half-life disintegration of a radioactive nuclei Technetium metastable (Tc-99m). Since it has a half-life of over six hours, it is mostly used in nuclear medicine for diagnosis of diseases, tissues and organs of human body disorder.

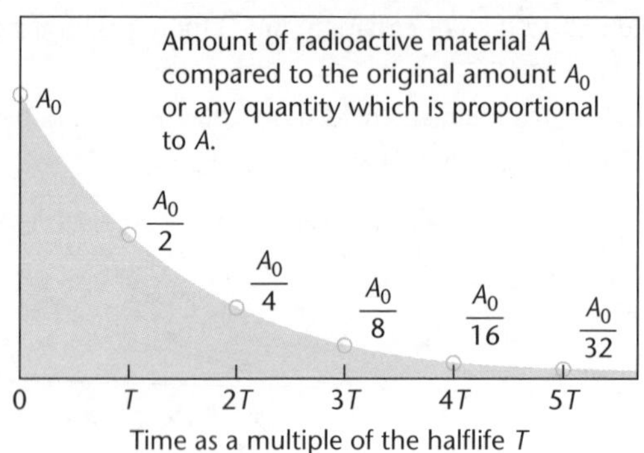

Decay of Technetium metastable (Tc-99m)

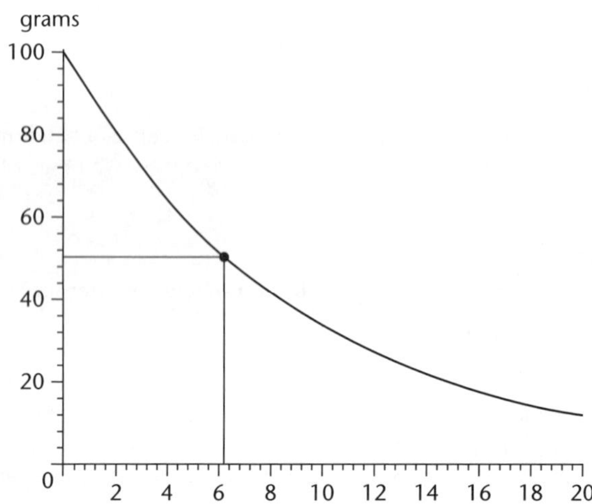

The decay formula

The number of atoms that disintegrate (*N*) over a certain period of time can be calculated using the mathematical expression called decay equation.

$$N = N_0 e^{-\lambda t}$$

Where N = number of atoms disintegrated over a period t,

N_0 = original number of atoms at $t = 0$ units.

λ = disintegration constant,

–ve sign indicates a decrease in the original value of the radioisotope.

Application of radioactive isotopes includes:

a. Using Carbon-14 to determine the age of an artefact to measure the amount of radiocarbon left in an artefact.

b. Producing a radiographic image in medical X-rays, MRI and nuclear medicine.

c. Radioactivity tracers are commonly used in the medical field in the study of plants and animals.

d. Radiation is used and produced in nuclear reactors, which controls fission reactions to produce energy and new substances from the fission products.

e. Radiation is also used to sterilise medical instruments and food.

f. Radiation is used by test personnel who monitor materials and processes by nondestructive methods such as X-rays.

Nuclear reactions

Radioactivity is caused by changes in the nucleus of an atom. Any change or rearrangement of nucleons is called a **nuclear reaction**.

In nuclear reactions:

- The *atomic number* is **conserved**. This means that the sum of the atomic numbers before the reaction equals the sum of the atomic numbers after the reaction.
- The *mass number* is conserved. The sum of the mass numbers before the reaction equals the sum of the mass numbers after the reaction.

Alpha decay

This occurs when a nucleus breaks up and releases an alpha particle with some kinetic energy. The nucleus remaining is now a different element. Sometimes a photon of gamma radiation is also released.

Example C

Radium-226 decays to Radon-222 by alpha decay.

Before	After
(Ra) 88 protons, 138 neutrons } $^{226}_{88}Ra$	2 protons, 2 neutrons (α) + (Rn) 86 protons, 136 neutrons } $^{222}_{86}Rn$

The nuclear reaction equation is:

$$^{226}_{88}Ra \rightarrow ^{226}_{88}Ra + ^{4}_{2}He \qquad \left\{\begin{array}{c} 226 = 222 + 4 \\ 88 = 86 + 2 \end{array}\right\}$$

Beta decay

Beta decay takes place when a neutron in the nucleus decays to a proton plus an electron which has some kinetic energy. Sometimes this is accompanied by the release of gamma radiation as well.

Example D

Cobalt-60 is used as a source of beta and gamma radiation. The cobalt decays to Nickel-60.

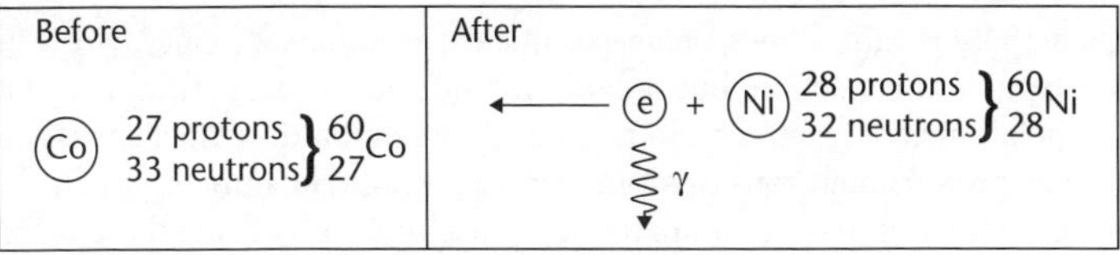

The nuclear reaction equation is

$$^{60}_{27}\text{Co} \rightarrow\ ^{60}_{28}\text{Ni} +\ ^{0}_{-1}\text{e} + \gamma \qquad \begin{Bmatrix} 60 = 60 + 0 \\ 27 = 28 - 1 \end{Bmatrix}$$

Note: 1. The beta particle is given a mass number of zero, since its mass is about $\frac{1}{2000}$th that of a neutron or proton, and an atomic number of –1.

2. The gamma ray has neither mass nor charge since it is a photon of electromagnetic radiation.

Unit 12.5 Activity 3A

1. How many protons, neutrons and electrons are in each of the following atoms
   a. $^{238}_{92}\text{U}$ **b**. $^{56}_{26}\text{Fe}$ **c**. $^{141}_{56}\text{Ba}$ **d**. $^{14}_{6}\text{C}$
2. Give *two* ways to distinguish between alpha, beta, and gamma radiation.
3. Smoke detectors contain tiny amounts of americium-241, each atom of which gives out an alpha particle when it decays.
   a. Complete the equation for this nuclear reaction
   $^{241}_{95}\text{Am} \rightarrow$
   b. What else could also be emitted in this reaction?
4. In nuclear reactors $^{239}_{92}\text{U}$ is formed, which then decays by emitting a beta particle.
   a. Complete the equation for this reaction
   $^{239}_{92}\text{U} \rightarrow$
   b. The nucleus formed in the reaction in part (a) also emits a beta particle. Complete the equation for this second reaction.
   c. what is the name of the final product of these two reactions?

The following information is referred to in questions 5, 6 and 7.

Consider the following nuclear equation:

$$^{4}_{2}\text{He} +\ ^{9}_{4}\text{Be} \rightarrow\ ^{12}_{6}\text{C} + \text{Y}$$

5. The atomic number for Y is:
   A. -1 **B**. 0 **C**. 1 **D**. 2 **E**. 4
6. The mass number for Y is:
   A. -1 **B**. 0 **C**. 1 **D**. 2 **E**. 4
7. The particle Y is:
   A. a proton **B**. an electron **C**. a neutron
   D. a beta particle **E**. an alpha particle

Other nuclear reactions

Natural radioactivity is caused by *spontaneous* nuclear reactions. The decay of a nucleus is not caused by any direct action, but because the nucleus is just naturally unstable. However, artificial nuclear reactions can be caused by bombarding nuclei with high speed particles such as protons, neutrons or alpha particles. Cloud chambers, which show a trail of vapour where radiation passes, are particularly useful for observing these reactions and identifying the products. This is an important method of studying the structure of the nucleus and the types of particles which are found in it.

The first artificial nuclear reaction was performed in 1919 by Rutherford. He placed a source of alpha particles in a container of nitrogen gas. He found that oxygen and protons were produced by the alpha particles colliding with nitrogen nuclei. The reaction is:

$$^{4}_{2}\text{He} + ^{14}_{7}\text{N} \rightarrow ^{17}_{8}\text{O} + ^{1}_{1}\text{H}$$

The alchemist's dream of the transmutation of elements had been achieved, an element was changed into a different element.

One of Rutherford's students, P.M.S. Blackett, set to work with a cloud chamber to photograph the collision which produces this reaction. After some 23 000 photographs he discovered collisions in only 8 of them! Here is one such photograph.

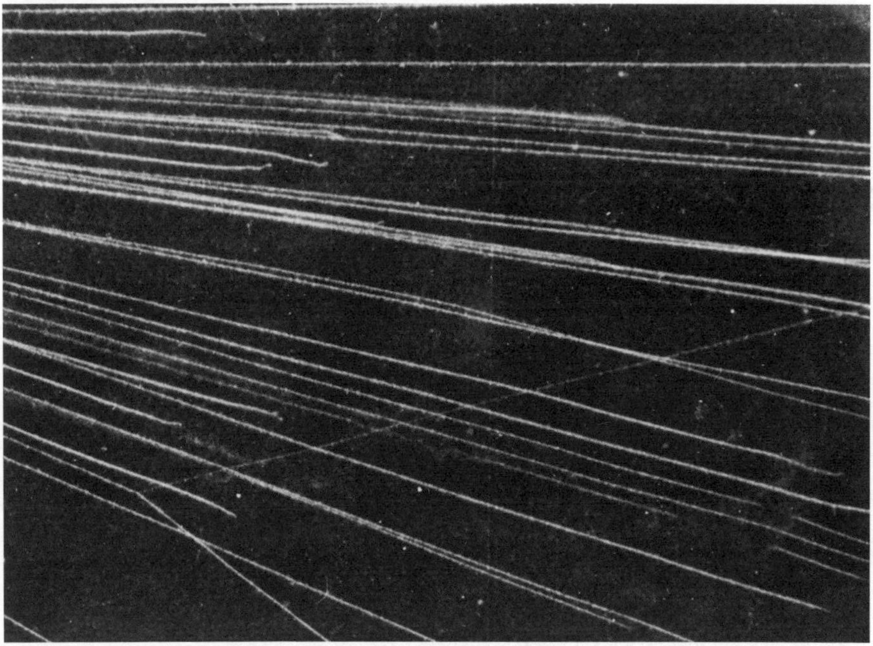

The speeds of the particles can be estimated from the lengths of the tracks.

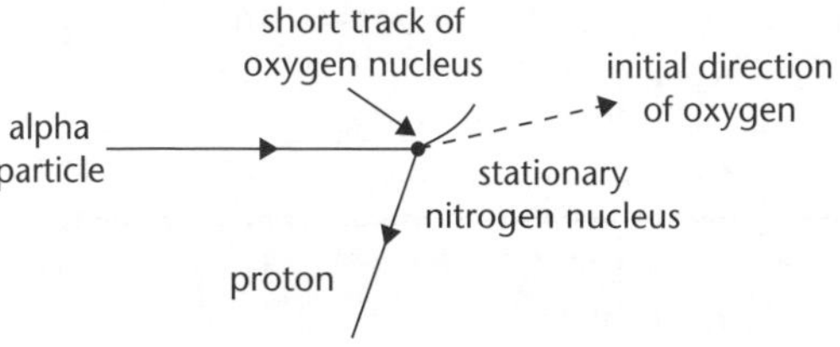

Multiplying the velocity of each particle by its mass gives the **momentum** p.

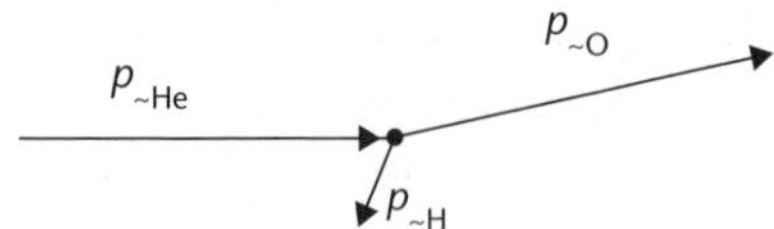

Since the nitrogen nucleus is stationary before the collision, the total momentum before the collision equals $\underset{\sim}{p}_{He}$.

Adding the momentum vectors after the collision shows that:

$$\underset{\sim}{p}_{He} = \underset{\sim}{p}_{O} + \underset{\sim}{p}_{H}$$

This shows that momentum is conserved in a nuclear reaction.

Mass-energy in nuclear reactions

One of the puzzles for the early discoverers of radioactivity was the source of the energy which is given out continuously by elements such as radium and uranium. The answer lies in Einstein's Theory of Relativity with the idea that energy is related to mass.

In a nuclear reaction the rest mass of the products can be greater or smaller than that of the reactants. The difference in mass m results in either a loss or gain of energy E given by:

$$E = mc^2$$

where c = the speed of light ie $c^2 = 9.0 \times 10^{16}\ m^2\ s^{-2}$

Example E

In the reaction between the nitrogen nucleus ${}^{14}_{7}N$ and an alpha particle, the alpha particle had an initial kinetic energy of 12.50×10^{-13} J. After the reaction the kinetic energies are:

${}^{1}_{1}H$: 9.92×10^{-13} J ${}^{17}_{8}O$: 0.64×10^{-13} J

$\therefore$ Total E_k after reaction = 10.56×10^{-13} J.

In this reaction there is a loss of kinetic energy of 1.94×10^{-13} J.

To explain this loss of energy, examine the rest masses of the particles involved.

${}^{4}_{2}He$: 6.64591×10^{-27} kg ${}^{14}_{7}N$: 23.25069×10^{-27} kg

${}^{1}_{1}H$: 1.67338×10^{-27} kg ${}^{17}_{8}O$: 28.22537×10^{-27} kg

Total mass before the reaction = 29.89660×10^{-27} kg

Total mass after the reaction = 29.89875×10^{-27} kg

There is an *increase* in rest mass during the reaction of 0.00215×10^{-27} kg.

This increase in mass is associated with a decrease in energy given by:

$E = mc^2$

$= 0.00215 \times 10^{-27} \times (3.0 \times 10^8)^2$

$= 1.94 \times 10^{-13}$ J

This is exactly equal to the lost kinetic energy of the particles.

So, during this reaction, some energy has been converted into an increase in mass.

In a nuclear reaction mass can be considered to be a form of energy. While the kinetic energy and mass *separately* are *not* conserved, the total of **mass-energy** [ie energy and its associated mass] is conserved.

The conservation laws

In nuclear reactions these quantities remain constant:

- mass number
- atomic number
- momentum
- mass-energy

Unit 12.5 Activity 3B

1. Polonium $^{212}_{84}Po$ can decay by emitting an alpha particle $^{4}_{2}He$, to form $^{208}_{82}Pb$.

The polonium nucleus is initially at rest 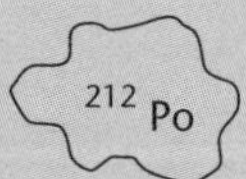

After decay the two products move in opposite directions.

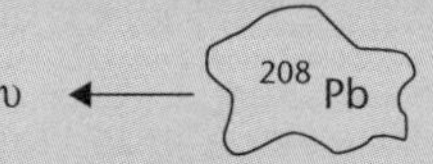

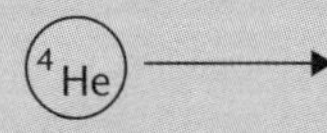

a. What is the total momentum
i. before decay? **ii**. after decay?

b. Calculate the speed of the alpha particle after decay in terms of v. (Use the mass number for mass.)

c. Calculate the total kinetic energy of the particles
i. before decay. **ii**. after decay.
(Use 'energy units' based on the speed v, and the mass number.)

d. Where has this kinetic energy come from?

2. The diagram shows a proton moving with a speed v towards a stationary nucleus X. The proton hits X and is absorbed.

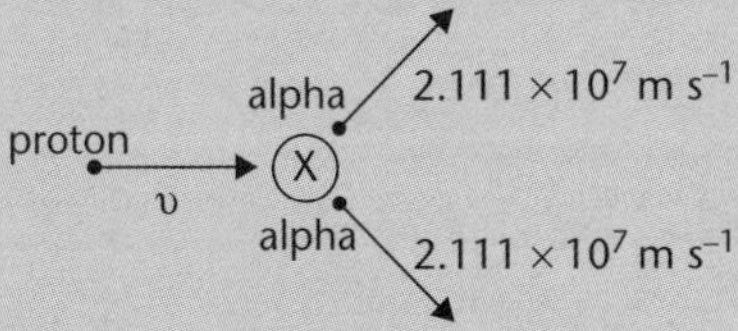

The resulting nucleus splits into two alpha particles, which move off in the directions shown with equal speeds of 2.111×10^{7} m s^{-1}.

a. The nuclear equation for the collision is: $^{1}_{1}H + X \rightarrow 2^{4}_{2}He$
How many protons and how many neutrons are in the nucleus of X?

b. Show that the kinetic energy of one of the emitted alpha particles is 1.481×10^{-12} J. The mass of an alpha particle is 6.646×10^{-27} kg.

c. Given that the energy released in the collision is 2.783×10^{-12} J, calculate the mass change ($c = 3.00 \times 10^{8}$ m s^{-1}).

d. Determine the speed v of the proton before collision, showing your working clearly. The mass of a proton is 1.673×10^{-27} kg.

e. State a quantity other than energy which is conserved in the above nuclear reaction.

3. In 1934 Irene Curie and Frederic Joliot produced the first artificial radioisotope using the following nuclear reactions:

$$^{4}_{2}\text{He} + ^{27}_{13}\text{Al} \rightarrow ^{1}_{0}\text{n} + ^{r}_{s}\text{Y} + Q$$

where $^{r}_{s}\text{Y}$ represents the radioactive isotope they produced and Q is the energy released.

a. Establish the numerical values of r and s.

b. The accurate masses of the nuclei involved are as follows:
$^{4}_{2}\text{He} = 6.6463 \times 10^{-27}$ kg
$^{27}_{13}\text{Al} = 44.8030 \times 10^{-27}$ kg
$^{1}_{0}\text{n} = 1.6749 \times 10^{-27}$ kg
If mass were conserved in the reaction, what would be the mass of $^{r}_{s}\text{Y}$?

c. Actually the mass of $^{r}_{s}\text{Y}$ is 49.7791×10^{-27} kg. Account for this mass difference and calculate the value of Q in joules (speed of light $c = 3.00 \times 10^{8}$ m s^{-1}).

4. The mass of the neutron was first measured accurately in 1934 by Chadwick and Goldhaber. They used a nuclear reaction in which a gamma ray is absorbed by a deuteron, which then splits up into a neutron and a proton.

$$\gamma + ^{2}_{1}\text{H} \rightarrow ^{1}_{1}\text{H} + ^{1}_{0}\text{n} + E_K$$

a. Explain what is meant by
i. gamma-ray ii. deuteron iii. neutron iv. proton

b. Use the energies and particle masses given below to calculate the mass of the neutron produced in this reaction.

Particle	Mass
$^{2}_{1}\text{H}$	3.3436×10^{-27}
$^{1}_{1}\text{H}$	1.6726×10^{-27}

The gamma ray energy was 4.20×10^{-13} J.
The kinetic energy E_K released was 6.25×10^{-14} J
Speed of light in vacuum = 3.00×10^{8} m s^{-1}

5. In the nuclear reaction

$$^{13}_{6}\text{C} + ^{4}_{2}\text{He} \rightarrow ^{16}_{8}\text{O} + ^{1}_{0}\text{n} + 2\text{MeV}$$

stationary carbon atoms are bombarded by helium nuclei. Which of the following is *not* conserved in the reaction?

A. The number of protons. **B**. The number of neutrons.
C. The kinetic energy. **D**. The angular momentum.
E. The linear momentum.

Holding the nucleus together

If a nucleus was put together from protons and neutrons, the mass of the resulting nucleus would be *less* than the total mass of the individual nucleons. This reduction in mass is called a **mass deficit**.

In order to break up a nucleus into separate nucleons the mass deficit must be restored by adding extra energy. This energy E changes into mass m according to Einstein's equation.

$$E = mc^2$$

This energy shortage E has the effect of holding the nucleus together, so it is called the binding energy of the nucleus. **Binding energy** is the amount of energy which must be supplied in order to make up the mass deficit, ie to separate the nucleus into individual nucleons. So binding energy represents the amount of 'glue' holding the nucleus together.

Example F

Using the masses given, find the binding energy of a tritium nucleus, ${}^{3}_{1}\text{H}$.

Rest Masses:	Protons	1.67252×10^{-27} kg
	Neutrons	1.67483×10^{-27} kg
	Tritium nucleus	5.0066×10^{-27} kg

Solution:

Mass of 1 proton + 2 neutrons

$= (1 \times 1.67252) \times 10^{-27} + (2 \times 1.67483) \times 10^{-27}$

$= 5.02218 \times 10^{-27}$ kg

Mass deficit when forming ${}^{3}_{1}\text{H} = (5.02218 - 5.0066) \times 10^{-27}$

$= 0.01558 \times 10^{-27}$ kg

Binding Energy $E = mc^2$

$= 0.01558 \times 10^{-27} \times (3 \times 10^8)^2$

$= 1.4022 \times 10^{-12}$ J

This is the energy needed to break up a tritium nucleus into its separate nucleons.

Nuclear units

Since the mass and energy values associated with the nucleus are very small, special units are sometimes adopted.

For mass, the average mass of a particle in the carbon-12 nucleus is used. This is called the **atomic mass unit** (symbol u).

The mass of ${}^{4}_{2}\text{He} = 4.0015$ u

$1 \text{ u} = 1.660 \times 10^{-27}$ kg

For energy, the **electron-volt** (symbol eV) is used. (See Page 366.) For larger energies the megaelectron-volt (symbol MeV) is also used.

$$1 \text{ MeV} = 10^6 \text{ eV}$$

Binding energy per nucleon

Obviously the larger the nucleus the greater the binding energy (or 'glue'). However, a useful way to show the 'strength' of the nuclear glue is the **binding energy per nucleon** in the nucleus.

$$\text{Binding Energy per Nucleon} = \frac{\text{Total Binding Energy of the Nucleus}}{\text{Number of Nucleons}}$$

Graph of Binding Energy per Nucleon against Mass Number

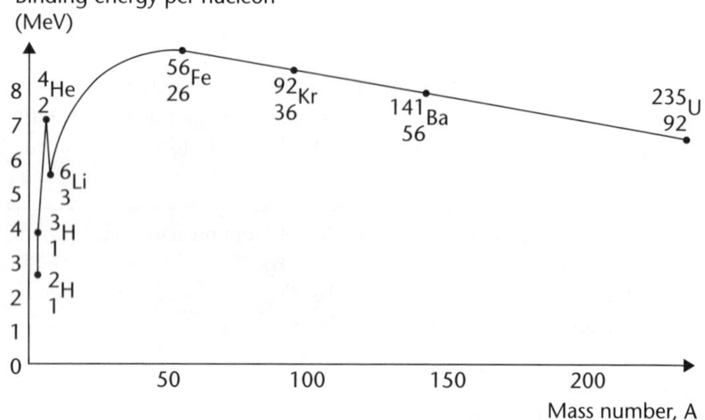

The more binding energy per nucleon, the more stable that nucleus will be. The graph shows that mid sized nuclei are most stable, with iron-56 the most stable nucleus of all.

Nuclear fission

Another way of thinking about binding energy is to consider the average **mass per nucleon** in different nuclei.

Graph of Average Mass per Nucleon against Mass Number

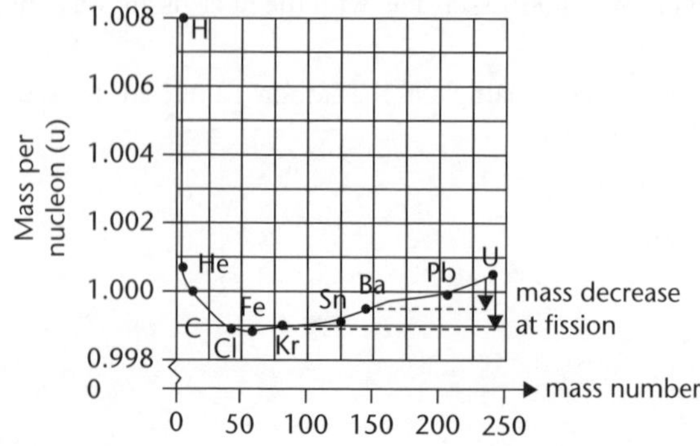

A **fission** reaction occurs when a large nucleus splits into smaller fragments. When this happens each fragment has less mass per nucleon. This lost mass is released as energy in the form of kinetic energy and gamma rays. This happens in a controlled way in a nuclear reactor or in the **explosion** of an atomic bomb.

Example G

Uranium-235 is an isotope of uranium which was used in the atomic bomb which exploded over Hiroshima, Japan, on 6 August 1945. If a slow moving neutron collides with the uranium nucleus, it causes a fission reaction.

$$^{235}_{92}\text{U} + ^{1}_{0}\text{n} \rightarrow ^{141}_{56}\text{Ba} + ^{92}_{36}\text{Kr} + 3^{1}_{0}\text{n} + \gamma$$

Note: There are some 30 different ways in which the uranium-235 nucleus can split up, but all involve a reduction in mass, and a release of energy.

The fission reaction of uranium was first discovered in 1938 by German scientists Otto Hahn and Fritz Strassmann, and in 1939 Lise Meitner and Otto Frisch, two refugees from Nazi Germany, calculated the amount of energy released per atom in the fission reaction.

Example H

In the fission reaction of Example F, the rest masses are:

$^{235}_{92}\text{U}$: 390.2480×10^{-27} kg $^{141}_{56}\text{Ba}$: 233.9616×10^{-27} kg

$^{92}_{36}\text{Kr}$: 152.5794×10^{-27} kg $^{1}_{0}\text{n}$: 1.6747×10^{-27} kg

How much energy is released for each uranium atom which undergoes a fission reaction?

Solution:

Before the reaction, rest mass of $^{1}_{0}\text{n} + ^{235}_{92}\text{U} = 391.9227 \times 10^{-27}$ kg

After the reaction, rest mass of $^{141}_{56}\text{Ba} + ^{92}_{36}\text{Kr} + 3^{1}_{0}\text{n} = 391.5651 \times 10^{-27}$ kg

$\therefore$ Decrease in rest mass = 0.3576×10^{-27} kg

The energy released by this reduction in mass is:

$E = mc^2$

$= 0.3576 \times 10^{-27} \times (3.00 \times 10^8)^2$

$= 3.22 \times 10^{-11}$ J

The chain reaction

In a fission reaction one neutron is needed to start the reaction, but two or three neutrons are produced as products (see diagram on next page). If these neutrons then collide with more nuclei, the number of reactions will suddenly grow very large – a **chain reaction** will occur.

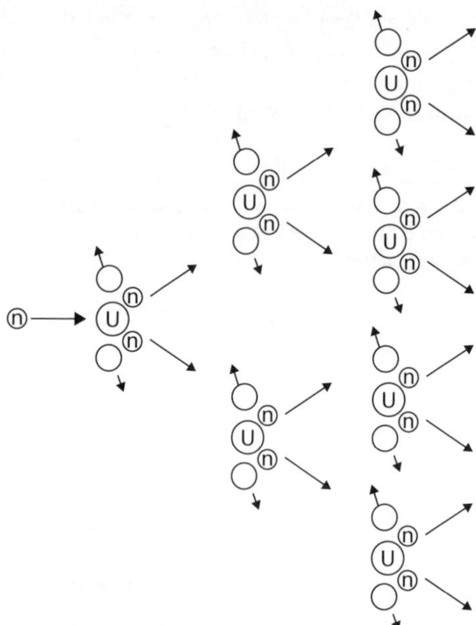

It was quickly realised that this chain reaction could be used to build an atomic bomb which would release huge amounts of energy. During World War II vast amounts of money and thousands of scientists were used in a race to build the bomb before Nazi Germany.

The main difficulty was not in the construction of the bomb itself, which is fairly straightforward, but in separating enough of the isotope uranium-235. Natural uranium is mainly $^{238}_{92}U$, and only about 0.7% of atoms are the $^{235}_{92}U$ nuclei needed to provide fuel for the fission reaction.

Because isotopes are chemically identical, it was a very complex, difficult and expensive task, but by 1945 a few kilograms of fairly pure uranium-235 were produced.

The atomic bomb

Once the enriched uranium-235 had been purified, all that was needed for an atomic explosion was to bring together a large enough amount (called the **critical mass**) and the chain reaction would cause the explosion. The chain reaction cannot build up in a small mass, since too many neutrons escape. This is called a **sub-critical mass**.

Cross-section of an atomic bomb

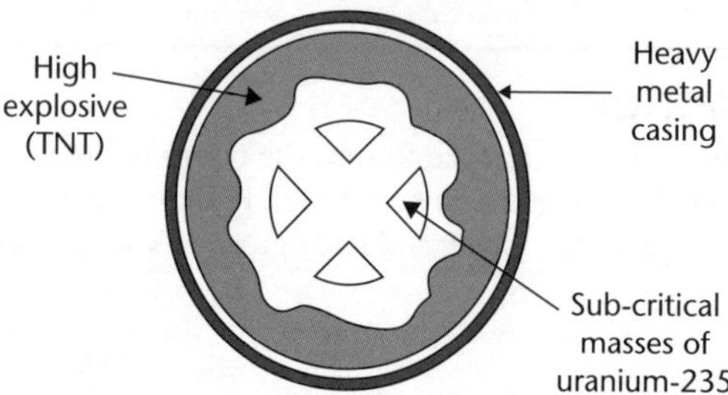

When the TNT is detonated, the sub-critical masses are forced together and form a critical mass, which undergoes a rapid chain reaction.

The first atomic explosion was a secret test carried out in the desert of New Mexico in July 1945. A few weeks later the second and third bombs were dropped on the Japanese cities of Hiroshima and Nagasaki. Japan surrendered a few days later and World War II ended.

The damage caused by an atomic explosion comes from several factors:

- an intense burst of radiation at the time of the explosion.
- the release of **radioactive fallout**, the highly radioactive products of the fission reactions.
- the intense heat and destructive force of the explosion itself.

Note: The size of a fission bomb is limited by the number of subcritical masses which can be brought together at once.

Nuclear power

Even before the first atomic explosion the potential for nuclear reactions to become a power source was realised.

The world's first **nuclear reactor**, where a controlled fission reaction took place, was built in 1942 in a disused squash court at the University of Chicago.

The core of a nuclear reactor has three main parts:

- *fuel rods*: made usually of enriched uranium or plutonium.
- *control rods*: made of boron or cadmium to absorb neutrons. The control rods are inserted in the core to slow down the reaction or removed to speed it up.
- *moderator*: usually graphite or water. The moderator slows down the neutrons released in the fission reaction so they are more likely to collide with another nucleus.

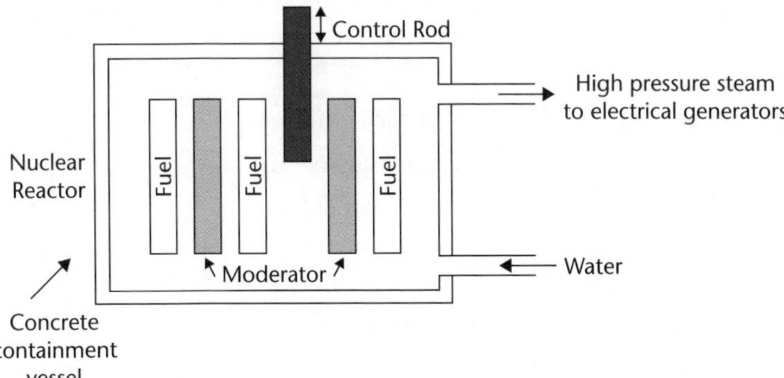

Some reactors also produce plutonium-239 as a product of the reactions in the core. The plutonium can be separated from the used fuel rods and is itself used as a fuel in reactors and atomic bombs.

The two main concerns about nuclear power are:

- safety: accidents at nuclear power plants have released radiation into the environment – the worst, by far, being in 1986 at Chernobyl. Nuclear plants in Western countries have better safety records.

- nuclear wastes: the highly radioactive products of nuclear reactors often have very long half-lives. The disposal of these wastes is a major problem, as they must be stored safely for thousands of years. No long term solution has been found to this storage problem.

Fusion reactions

Another kind of nuclear reaction is a **fusion** reaction, where two small nuclei join together to form a larger nucleus.

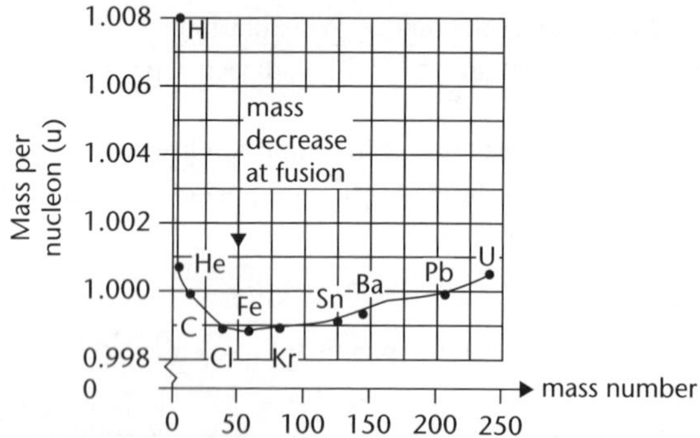

The graph of mass per nucleon against mass number shows that fusion reactions also result in a decrease in mass per nucleon. Again this loss of mass results in a release of energy. A fusion reaction requires a very high temperature (millions of degrees centigrade), in order to make the nuclei move fast enough to collide and stick together.

Example I

In a hydrogen bomb the nuclei of two hydrogen isotopes, deuterium ${}^{2}_{1}H$ and tritium ${}^{3}_{1}H$, join together to form a helium nucleus.

$$ {}^{2}_{1}H + {}^{3}_{1}H \rightarrow {}^{4}_{2}He + {}^{1}_{0}n $$

The rest masses are:

${}^{2}_{1}H$: 3.3434×10^{-27} kg — ${}^{3}_{1}H$: 5.0066×10^{-27} kg

${}^{4}_{2}He$: 6.6443×10^{-27} kg — ${}^{1}_{0}n$: 1.6747×10^{-27} kg

How much energy is released by each fusion reaction?

Solution:

Before reaction, rest mass of ${}^{2}_{1}H + {}^{3}_{1}H = 8.3500 \times 10^{-27}$ kg

After reaction, rest mass of ${}^{4}_{2}He + {}^{1}_{0}n = 8.3190 \times 10^{-27}$ kg

$\therefore$ Decrease in rest mass $= 0.031 \times 10^{-27}$ kg

The energy release by this reduction in mass is

$E = mc^2$

$= 0.031 \times 10^{-27} \times (3.00 \times 10^{8})^2$

$= 2.79 \times 10^{-2}$ J

The first Hydrogen (or Thermonuclear) bomb was exploded in 1952. Hydrogen bombs need a small fission bomb to start the fusion reaction by raising the temperature to several million degrees Celsius.

A basic hydrogen bomb

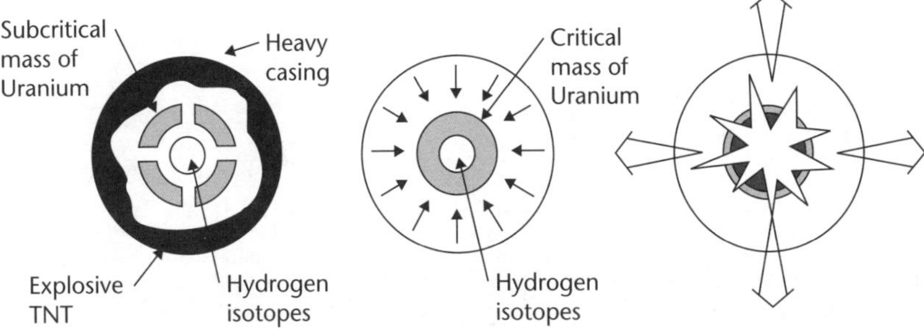

Note: 1. Hydrogen bombs are not limited in size since the amount of hydrogen fuel can be very large.

2. Fusion reactions are 'clean', since they don't produce radioactive products as fission reactions do.

Fusion reactions are the source of energy in the sun and other stars. In the future they may provide a clean form of energy on earth, if a thermonuclear reactor can be built. With the hydrogen fuel readily available and no problems of radioactive wastes to dispose of, a thermonuclear reactor could be the answer to the energy needs of the future. Much research is taking place at present, but the problems of starting up a reaction and extracting the energy from it at such high temperatures are making progress very slow.

Unit 12.5 Activity 3C

1. The diagram shows a graph of the binding energy per nucleon *against* the mass number for the elements.

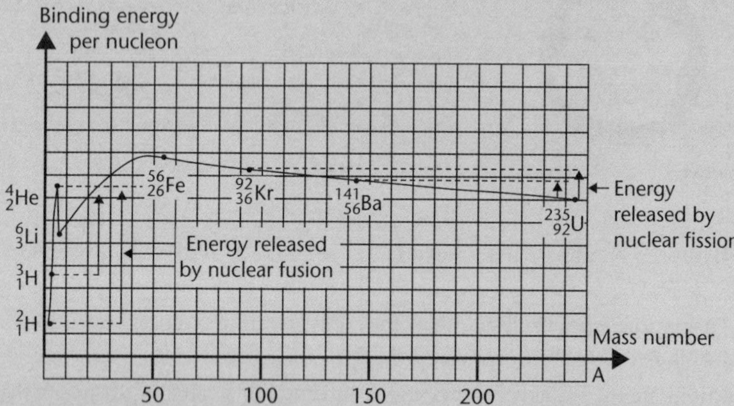

   a. Why is iron (Fe) regarded as being the most stable element?
   b. Explain what is mean by
      i. nuclear fission ii. nuclear fusion
   c. A helium nucleus is represented by $^{4}_{2}He$:
      i. How many protons are there in a helium nucleus?
      ii. How many neutrons are there in a helium nucleus?
   d. You are given the following masses:
      Mass of a helium nucleus = 6.6463×10^{-27} kg
      Mass of a proton = 1.6725×10^{-27} kg
      Mass of a neutron = 1.6748×10^{-27} kg
      i. Determine the value of the mass loss in forming a helium nucleus from its component nucleons. Show your working clearly.
      ii. Determine the energy gained in forming the helium nucleus due to the mass loss. Show your working clearly.
      [Velocity of light = 3.00×10^{8} m s^{-1}]
      iii. Determine the value of the binding energy per nucleon for helium, showing your working clearly.
2. Is the statement 'matter can be neither created or destroyed' true or false? Explain.
3. Why are neutrons so often used to cause nuclear reactions by being fired at the nucleus?
4. In an atomic bomb explosion, the total amount of mass which changes to energy is about 0.2 g. How many joules of energy are released?
5. Plutonium-239 can undergo a fission reaction when a neutron collides with the nucleus.
   a. Write an equation for the fission reaction which has strontium-93 and barium-142 as products.

b. Use the information below to calculate

i. the reduction in mass in this reaction.

ii. the energy released.

Reset masses:	Plutonium-239	$396.929\ 35 \times 10^{-27}$ kg
	Strontium-93	$154.278\ 37 \times 10^{-27}$ kg
	Barium-142	$235.642\ 16 \times 10^{-27}$ kg
	Neutron	$1.674\ 83 \times 10^{-27}$ kg

6. 'Cold fusion' hit the newspaper headlines in 1989 when it was claimed that the nuclear reaction ${}^{2}_{1}H + {}^{3}_{1}H \rightarrow {}^{4}_{2}He + {}^{1}_{0}n$ was detected when electric current was passed between metal plates in a beaker of heavy water.

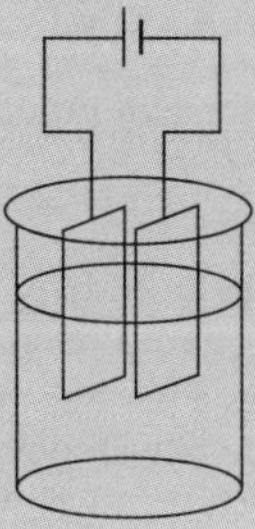

This was greeted with disbelief because:

A. there is no deuterium in heavy water.

B. the deuterium atoms would move too slowly.

C. the deuterium atoms would move too fast.

D. deuterium is radioactive.

E. deuterium nuclei are uncharged.

Uses of radioactivity

Carbon dating

The age of an archaeological specimen which was once alive and thus contains carbon can be established using the half-life of the radioactive carbon-14 isotope. While the specimen was alive, the carbon atoms in it comprised a certain proportion of carbon-14. At death, the decay of the carbon-14 means that the proportion of carbon-14 begins to reduce. Later, by establishing the proportion of carbon-14 present in a small sample and knowing that the half life is 5 730 years, the time since death can be calculated. Carbon dating works well on organic specimens such as wood, bone and cloth that are between 2 000 and 20 000 years old.

Radiotherapy

Gamma rays from a radioactive cobalt-60 source are directed from various angles at a cancerous tumour in a patient's body for a short time. The gamma rays kill the cancerous cells while doing only little damage to healthy cells because cancerous cells are more susceptible to damage from gamma rays .

Sterilisation

Gamma radiation from cobalt-60 can sterilise medical products such as syringes, dressings and needles. The radiation kills micro-organisms in the packaging without damaging the medical instruments and dressings.

Nuclear medicine

Radioactive isotopes are used in medicine as tracers to show lung function and blood flow. Air containing xenon-133 is breathed in and a 'gamma camera' is used to take a moving picture. Most of the xenon-133 is breathed out and the patient receives only a small dose of radiation.

Technetium-99, with a half-life of 6 hours, is injected into a patient's bloodstream and a gamma camera monitors its flow around the body.

Bone imaging is an extremely important use of radioactive properties.

Example J

An athlete is experiencing severe pain in both shins. A possible cause is a stress fracture in either tibia. The athlete is given an injection containing technetium.

After a short wait the patient undergoes a bone scan. At this point, any area of the body that is undergoing unusually high bone growth will show up as a stronger image on the screen. Therefore, if the athlete has a stress fracture, it will show up on the bone imaging scan.

Bone imaging is also useful in diagnosing arthritic problems, bone abnormalities, and various other conditions.

Nuclear radiation is an essential tool of nuclear medicine – it is used to measure:

- Uptake of radioactive iodine by the thyroid.
- Uptake of radioactive technetium by the brain, liver, bone or spleen.
- Uptake of radioactively-labelled drugs or antibodies by tumours or target tissues.

Radioactive isotopes, either alone or incorporated into other compounds (such as drugs), can be administered to the body, and the ability of specific tissues or organs to take up the isotopes evaluated.

Useful isotopes include the following:

Element	Radioisotope	Accumulation
Iodine	$^{131}_{53}I$	Thyroid, lung, kidney
Chromium	$^{51}_{24}Cr$	Spleen
Selenium	$^{75}_{24}Se$	Pancreas
Technetium	$^{99}_{43}Tc$	Brain, lung, liver, spleen, bone, kidney, thyroid, heart, skeletal muscle
Gallium	$^{67}_{31}Ga$	Lymphomas
Hydrogen	$^{3}_{1}H$	A variety of clinical assays

Smoke detectors

Smoke detectors are widely used in homes as an early warning system in case of fire. Many of these units contain a small amount of americium-241. Americium is an alpha-particle emitter. By utilising the radioactive properties of this material, smoke from a fire can be detected at a very early stage. This early warning capability has saved many lives (80% of fire injuries and 80% of fire fatalities occur in homes *without* smoke detectors).

Agricultural applications – radioactive tracers

Radioisotopes can be used to help understand chemical and biological processes in plants and animals. This is possible because:

- Radioisotopes are chemically identical with other isotopes of the same element and will be substituted in chemical reactions.
- Radioactive isotopes of the element can be easily detected with a Geiger counter.

Example K

A solution of phosphate, containing radioactive phosphorus-32, is injected into the roots of a plant. Since phosphorus-32 behaves identically to phosphorus-31 (the more common and non-radioactive form of the element), it is used by the plant in the same way. A Geiger counter detects the movement of the radioactive phosphorus-32 throughout the plant. This information helps to understand the way plants use phosphorus to grow and reproduce.

Carbon-11 has a half-life of about 20 minutes. Scientists have used it to follow the movement of carbon dioxide in plants.

Food irradiation

Food is irradiated by exposing it to the gamma rays from cobalt-60. The energy from the gamma rays passing through the food destroys many disease-causing bacteria as well as those that cause food to spoil, but is not strong enough to change the quality of the food. The food never comes in contact with the radioisotope and is therefore never at risk of becoming radioactive.

Detecting cracks in pipes

A radioactive isotope that emits gamma rays is placed inside a pipe and a detector is placed on the outside. If a crack or flaw in a welded joint is present, less gamma rays will be absorbed than the surrounding metal and the detector will give a higher reading. Beta particles and a detector can be used to monitor the thickness of plastic or paper when they are being manufactured. If the plastic is too thick, the detector will count less beta particles per second.

Living with radiation

Radiation is dangerous even in small amounts. The nuclear accident at Chernobyl power station in the Ukraine in April 1986, when the cooling system failed, heightened debate about the wisdom of using nuclear power and nuclear radiation. Some of the radioactive isotopes released into the atmosphere from Chernobyl were iodine-131 and worse, caesium-137, which both emit beta-particles and gamma rays. After spreading quickly to mainland Europe, the radioactive cloud reached Scotland a few days later. Iodine-131 initially caused concern since it found its way into cows' milk. Its half-life is only 8 days, so its activity became very low after a few weeks. Caesium-137, however, has a half-life of 30 years and so causes continued health problems.

The cancers and births of deformed babies caused by the two atomic bombs dropped on Japan during World War II have been well documented. There is also a continuing problem of what to do with the radioactive waste material produced in nuclear power stations.

There are several points in favour of continuing with nuclear power stations. The world's supplies of oil, coal and natural gas are running out. Some countries in the world do not have river systems that can be used to generate hydroelectric power. (New Zealand is very fortunate in this respect and the prospect for nuclear power stations is remote.) More people are killed each year by mining coal, drilling for oil and by breathing in pollution from their combustion than by using nuclear fuel. Impressive safety designs are built into nuclear power stations and nuclear accidents are far fewer than air crashes, fires, or dam collapses.

Nowdays, much effort and research are being directed into alternative sources of energy production. Wave generators, hydrogen fuel cells, wind generators, solar energy converters, nuclear fusion reactors that produce non-radioactive waste are a few of the ideas being developed.

Unit 12.5 Activity 3D: Fundamantal particles, isotopes

1. How many protons, neutrons, and electrons does an atom of ${}^{235}_{92}U$ have?
2. ${}^{238}_{92}U$ is the most common isotope of uranium. How many neutrons are in the nucleus of uranium-238?
3. A beam of α-particles emitted by a radioactive source is directed towards a very thin sheet of gold foil in a vacuum. State the results that would be observed.
4. Protons are fired into a sample of radon gas. Draw a diagram that best represents the path of a proton as it passes near the nucleus of a radon atom.
5. State what is meant by the term *isotope* and give an example of two nuclei which are isotopes.
6. Workers in a nuclear power station wear badges to find out how much nuclear radiation they have been exposed to. The badge contains a photographic film in a light-proof packet. After being worn for several weeks, the film is taken from the badge and developed. The parts of the film which received radiation will then show up as dark areas.

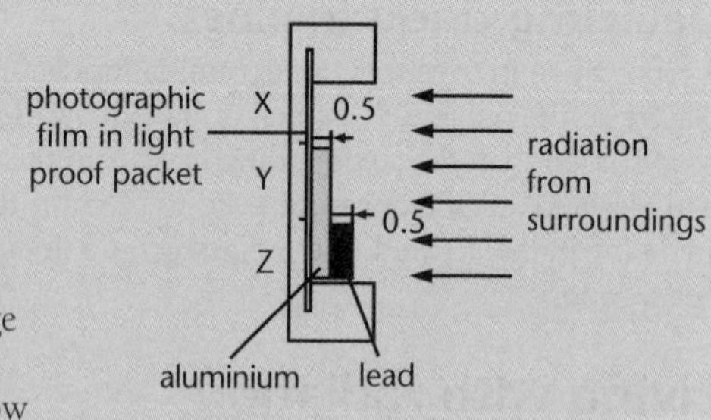

   a. Why would it be wrong to keep the badge in a pocket?
   b. How would the developed film show a worker had received a large dose of alpha radiation?
   c. How would the developed film show a worker had not been exposed to gamma radiation?
   d. Why is the film in a light-proof packet?

7. Uranium-238 can decay in a series of stages. The first five steps in the decay chain are:

$$^{238}_{92}\text{U} \xrightarrow{1} {}^{234}_{90}\text{Th} \xrightarrow{2} {}^{234}_{91}\text{Pa} \xrightarrow{3} {}^{234}_{92}\text{U} \xrightarrow{4} {}^{230}_{90}\text{Th} \xrightarrow{5} {}^{226}_{88}\text{Ra}$$

a. What particles are emitted at each stage?

b. What else might be emitted during some (or all) of the stages?

8. Radium, Ra, atomic number 88, and mass number 226, is a radioactive chemical element discovered by Marie Curie. It emits an alpha-particle and gamma-rays, and thereby changes into another element radon, Rn. State the formula for the radon isotope formed by this process.

9. Radioactive neptunium, $^{239}_{93}\text{Np}$, emits a β-particle. The heavy nucleus remaining is also radioactive, and decays to $^{235}_{92}\text{U}$. What small particle is emitted simultaneously with the formation of uranium-235?

10. Complete the following equations:

a. $^{23}_{11}\text{Na} + {}^{4}_{2}\text{He} \longrightarrow {}^{26}_{12}\text{Mg} + ?$

b. $^{10}_{5}\text{B} + {}^{4}_{2}\text{He} \longrightarrow {}^{13}_{7}\text{N} + ?$

c. $^{106}_{47}\text{Ag} \longrightarrow {}^{106}_{48}\text{Cd} + ?$

d. $^{238}_{92}\text{U} \longrightarrow {}^{234}_{90}\text{Th} + ?$

11. A sample of pure radon gas, $^{222}_{86}\text{Rn}$, is sealed in a glass container. Radon is radioactive. The sample has a half-life of 4 days.

a. If the pressure inside the glass container is doubled, state what will happen to the half-life of the radon.

b. Calculate the fraction of radioactive radon remaining twelve days after the original sample of radon gas was sealed in the glass container.

12. The diagram shows a graph of counts/second against time (minutes) for a sample of radioactive material placed near a suitable detector. The decay rate of a radioactive sample is proportional to the number of radioactive atoms remaining.

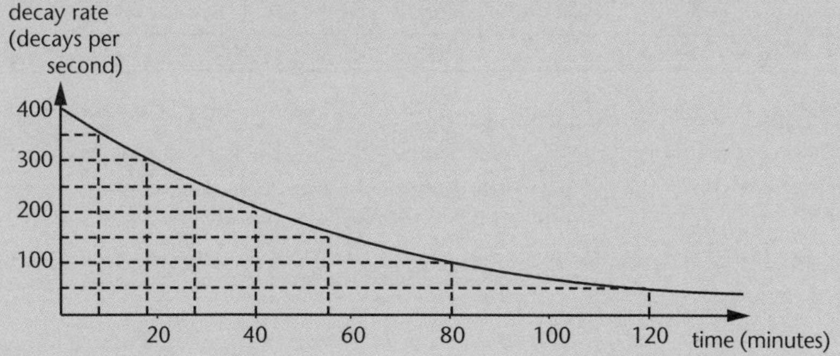

a. Calculate the half-life (in minutes) of the radioactive sample.

b. What count rate (in counts/second) would you have expected if the detector had been turned on one half-life before the start?

13. A radioactive source has a half-life of 80 hours. How long will it take to decay to $\frac{1}{16}$ of its original amount?

14. A radioactive source with an activity of 400 counts per minute is used for industrial purposes.

a. If the half-life of the source is 20 days, sketch a graph showing how its activity will change over a period of 60 days.

b. From the graph, estimate its activity after 5, 10, 25 and 50 days.

c. This source is useful for as long as its activity remains above 50 counts per minute. What is the useful life of the source?

15. The thickness of paper is measured using a radioactive source which emits beta particles. The source is put on one side of the paper and a Geiger counter on the other side of the other. The paper rolls from the papermaking plant onto a roller.

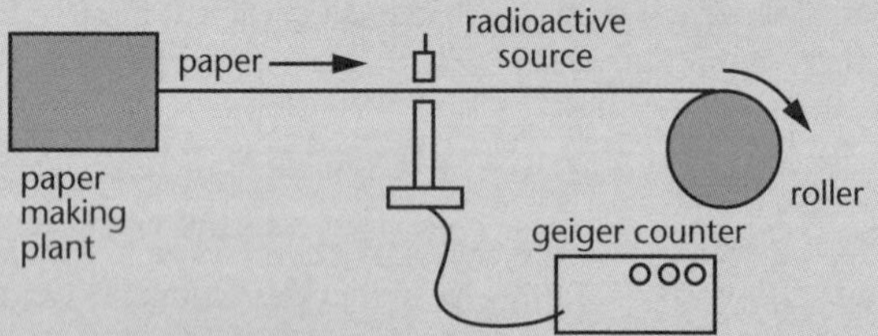

a. Why are beta particles more suitable than alpha particles or gamma rays for this job?

b. Write down one precaution you would take when handling the radioactive source.

c. Chris is worried that the paper will end up being radioactive. What would you say in answer to this?

The table shows the reading on the counter during 70 s.

time in seconds	10	20	30	40	50	60	70
total count since the start	50	100	150	195	235	275	315
count in 10 seconds	50	50	50				

d. Copy and complete the table to show the count in each 10 second time period.

e. Look at the table of results. What happened to the thickness of the paper? Why do you say this?

f. **i.** At what time did the paper begin to change thickness?

ii. The paper was moving at 3 m s^{-1}. What length of paper passed the source before it changed thickness?

16. Tracy carries out experiments with three radioactive sources. She puts different absorbing materials in front of each, and measures the radiation passing through. These are her readings:

Radiation received (counts per second)				
Source	No absorber	Thick cardboard	2 cm of aluminium	2 cm of lead
A	8	8	7	5
B	12	5	4	3
C	10	9	0	0

Tracy knows that: one source gives out beta particles; another source gives out gamma rays; the third source gives out two types of radiation. Work out what is coming from each source.

Appendix
List of constants

Acceleration due to gravity, $g = 9.81\ m\ s^{-2}$ ($10\ m\ s^{-2}$) on earth's surface

Pi, $\pi = 3.142$ approximately

Elementary electric charge, $e = 1.6 \times 10^{-19}$ C

Number of free electrons per cubic metre for copper, $n = 10^{29}$

Magnetic field constant, $k = 2 \times 10^{-7}\ N\ A^{-2}$

Mass of an electron, $m_e = 9 \times 10^{-31}$ kg

Speed of light, $c = 3.0 \times 10^{8}\ m\ s^{-1}$

Refractive index, n: table of values

Wavelengths of colours: table of values

Speed of sound in air at 0°C, $v = 332\ m\ s^{-1}$

Speed of sound in water, $v = 1\ 410\ m\ s^{-1}$

Speed of sound in iron at 20°C, $v = 5\ 000\ m\ s^{-1}$

Half-life: table of values

Answers

Answers for many questions include a 'Marking Guide':

- **A** ('Achievement', meaning 'satisfactory achievement').
- **M** ('Merit', meaning 'high achievement').
- **E** ('Excellence', meaning 'very high achievement').

The Marking Guide has been made by the authors and the publishers and is not an official guide but we hope it will be a help to students who are striving for the best possible results.

Preliminary Unit

The importance of physics

Preliminary Unit Activity A: Physics quantities and units (page 3)

1. **a.** A s (**A**) **b.** m s^{-2} (**A**) **c.** kg m s^{-2} (**A**)
2. **a.** 10^3 cm^3 (**A**) **b.** 10^3 L (**A**) **c.** 10^3 mL (**A**) **d.** 1 mL (**A**)
3. **a.** 2h 47 min (**A**) **b.** 16 min 40 s (**A**) **c.** 11 days 14 h (**A**)
   d. 31 years 251 days 1 year = 365.25 days (**A**)
4. **a.** 10^5 (**A**) **b.** 10^3 (**A**) **c.** 10^6 (**A**) **d.** 10^{-3} (**A**)
5. **a.** 10^3 km (**A**) **b.** 10^5 cm (**A**) **c.** 10^3 µA (**A**) **d.** 10^{-2} Ms (**A**)
6. **a.** 5.2×10^5 t (**A**) **b.** 3.8×10^4 t (**A**) **c.** 0.5 t (**A**) **d.** 5×10^{-4} t (**A**)

Preliminary Unit Activity B: Useful mathematics (page 12)

1. **a.** $13\frac{1}{2}$ (**A**) **b.** $1\frac{2}{3}$ (**A**) **c.** 0.636 (**A**) **d.** 81.7 (**A**) **e.** 23.0 (**A**)
2. **a.** –57.7 (**A**) **b.** 8.35×10^9 (**A**)
3. **a.** $x = 12$ (**A**) **b.** $x = 8$ (**A**) **c.** $x = 3.33$ (**A**) **d.** $x = 4$ (**A**)
4. **a.** 5 : 3 (**A**) **b.** 8 : 3 (**A**) **c.** 3 : 2 (**A**) **d.** 20 : 9 (**A**)
5. **a.** $f = \frac{v}{\lambda}$ (**A**) **b.** $I = \sqrt{\frac{P}{R}}$ (**A**) **c.** $m = \frac{2E_K}{v^2}$ (**A**) **d.** $P_1 = \frac{P_2V_2}{V_1}$ (**A**) **e.** $V_2 = \frac{P_1V_1}{P_2}$ (**A**)
   f. $t = \frac{2d}{v_i + v_f}$ (**A**) **g.** $T_1 = \frac{V_1T_2}{V_2}$ (**A**)
6. **a.** $x = 15, \theta = 53.1°$ (**A**) **b.** $x = 26, \theta = 22.6°$ (**A**) **c.** $x = 1.41, \theta = 45°$ (**A**)
   d. $x = 1.73, \theta = 30°$ (**A**) **e.** $x = 2.24, \theta = 63.4°$ (**A**) **f.** $x = 4.58, \theta = 23.6°$ (**A**)
7. **a.** Δ height = 10.9 cm (**A**) **b.** Δ temperature = –17.3 °C (**A**) **c.** Δ time = 6 h 40 min (**A**)

Unit 12.1 Fluids

Topic 1: Fluid statics

Unit 12.1 Activity 1A (page 14)

1. b, d, a, c, e.
2. $\rho = \frac{\text{mass}}{\text{volume}}$, mass $= \rho \times$ volume $= 1.2\frac{\text{kg}}{\text{m}^3}(4 \times 5 \times 3)\text{m}^3 = 72\frac{\text{kg}}{\text{m}^3}$
3. Pressure $= \frac{\text{Force}}{\text{Area}}$, Force = Pressure × Area $= 1.01 \times 10^5 (4.0 \times 5.0) = 2.02 \times 10^6$ Pascals

Unit 12.1 Activity 1B: Research activity (page 15)

Discuss with teacher.

Unit 12.1 Activity 1C (page 17)

1. The solution will rise the highest in Tube A due to its smaller diameter. Capillary rise depends on the surface tension $\gamma = \frac{F}{L}$. For a cylindrical tube $L = 2\pi r$. For small value of "r" F, which is the force, tension is large and therefore pulls on the whole surface of the liquid up through the narrow tube height, which is called a capillary rise.
2. From Equation 6: $P_{\text{INSIDE}} - P_{\text{OUTSIDE}} = \rho gh = \frac{2\gamma}{r}\cos\theta$

$$= (1040\ \text{kg/m}^3)(9.81\ \text{m/s}^2)(0.20\ \text{m}) = \frac{2(0.0728)}{0.0000030}\cos\theta$$

$$\cos\theta = \frac{2{,}040.48 \times 0.0000030}{2(0.0728)} = 0.42,\ \theta = \cos^{-1}(0.42) = 65^0.$$

Unit 12.1 Fluids

Topic 2: Fluid dynamics

Unit 12.1 Activity 2A (page 20)

1. Both A and B are correct. Bernoulli's principle states that when fluid (liquids and gases) moves its ability to produce pressure is lowered, and the faster the movement the lower the pressure. Examples include water flowing through a pipe with a uniform cross-section and an aeroplane building up pressure difference between its wings which causes lift.
2. A. For flows in which the density does not vary significantly as the traffic slows down more cars will be behind one another in one lane which implies a constant density along the lane. This is incompressible.
3. Data: $\rho_{oil} = 850\ \text{kg/m}^3$, diameter 8.0 cm, flow rate = 9.5 litres/second.

   a. From equation (7). $\frac{\text{Flow.rate}}{\text{cross} - \text{section.area}} = \text{velocity}$

   $$\frac{9.5\ \text{l/s}}{\pi(0.08)^2} = 472.5\frac{\text{l/s}}{\text{m}^2} = 0.4725\frac{\text{m}}{\text{s}}\left(1000\ \text{l} = 1\ \text{m}^3\right)$$

   b. Mass flow rate = mass/time mass = $\rho V = \rho A v = (850\ \text{kg/m}^3)(\pi \times 0.08^2)(0.4725) = 8\ \text{kg/s}$

Unit 12.1 Activity 2B (page 21)

1. The hose end exposed to air may go under water causes suffocation and drowning.
2. (iii) The air pressure is 1 atmosphere at sea level so all objects should experience the same air pressure sitting on top of them at sea level. The human body and other ridged bodies are built to withstand that amount of pressure hence will not shrink under it. The fact that barometer reading remains the same implies air pressure is the same.
3. Absolute pressure at the bottom of the tank is
   $P_{abs} = P_{atm} + \rho gh = 1.01 \times 10^5 + 1000 \times 9.81 \times 12 = 218{,}720.Pa$
   The gauge Pressure = $P_{abs} - P_{atm} = \rho gh = 1000 \times 9.81 \times 12 = 117{,}720.Pa$
4. B; it decreases due to buoyant force (up-trust) acting upwards on the statue sinking in it.
5. $P = \frac{F}{4A},\ A = \frac{F}{4 \times P} = \frac{12{,}000}{4 \times 200{,}000} = 1.5 \times 10^{-2}\ \text{m}^2$

6. The air trapped inside the funnel has more pressure hence pushes the water up the hose, then it is supported by the buoyant force of the water in the funnel to sustain it.

7. Pascals Law which states that any pressure applied to an enclosed fluid is transmitted undiminished to every point of fluid.

$$P_1 = \frac{F_1}{A_1} = P_2 = \frac{F_2}{A_2},\ F_2 = \frac{F_1 A_2}{A_1} = \frac{F_1(4 \times A_1)}{A_1} = 4F_1 = 4(1.2 \times 10^4) = 4.8 \times 10^4\,\mathrm{N}$$

8. **a**. Density, $\rho = \frac{\text{mass}}{\text{volume}}$, volume $= \frac{\text{mass}}{\text{density}} = \frac{27}{2700} = 0.01.\mathrm{m}^3$

b. Tension,

$$T = F_{\text{GRAVITY}} - F_{\text{BUOYANT}} = mg - \rho Vg = 27 \times 9.81 - (1{,}000)(0.01)(9.81) = 264.87 - 98.10 = 166.77.\mathrm{N}$$

9. $F_{\text{BUOYANT}} = F_{\text{AIR}} - F_{\text{WATER}} = 0.98 - 0.88 = 0.1.\mathrm{N}.$

$$F_{\text{BUOYANT}} = \rho Vg,\ V_{\text{OBJECT}} = \frac{F_{\text{BUOYANT}}}{\rho g} = \frac{0.1}{1000 \times 9.81} = 1.02 \times 10^{-5}\,\mathrm{m}^3$$

$$\rho_{\text{OBJECT}} = \frac{\text{mass}}{\text{volume}} = \frac{(0.98/9.81)}{1.02\times10^{-5}} = 9{,}794\frac{\mathrm{kg}}{\mathrm{m}^3}$$

10. Assuming that the volume of water in lagoon is fixed. The volume of water the fish would have displaced when it was in the lagoon would be $V_{\text{WATER}} = \frac{\text{mass}_{\text{FISH}}}{\rho_{\text{WATER}}} = \frac{10\ \mathrm{kg}}{1000(\mathrm{kg/m^3})} = 0.01\ \mathrm{m}^3$.

Therefore the water level would have fall by 0.01 m^3 in volume.

11. The buoyant force exerted upwards on the surfer and the surfing board by the sea water is the same. It is greater than the weight of the surfer and board hence keeps the surfer floating like ship with tonnes of cargo floating in the ocean. However the surfer moves by balancing himself on the surfing board and is carried by wave energy as he rides with the wave.

12. **a**. $\text{Flow.Rate} = Av,\ v = \frac{\text{Flow.Rate}}{A} = \frac{(20\ \mathrm{L}/60\ \mathrm{s})(1 \times 10^{-3}\,\mathrm{m^3/L})}{\pi r^2} = \frac{3.33 \times 10^{-4}}{\pi(0.01)^2} = 1.06\frac{\mathrm{m}}{\mathrm{s}}$

b. $\text{Flow.Rate} = Av,\ v = \frac{\text{Flow.Rate}}{A} = \frac{(20\ \mathrm{L}/60\ \mathrm{s})(1 \times 10^{-3}\,\mathrm{m^3/L})}{\pi r^2} = \frac{3.33 \times 10^{-4}}{\pi(0.0025)^2} = 16.96\frac{\mathrm{m}}{\mathrm{s}}$

13. **a**. $P_1 - P_2 = \frac{1}{2}\rho_{\text{AIR}}(v_2^2 - v_1^2) = \frac{1}{2}(1.29)\left(30^2 - 0^2\right) = 580.5.Pa$

b. $P_1 - P_2 = \frac{F}{A},\ P_2 = P_1 + 580.5 = 101000 + 580.5 = 101{,}580.5.Pa$

$$F_{\text{ROOF}} = P_2(A) = 101{,}580.5(175) = 1.8 \times 10^7.\mathrm{N}$$

14. $P_1 + \frac{1}{2}\rho v_1^2 + \rho g y_1 = P_2 + \frac{1}{2}\rho v_2^2 + \rho g y_2$

$$A_1 v_1 = A_2 v_2, v_2 = \frac{A_1 v_1}{A_2} = \frac{\pi(0.2)^2(0.1)}{\pi(0.1)^2} = 0.40\frac{\mathrm{m}}{\mathrm{s}}$$

$$P_1 - P_2 = \frac{1}{2}\rho v_2^2 + \rho g y_2 - \frac{1}{2}\rho v_1^2 - \rho g y_1 = \frac{1}{2}\rho\left(v_2^2 - v_1^2\right) - \rho g(y_1 - y_2)$$

$$P_1 - P_2 = \frac{1}{2}\rho\left(v_2^2 - v_1^2\right) + \rho g(y_2 - y_1) = \frac{1}{2}(1000)\left(0.4^2 - 0.1^2\right) - (1000)(9.81)(0.0 - 3.0)$$

$$= 75 + 29{,}430.Pa$$

$$P_1 - P_2 = 75 + 29,\ 430 = 29{,}505.Pa$$

15. Flow speed at the second floor $A_1v_1 = A_2v_2,\ v_2 = \frac{A_1v_1}{A_2} = \frac{\pi(0.01)^2(1.5)}{\pi(0.005)^2} = 6.0\frac{\text{m}}{\text{s}}$

Pressure at the second floor. Apply equation 10:

Apply Equation 10: $P_1 + \frac{1}{2}\rho v_1^2 + \rho g y_1 = P_2 + \frac{1}{2}\rho v_2^2 + \rho g y_2$

$P_1 + \frac{1}{2}\rho v_1^2 + \rho g y_1 - \frac{1}{2}\rho v_2^2 - \rho g y_2 = P_2$

$P_2 = P_1 + \frac{1}{2}\rho(v_1^2 - v_2^2) + \rho g(y_1 - y_2) = 400{,}000 + \frac{1}{2}(1000)\left(1.5^2 - 6.0^2\right) + (1000)(9.81)(0-5)$

$P_2 = 400{,}000 + \frac{1}{2}(1000)\left(1.5^2 - 6.0^2\right) + (1000)(9.81)(0-5) = 400{,}000 - 16{,}875 - 49{,}050 = 334{,}075.$

$P_2 = 334{,}075.Pa$

Volume flow rate in the bath room of the second floor = $Av = \pi(0.005)^2.(6.0) = 4.7 \times 10^{-4}\ \text{m}^3/\text{s}$

16. **a**. To push the blood to the heart and then the heart will pump the blood to the lungs via arteries for filtering of carbon dioxide. The filtered blood is then transported by veins back to the heart which is then pump to the rest of the body including the brain.

b. Small surface area, greater surface tension, will not wet too much area the water molecule is sitting on.

17. The brachial artery is located in the upper arms. The brachial artery gives the mean arterial pressure which represents the average arterial pressure throughout the cardiac cycle, and is the force that drives blood through the vasculature.

18. $P_2 - P_1 = \rho g h,\ h = \frac{P_2 - P_1}{\rho g} = \frac{1500}{1060 \times 9.81} = 0.144\ \text{m}$ (14.4 cm in real case is too close)

19. Adding detergent to water reduces surface tension which allow water to wet a surface that is not ordinarily wettable like the grease and oil and dissolve them.

20. Less surface tension as liquid droplets or bubbles will encounter with air resistance as it falls or rises in air, hence keeping its shape.

21. Research question. Task for the teacher and students to carry out for themselves.

Unit 12.2 Temperature and Heat

Topic 1: Temperature, thermal expansion and heat transfer

Unit 12.2 Activity 1A (page 27)

1. c, d, b, e, a, f
2. B and C are currently most suitable.

Unit 12.2 Activity 1B (page 29)

1. $L = \Delta L + L_O = (\alpha)(\Delta T)(L_O) + L_O = (1.2 \times 10^{-5})(35-20)(50) + (50) = 50.009\ \text{m}$

2. Increase in diameter would result in an increase in volume. Therefore:

$\Delta d = \alpha(\Delta T)d_O = \alpha(\Delta T)d_O = 3(2.4 \times 10^{-5})(240-20)(400) = 2.112\ \text{mm}$

3. **a**. Final steady state temperature gradient along the rod.

$\frac{\Delta T}{L} = \frac{T_{H-}T_L}{L} = \frac{(100-0)°\text{C}}{20\ \text{cm}} = 5°\text{C/cm}(500°\text{C/m})$

b. Heat current in the rod during the final steady state.

$$H = \frac{kA(T_{H-}T_L)}{L} = \frac{385(1.2 \times 10^{-4})(100-0)°C}{0.20} = 23.1 \text{ J/s(Watts)}$$

$$H = \frac{kA(T_{H-}T_{L1})}{L_1} = \frac{385(1.2 \times 10^{-4})(100-T_{L1})°C}{0.14} = 23.1 \text{ J/s}$$

c. $$T_{L1} = T_H - \frac{H \times L_1}{kA} = 100 - \frac{23.1(0.14)}{385(1.2 \times 10^{-4})} = 100 - 70 = 30°C$$

4. Volume expansion for glass: $\Delta V_{GLASS} = 3\alpha(\Delta T)V_O = 3(8.5 \times 10^{-6})(100-20)(100) = 0.204 \text{ cm}^3$
   Volume expansion of mercury: $\Delta V_{MERCURY} = \beta(\Delta T)V_O = (181 \times 10^{-6})(100-20)(100) = 1.448 \text{ cm}^3$
   Overflow = 1.448 – 0.204 = 1.224 cm^3
5. Research question for the teacher and the students to carry out.

Unit 12.2 Activity 1C (page 32)

1. B
2. D

Unit 12.2 Activity 1D (page 32)

Research questions for the teacher and the students to carry out.

Unit 12.2 Temperature and Heat

Topic 2: Heat and energy

Unit 12.2 Activity 2A: Heat and temperature (page 37)

1. The matt black cube is a good absorber and also a good radiator of heat; this property causes it to cool down quickly. (**M**)
2. Helium becomes less dense and balloon rises. (**M**)
3. A shiny surface is a poor absorber of radiation. (**M**)
4. By radiation only because outer space is a vacuum. (**M**)
5. Trapped air is a bad conductor of heat. (**M**)
6. Energy gained by water = Energy lost by sand.
   $mc_W\Delta T = mc_{sand}\Delta T$
   $1 \times c_W (40 - 20) = 1 \times 0.2\, c_W (T - 40)$
   $c_{sand} = \frac{1}{5} c_{water}$, therefore $20\, c_W = 0.2\, c_W (T - 40)$
   $20 = 0.2 (T - 40)$
   $T = 60°$ (**E**)
7. Because brass expands more than steel, the strip curves when heated. (**M**)
8. The metal lid expands more than the glass. (**M**)
9. The shiny surface reflects heat back to the body. (**A**)
10. I and II. (**A**)
11. Water colder than 4°C will rise to the top. (**A**)

12. **a**. It transmits thermal radiation *or* it stops air currents from cooling water pipes *or* it traps a layer of air behind the cover. (***A***)

b. Black. [It is a good absorber of radiation.] (***A***)

c. Clockwise. [It takes cold water from the bottom of the cylinder.] (***A***)

d. B, because hot water rises in the cylinder. (***A***)

e. **i**. Copper is a good conductor of heat. (***A***)

ii. The spiral maximises the surface area of coppper in contact with the water.

f. **i**. Foam. (***M***)

ii. To prevent heat loss and save on heating costs. (***M***)

g. Raise the cylinder to produce a gravity feed. (***M***)

h. A dark, heat-absorbent material. (***A***)

i. To get maximum exposure from sunlight. (***A***)

j. $Q = mc\Delta T$

$= 200 \times 4\ 200 \times (55 - 15)$

$= 200 \times 4\ 200 \times 40$

$= 33\ 600\ 000$

$= 3.4 \times 107$ J (***A***)

13. **a**. To reflect heat. (***A***)

b. Silver reduces radiation. (***A***)

c. Layers of newspaper act as an insulator. (***A***)

d. Move warm air away from the ceiling. (***A***)

e. Stone conducts heat away from feet. (***A***)

14. **a**. To cause convection currents. (***A***)

b. Reduces radiation. (***A***)

c. Plastic is an insulator. (***A***)

15. I and II (***A***)

16. Sweat absorbs heat from the body as it evaporates. (***A***)

17. The glass heats up rapidly and expands rapidly, causing the glass to crack. The cork coaster insulates the glass and keeps the base at a higher temperature, thus preventing sudden large temperature changes. (***M***)

18. **a**. The white-hot sparks have a high temperature but do not contain much heat energy. (***M***)

b. The steel head of the hammer conducts heat away from the hand faster than the rubber handle. (***M***)

c. This is due to the sudden expansion of the glass. (***M***)

d. The copper base conducts heat from the element to the contents of the pan. The insulating handle protects the user from burning their hands by slowing and preventing heat conduction. (***M***)

e. This is due to expansion of the metal. (***A***)

f. This is to enable expansion to occur without the road surface cracking. (***A***)

19. Allows air to remove heat rapidly by increasing the surface area. (***A***)

20. Condensing. (***A***)

21. Hot air is less dense than cold air and rises. Increasing the temperature of the air inside the balloon will cause it to rise. (***M***)

22. a. A thermostat contains a bimetallic strip. When this is heated, the two metals expand at different rates, causing the strip to bend and switch off the heating element – this keeps the oven at a constant temperature.

b.

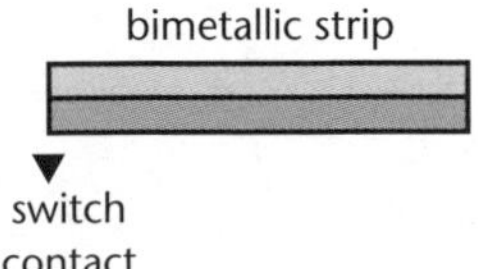

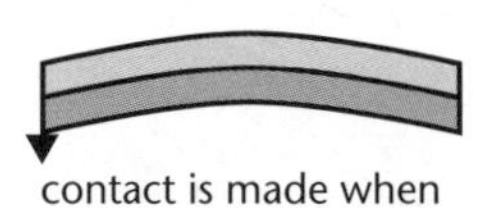

(***M***)

23 a.

Heat Transfer		Percentage of total heat loss
By	**To**	
radiation	surroundings	65%
conduction	*bench*	5%
evaporation	*air*	20%
convection	air currents	10%

(***A***)

b. **i.** evaporation (***A***) **ii.** convection (***A***) **iii.** conduction (***A***)

c. **i.** A (***A***)

ii. Cold water enters the tank at the bottom. Heating water at this point enables convection currents to occur. (***M***)

d. Rate = slope of graph = $\frac{40-10}{20-0} = \frac{20}{30}$ = 1.5°/min (***M***)

e. $Q = mc\Delta T = 40 \times 4200 \times 30 = 5 \times 106$ J OR 5 MJ (***M***)

f. $P = \frac{E}{t} = \frac{5\,040\,000}{20 \times 60}$ = 4200 W or 4.2 kW (***M***)

g. **i.** Lagging of pipes – cylinder wrap – enclosed cupboard. (***A***)

ii. It saves on electricity costs. (***A***)

Unit 12.2 Activity 2B: Multiple choice questions (page 42)

1. B
2. C
3. C
4. D
5. A
6. A
7. B
8. C
9. B
10. C

Unit 12.3 Waves

Topic 1: Types of waves

Unit 12.3 Activity 1A: Waves (page 50)

1. **a.** longitudinal (***A***) **b.** transverse (***A***) **c.** transverse (***A***) **d.** longitudinal (***A***)
e. transverse (***A***)

2. **a.** 0.75 m (***A***) **b.** 1×10^6 m s^{-1} (***A***) **c.** 18.3. Hz (***A***) **d.** 5×10^{14} Hz (***A***)
e. 5×10^{-5} m (***A***) **f.** 52.5 m s^{-1} (***A***)

3. **a.** C, A, B, D. (***M***) **b.** C, D, B, A. (***M***)

4. X-rays, UV, visible light, infrared, radio (***A***)

5. $v = f\lambda$

$$\lambda = \frac{v}{f} = \frac{3 \times 10^8}{2\ 500 \times 10^6} \text{ [1 MHz = 1 000 000 Hz]}$$

$\lambda = 0.12$ m (***M***)

6. **a.** $\lambda = \frac{v}{f} = \frac{3 \times 10^8}{650 \times 103} = 461.5$ m (***A***)

b. $f = \frac{v}{\lambda} = \frac{3 \times 10^8}{315} = 952$ kHz (***A***)

7. $v = \frac{d}{t}; t = \frac{d}{v} = \frac{1.5 \times 10^8}{3 \times 10^5} = 500$ s (8.3 min) (***M***)

8. Common property: they can be reflected
Not common property: light can travel through a vacuum, sound can't. (***M***)

9. $v = \frac{d}{t}; d = vt = 330 \times 15 = 4\ 950$ m = 5 km (***A***)

10. **a.** $v = f\lambda; \lambda = \frac{v}{f} = \frac{1\ 570}{1 \times 10^6} = 1.57 \times 10^{-3}$ m = 1.6 mm (***A***)

b. X-rays can be detrimental to the developing baby. (***A***)

11. $v = \frac{d}{t}; t = \frac{d}{v}$

Time for *P*-waves to arrive = $\frac{600\ 000}{8\ 000} = 75$ s

Time for *S*-waves to arrive = $\frac{600\ 000}{4\ 500} = 133.3$ s

Time delay = 133.3 – 75 = 58.35 s (***M***)

12. $d = v \times t$
$= 8\ 000 \times 5 \times 60$ [convert to s]
$= 2\ 400\ 000$ m
$= 2\ 400$ km (***A***)

13. a.

(***A***)

b.

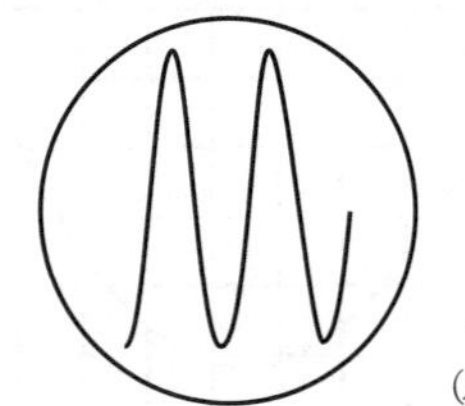

(***A***)

14. a. 3×10^8 m s^{-1} (***A***)

b. i. Infrared. (***A***)

ii. Radiowaves. (***A***)

iii. Film. (***A***)

iv. Eye, camera. (***A***)

15. $v = 2\frac{d}{t}$; $d = \frac{vt}{2} = \frac{330 \times 4}{2} = 660$ m (***M***)

16. $v = f\lambda = \frac{\lambda}{T} = \frac{3}{0.5} = 6$ cm s^{-1} (***A***)

17. Measured wavelength = 3 cm; scale = 1 : 5; actual wavelength = 15 cm. (***M***)

18. a.

Part of trace	P or S	Longitudinal or transverse	Slow or fast
A	P	longitudinal	fast
B	S	transverse	slow

(***A***)

b. $d = v \times t = 8\,000 \times 4 \times 60 = 1\,920$ km (***A***)

c. 8 divisions = $8 \times 2 = 16$ s (***A***)

d. $v = \frac{d}{t} = \frac{1\,920 \times 10^3}{256}$ [240 + 16 = 256 s]

$= 7\,500$ m s^{-1} (***M***)

e. By the amplitude of the trace. (***A***)

19. $v_{\text{sound}} = \frac{\text{extra distance travelled}}{0.25\text{ s}}$

$= \frac{80}{0.25} = 320$ m s^{-1} (***M***)

20. $v = f\lambda$; $\lambda = \frac{v}{f}$

$\lambda_{\text{long}} = \frac{330}{20} = 16.5$ m; $\lambda_{\text{short}} = \frac{330}{20 \times 10^3} = 0.0165$ m

range = 16.5 – 0.0165

= 16.48 m (***E***)

21. a. The loudspeaker must vibrate back and forth. (***A***)

b. Sound travels as a longitudinal wave. (***A***)

c. At 'A' the particles are close together while at 'B' they are further apart. (***A***)

d. i. Compressions. (***A***) **ii.** Rarefactions. (***A***)

e. i. electrical energy to kinetic energy (***A***) **ii.** sound energy (***A***)

22. a. i.

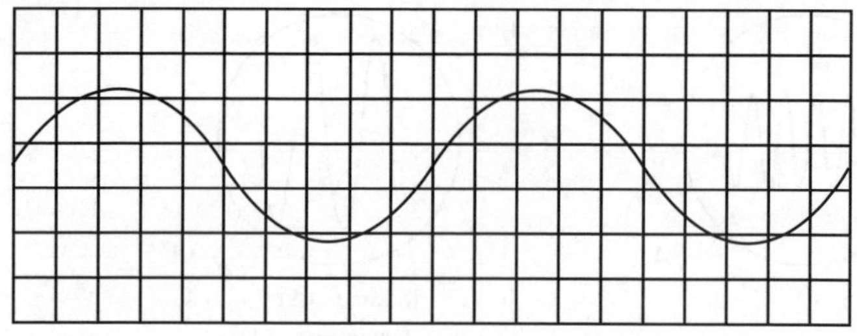

(A)

ii.

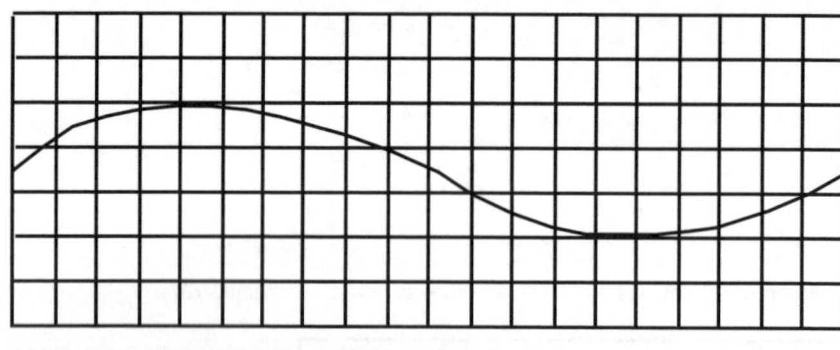

(A)

iii.

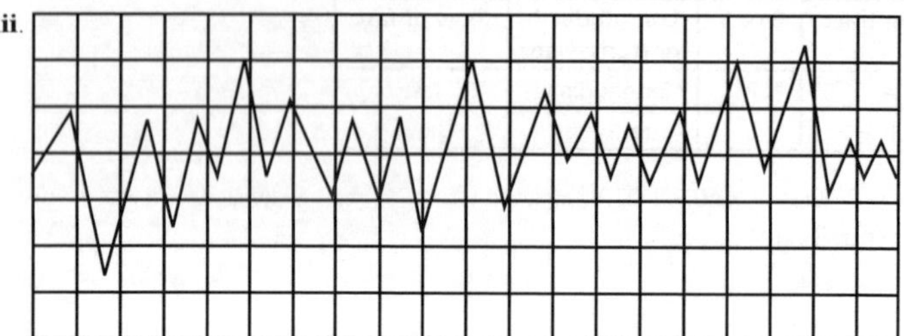

(A)

b. i. sound energy ii. electrical energy

23. a. i. 280 (**A**) ii. $1\,050 \times 10^3$ (**A**) iii. 2.9×10^8 (**A**) iv. 920×10^3 (**A**)
v. 3.0×10^8 (**A**) vi. 360 (**A**) vii. 820×10^3 (**A**) viii. 2.95×10^8 (**A**)
ix. 300 (**A**)

b. (Dial 1) × (Dial 2) = 3×10^8 m (**A**)

c. $v = f\lambda$; λ = = = 273 m (**A**)

24. $\lambda = 4$ m (**A**)

25. $f = \frac{v}{\lambda} = \frac{3 \times 10^8}{1\,100 \times 10^3} = 1.5$ Hz (**A**)

26. I. true II. false III. false (**M**)

27. Sound travels slower (330 m s^{-1}) than light (3×10^8 m s^{-1}). (**A**)

28. 576 = 512 × 1.125; 640 = 576 × 1.111 (**A**)

29. 128 ÷ 2 ÷ 2 = 32 Hz (**A**)

30. $v = f\lambda$; $f = \frac{v}{\lambda} = \frac{330}{1} = 330$ Hz; the closest is the 320 Hz 'E' note. (**M**)

31. 96 MHz = 96×10^6 Hz = 96 000 kHz (**A**)

32. **a**. $v = \dfrac{d}{t}$

$d = v \times t = 1\ 600 \times 0.1 = 160$ m (***A***)

b. The shoal of fish is = 80 m below the boat. (***A***)

c. Yes, because some ultrasound is reflected off the top of the shoal and some off the bottom, causing a time delay in the registering of the echoes. (***M***)

33. A false (***A***) B false (***A***) C correct (***A***) D false (***A***)

34. $\lambda = \dfrac{v}{f} = \dfrac{330}{1\ 650} = 0.2$ m; (***A***)

35. $\lambda = \dfrac{v}{f} = \dfrac{300}{600} = 5$ m (***A***)

36. $v = f\lambda$, the frequency remains the same, the speed decreases; therefore the wavelength decreases. (***M***)

37. Gamma. (***A***)

38. KN (***A***)

39. **a**. Longitudinal. (***A***)

b. $\lambda = \dfrac{v}{f} = \dfrac{1\ 500}{66\ 000} = 0.02$ m (***A***)

40. **a**.

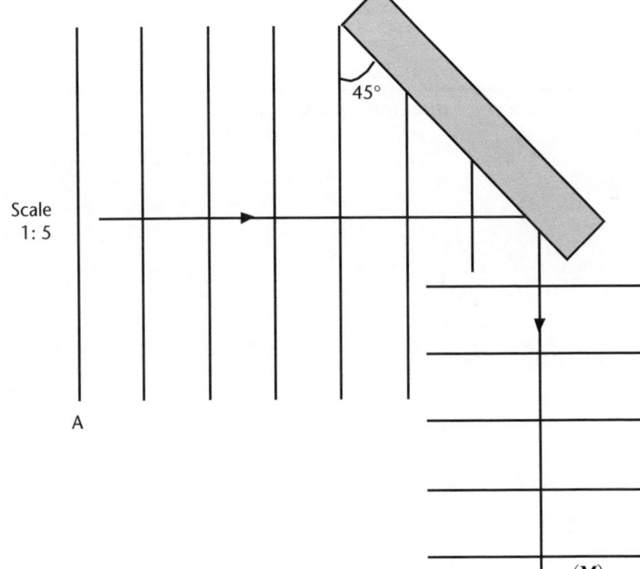

b. $1 \times 5 = 5$ cm (***A***) **c**. $f = \dfrac{5}{3} = 1.7$ Hz. (***A***)

41. longitudinal, transverse (***A***)

42. Distance *P* (***A***)

43. 1.5 waves/second. (***A***)

44. **a**. $1\ 000 \times 0.33 = 330$ m s^{-1} (***A***)

b. $165 \times 2 = 330$ m s^{-1} (***A***)

c. $8\ 000 \times 0.04 = 320$ m s^{-1} (***A***)

d. $330 \div 15 = 22$ m (***A***)

e. $1\ 500 \div 10\ 000 = 0.15$ m (***A***)

f. $3.8 \div 20\ 000 = 1.9 \times 10^{-4}$ m (***A***)

45. $\lambda = \frac{v}{f} = \frac{6}{2} = 3$ m (***A***)

46. $d = vt = 330 \times 5 = 1\,650$ m (***A***)

47. $v = \frac{2d}{t}$

$d = \frac{vt}{2} = \frac{1\,500 \times 0.4}{2} = 300$ m (***A***)

48. **a**. **i**. vibrations (***A***) **ii**. Q to R (Since air particles are compressed.) (***A***)
iii. longitudinal (***A***) **iv**. frequency (***A***)
v. wavelength (***A***) **vi**. audible (***A***)

b. $l = \frac{v}{f} = \frac{330}{220} = 1.5$ m (***A***)

49. **a**. The sound first heard was transmitted through the pipe. The later sound was transmitted through the air. (***M***)

b. By using a longer pipe. (***M***) **c**. $f = \frac{v}{\lambda} = \frac{2\,500}{2} = 1\,250$ Hz (***A***)

d. The Moon has no atmosphere. Sound cannot travel in a vacuum. (***M***)

50. $T = \frac{1}{f} = 0.33$ s

It takes the sound 0.33 s to travel 110 m.

$v = \frac{d}{t} = \frac{1}{f} = 333$ m s^{-1} (***E***)

Unit 12.3 Activity 1B: Multiple choice questions (page 59)

1. C
2. B
3. D
4. C
5. D
6. A
7. B
8. B
9. C
10. B
11. D
12. B
13. D
14. C
15. D
16. C
17. B
18. A
19. B
20. D

Unit 12.3 Waves

Topic 2: Reflection of light

Unit 12.3 Activity 2A: Reflection (page 67)

1. **a.** $c = 3 \times 10^8 \text{ m s}^{-1} = 300\,000 \text{ km s}^{-1}$ (**A**)

   b. $c = \frac{d}{t}$ $d = c \times t = 300\,000 \times 60 \times 60 \times 24 \times 365$
   $= 9.46 \times 10^{12}$ km (**A**)

   c. $4\frac{1}{2}$ light years $= 4.5 \times 9.46 \times 10^{12}$
   $= 4.3 \times 10^{13}$ km (**A**)

2. 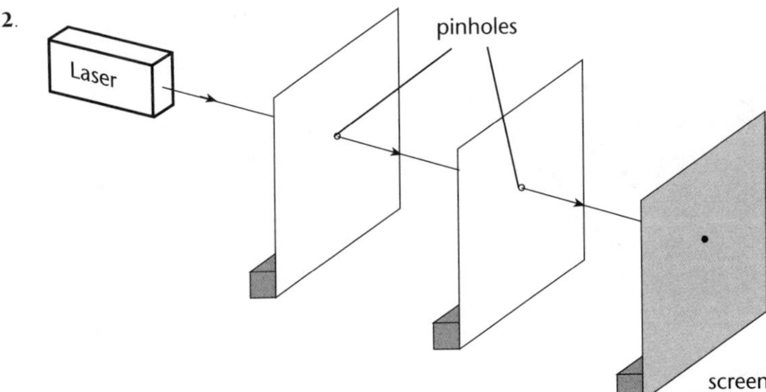

   When the pinholes are exactly lined up, the dot of laser light appears on the screen, indicating that the light must have travelled in a straight line. (**M**)

3. 
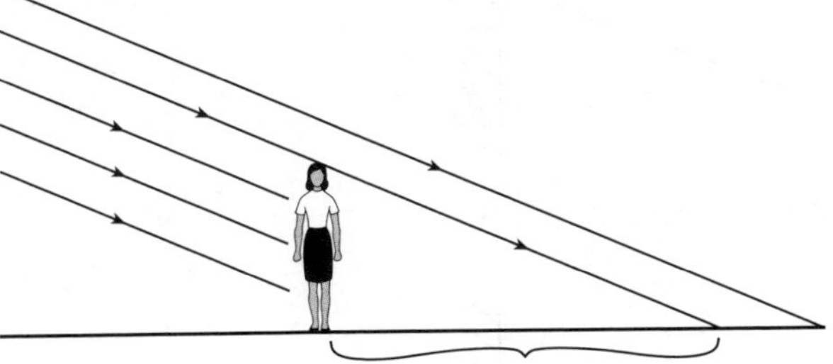

   Shadow occurs because rays of light that travel in a straight line are blocked off. (**M**)

4. 
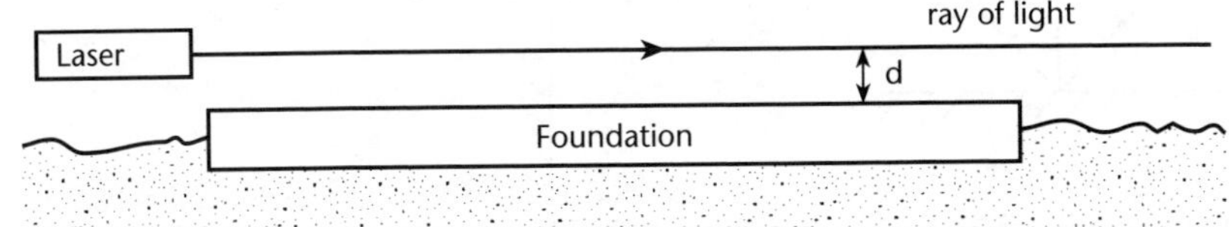

   The distance d. between the ray of light (laser beam) and the foundation must be the same everywhere. (**M**)

5. XO = incident ray; ON = normal; OY = reflected ray; XON = angle of incidence; YON = angle of reflection. (**A**)

6.

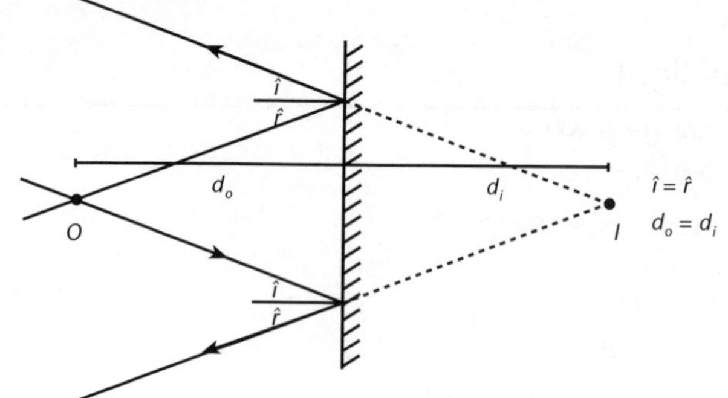

(**M**)

7.

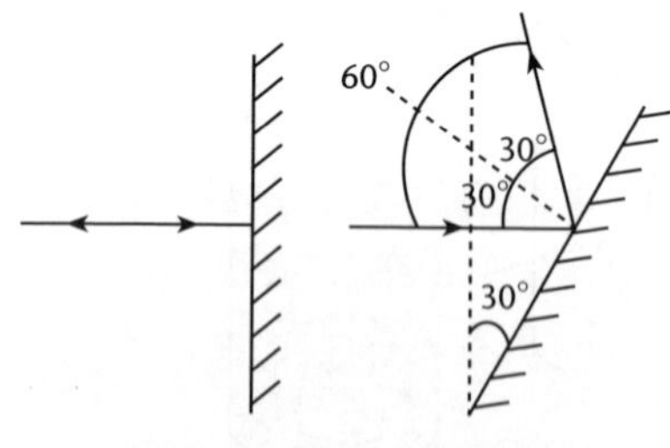

The reflected ray is rotated 60°

(**M**)

8.

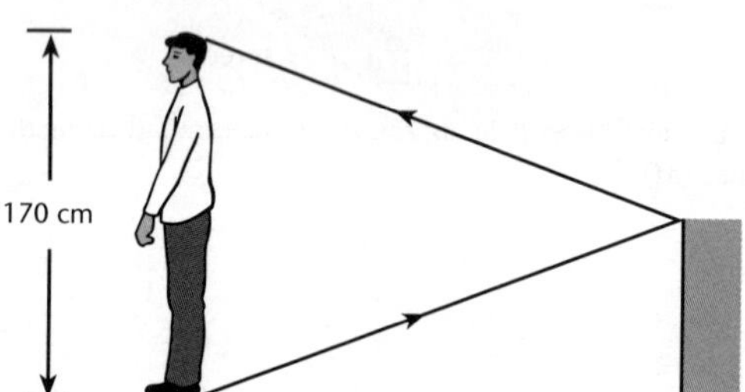

The minimum height of the mirror is $\frac{170}{2}$ = 85 cm. (**E**)

9.

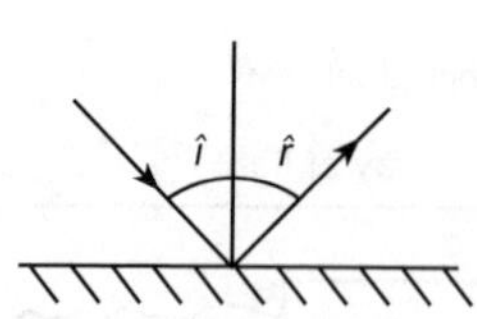

Measure various pairs of î and r̂ to show that they are always equal.

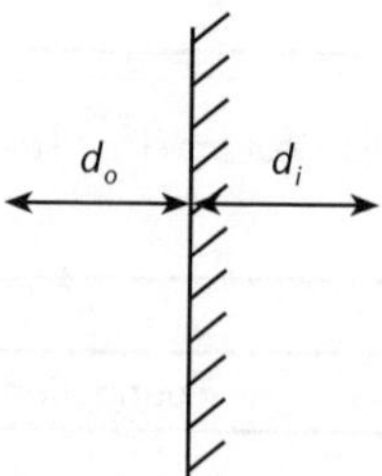

Measure various d_o and d_i to show that they are always equal.

(**A**)

10.

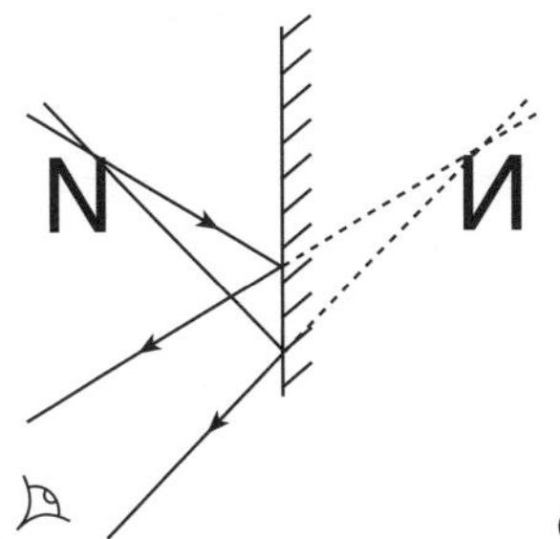

(***M***)

11.

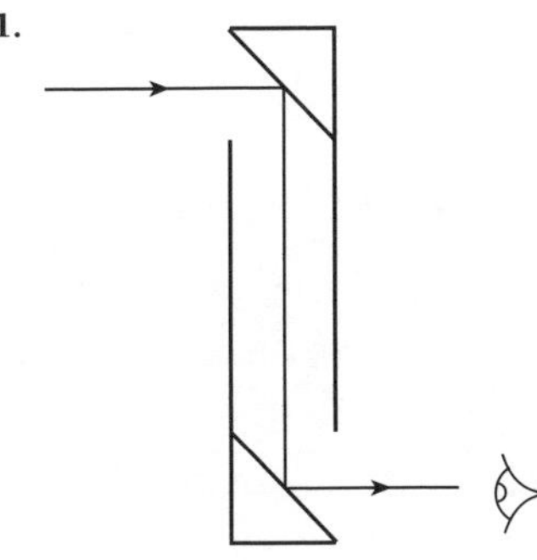

Because it could be used to look over the edge of a trench without exposing a soldier to enemy fire. (***M***)

12. **a**. 2.3 m. (***A***)

b. This is known as the rectilinear propagation of light, ie light travels in straight lines, unless it is disturbed. (***A***)

c. 2.0 m. (***A***)

d. Laterally inverted means that the right appears on the left and the left on the right, eg if you hold up your left hand your image appears to hold up the right hand. (***A***)

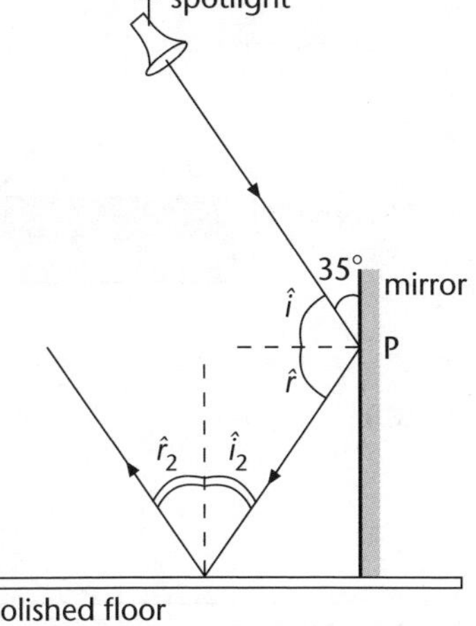

e. $\hat{i} = 90° - 35°$

$\therefore\ \hat{i} = 55°$ (***A***)

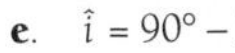

f. **i**. *see diagram* (***M***)

ii. $\hat{r} + \hat{i}_2 = 90°$ $\Rightarrow$ $\hat{i}_2 = 90° - 55° = 35°$

$\hat{r}_2 = \hat{i}_2$ $\Rightarrow \therefore\ \hat{r}_2 = 35°$ (***M***)

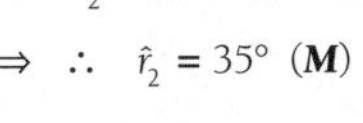

Unit 12.3 Waves

Topic 3: Refraction of light

Unit 12.3 Activity 3A: Refraction of light (page 76)

1. a.

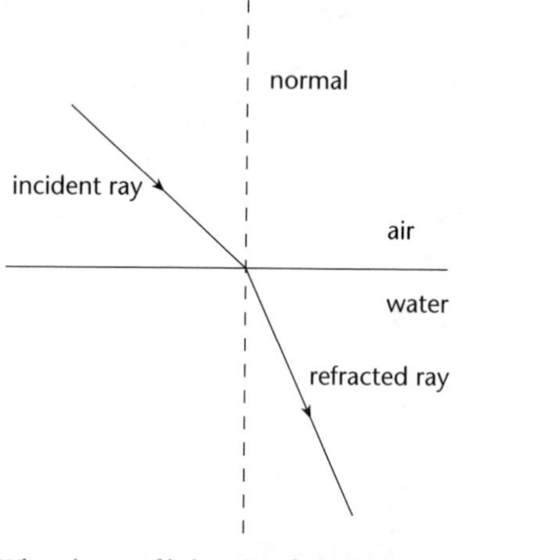

(***A***)

b. When the ray of light passes from air into water it slows down. Since water is optically more dense than air, the ray bends towards the normal. (***M***)

2. a.

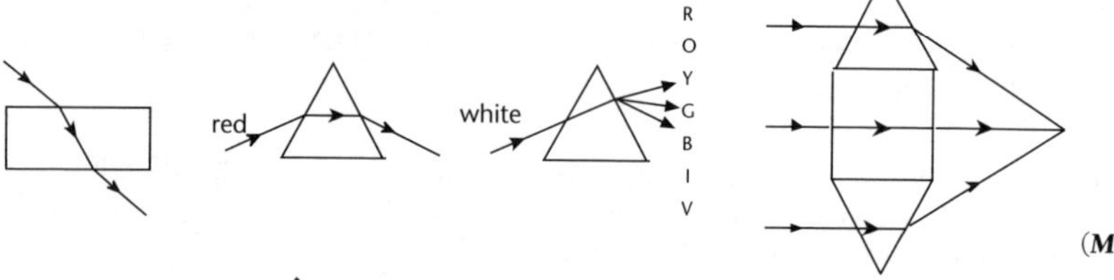

(***M***)

3.

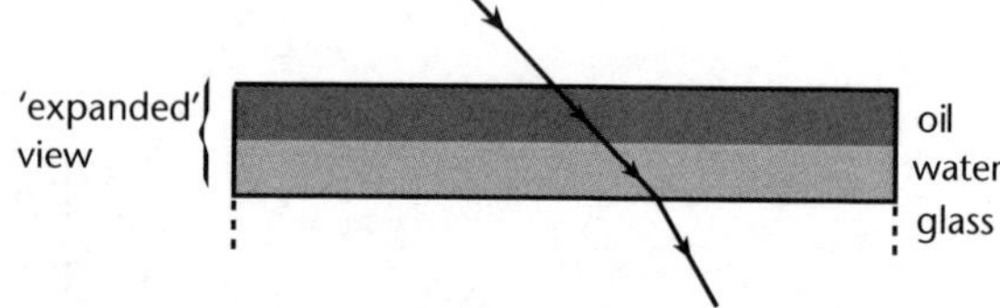

(***M***)

4. a., b.

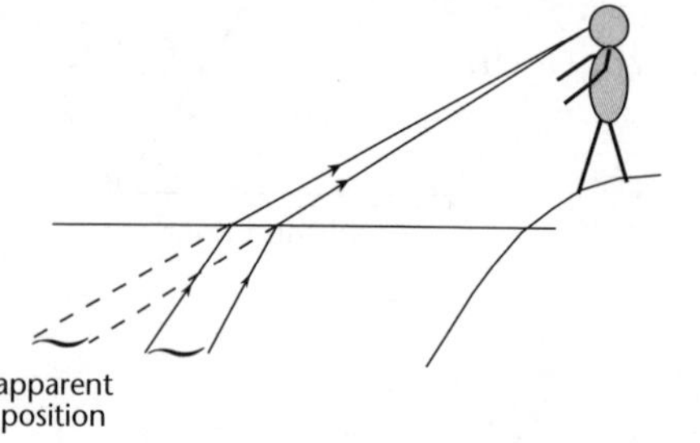

(***M***)

c. Rangi needs to direct his spear in front of, and deeper than, the position where he 'sees' the eel. (***M***)

5. The writing appears larger and closer to Rowena's eye. (***M***)

6.

red and blue ray

glass

blue

red

air

(***A***)

7. Green (***A***)

8.

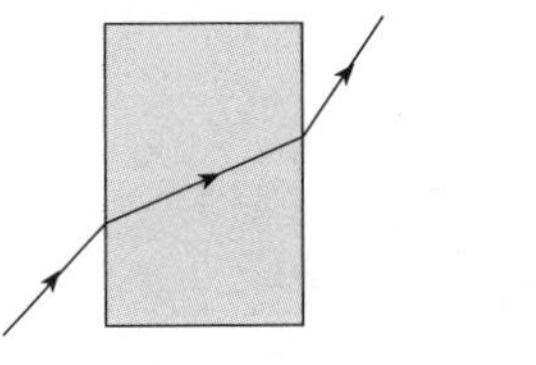

(***A***)

9.

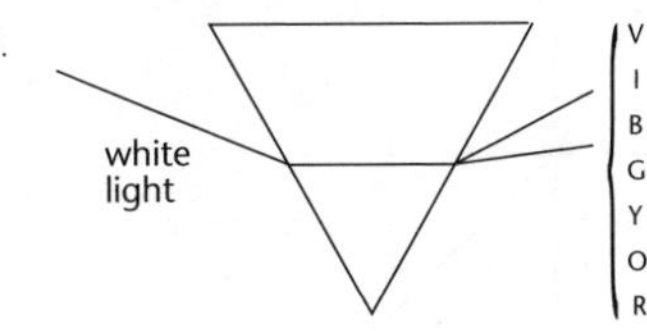

(***A***)

10. When a ray of light passes from air to glass the ratio of $\frac{\sin \hat{i}}{\sin \hat{r}} = 1.5$ (***M***)

11. **a**. **i**. 0.4695 (***A***) **ii**. 0.5446 (***A***) **iii**. 0.6157 (***A***) **iv**. 0.6560 (***A***)
v. 0.3090 (***A***) **vi**. 0.3584 (***A***) **vii**. 0.4067 (***A***) **viii**. 0.4373 (***A***)

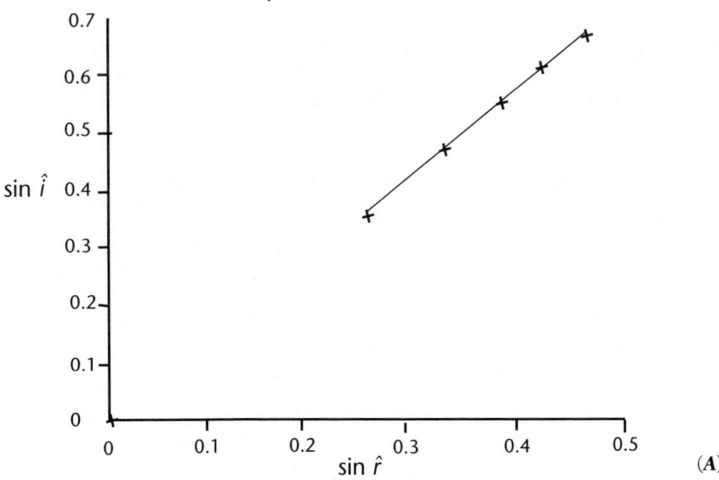

(***A***)

b. **i.** The refractive index of Perspex is equal to the slope of the graph = 1.5 (***A***)

ii. sin 45° = 0.7071; the corresponding value for

$\sin \hat{r} = 0.4695$

$\hat{r} = \sin^{-1} 0.4695$

$= 28°$ (***M***)

12. $v_{glass} = \dfrac{c}{1.5} = \dfrac{300\,000}{1.5} = 200\,000 \text{ km s}^{-1} = 2 \times 10^8 \text{ m s}^{-1}$ (***A***)

$v_{paraffin} = \dfrac{c}{1.4} = \dfrac{3 \times 10^8}{1.4} = 2.1 \times 10^8 \text{ m s}^{-1}$ (***A***)

$v_{diamond} = \dfrac{c}{2.4} = 1.3 \times 10^8 \text{ m s}^{-1}$ (***A***)

13. **a.**

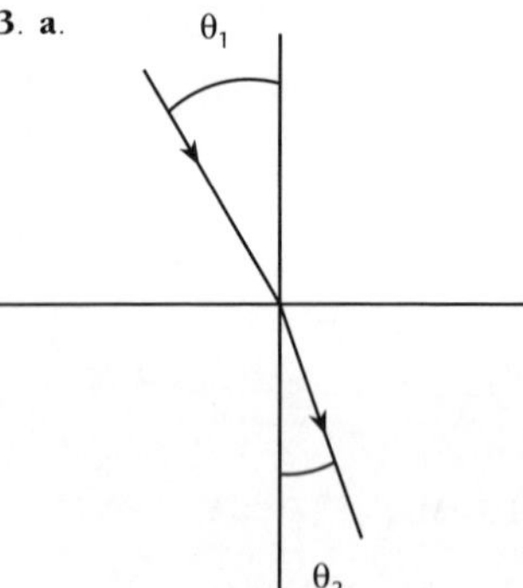

$n_2 > n_1$
$\theta_2 < \theta_1$ (***A***)

b.

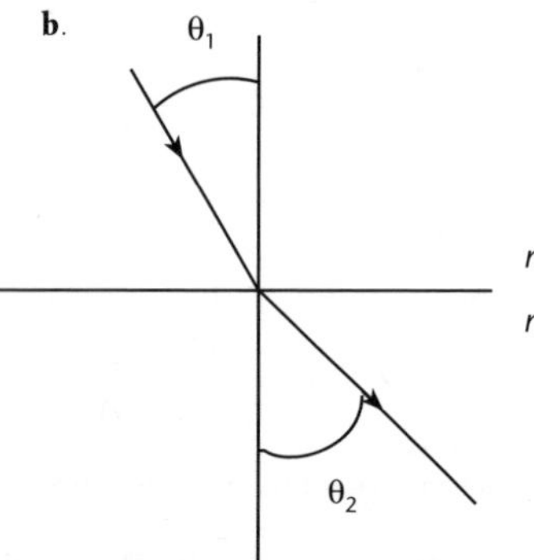

$n_2 > n_1$
$\theta_2 > \theta_1$ (***A***)

14. **a.** $n_1 \sin \hat{\imath} = n_2 \sin \hat{r}$

$1 \times \sin 24° = 1.33 \sin \hat{r}$

$\sin \hat{r} = \dfrac{\sin 24°}{1.33} = 0.3058$

$\hat{r} = \sin^{-1} 0.3058 = 17.8°$ (***M***)

b. $\sin \hat{r} = \dfrac{\sin 35°}{1.33} = 0.4313$

$\hat{r} = \sin^{-1} 0.4313 = 25.5°$ (***M***)

c. $\sin \hat{r} = \dfrac{\sin 53°}{1.33} = 0.6005$

$\hat{r} = \sin^{-1} 0.6005 = 36.9°$ (***M***)

15. **a.** $n_1 \sin \hat{\imath} = n_2 \sin \hat{r}$

$1 \sin \hat{\imath} = 1.5 \sin 24°$

$\sin \hat{\imath} = 0.6101$

$\hat{\imath} = \sin^{-1} 0.6101$

$\hat{\imath} = 37.6°$ (***M***)

b. $\sin \hat{\imath} = 1.5 \sin 35°$

$\sin \hat{\imath} = 0.8604$

$\hat{\imath} = \sin^{-1} 0.8604$

$\hat{\imath} = 59.4°$ (***M***)

16. Velocity decreases, wavelength decreases, frequency stays the same. (***A***)

17. **a.**, **b.**

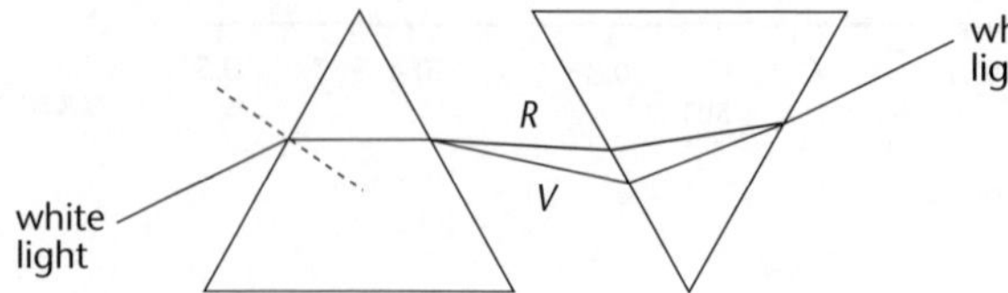

(***A***)

18.

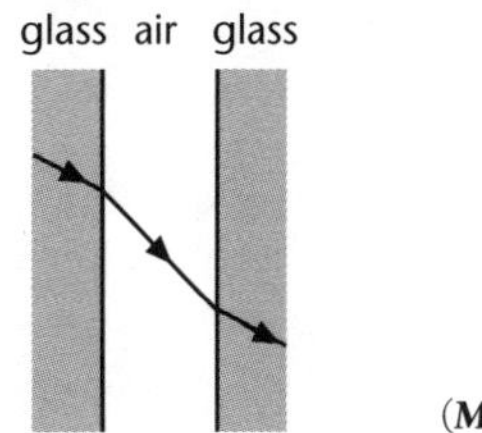

(***M***)

19. position *D*. (***A***)

20. a.

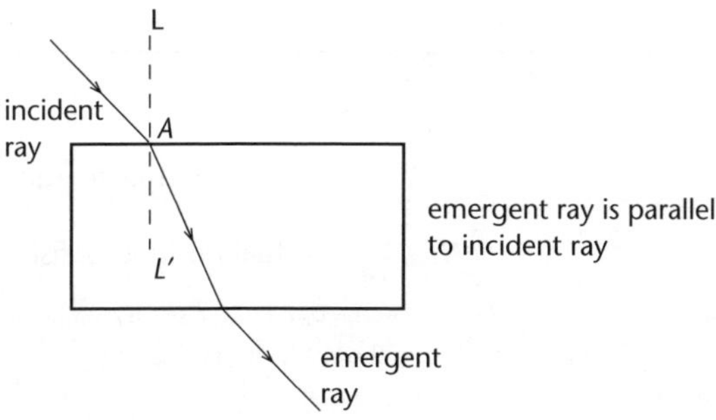

(***A***)

b. Normal. (***A***)

c. At A the light slows down and bends towards the normal. (***A***)

d.

(***A***)

21. a. i. (***A***)

ii. (***A***)

iii. (***M***)

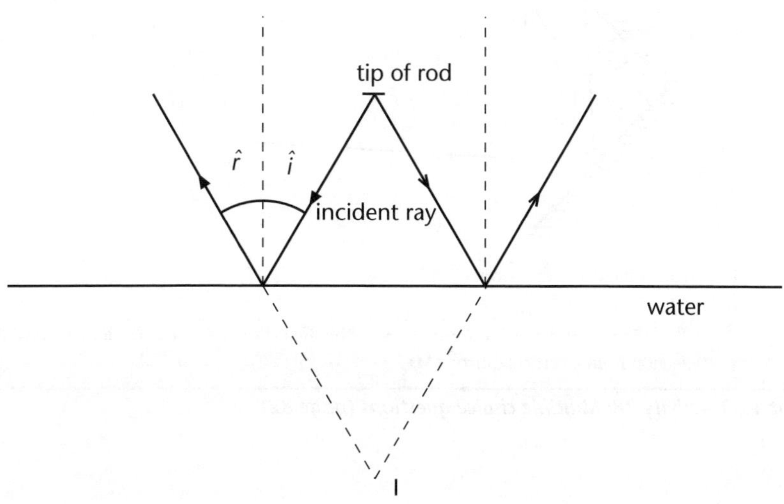

b. **i**.

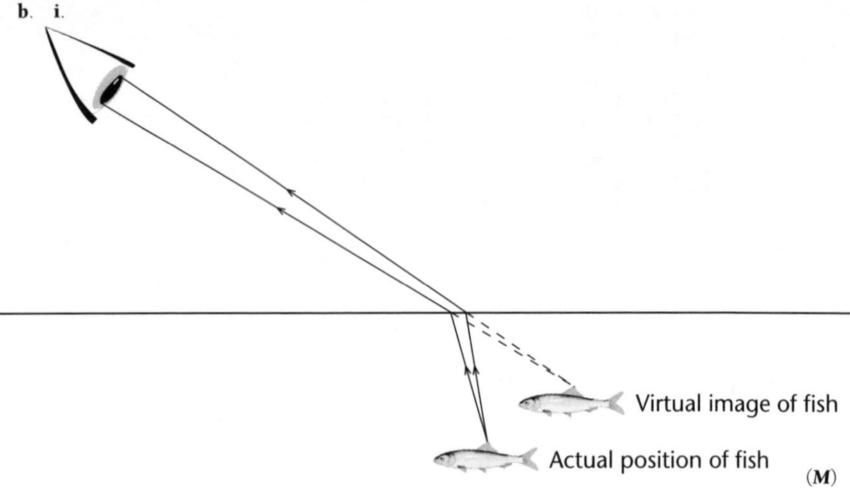

(***M***)

ii. The image is virtual because it only appears to be at the location shown. It cannot be projected onto a screen. The light rays do not originate from the virtual image. (***E***)

c. **i**. Total internal reflection. (***A***)

ii. Light rays pass from a more dense medium (water) into a less dense medium (air). The angle of incidence has to be greater than the critical angle. (***E***)

d. $n_{air} \sin \theta_{air} = n_{water} \sin \theta_{water}$

$1 \times \sin 40° = 1.33 \sin \theta_{water}$

$\theta_{water} = \sin^{-1}\left(\frac{\sin 40°}{1.33}\right)$

$\theta_{water} = 28.9°$ (***M***)

e.

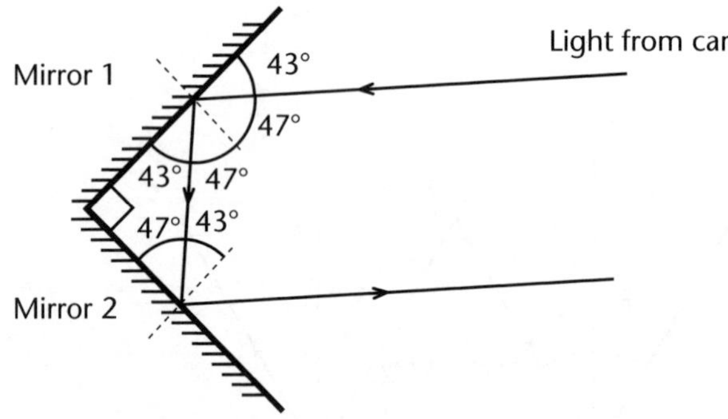

i. Angle of incidence at mirror 1 = 47°

ii. Angle of incidence at mirror 2 = 43°

iii. Reflected rays are parallel to incident rays. The light does not spread in all directions like reflection from a vertical mirror. (***M***)

Unit 12.3 Activity 3B: Multiple choice questions (page 82)

1. D

2. D

3. B

4. D
5. B
6. B
7. B
8. A
9. C
10. D
11. B
12. A
13. B
14. B
15. C

Unit 12.3 Waves

Topic 4: Waves and the electromagnetic spectrum

Unit 12.3 Activity 4A: Pulses and waves (page 91)

1. 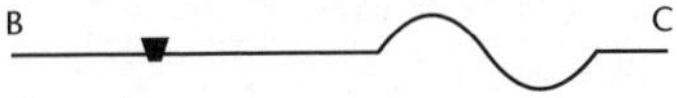 B ——▼—— C (**A**)

2. a. Wavelength. (**A**) b. Upwards. (**A**)

3. 80 m s^{-1}

4.

	Speed ($m\ s^{-1}$)	Frequency (Hz)	Wavelength (m)
a.	***60***	20	3
b.	***5***	10	0.5
c.	330	100	***3.3***
d.	3×10^8	10^6	***300***
e.	3×10^8	$\boldsymbol{2 \times 10^6}$	150
f.	350	***3 500***	0.1

(**A**)

5. 101 m (**A**)

6. 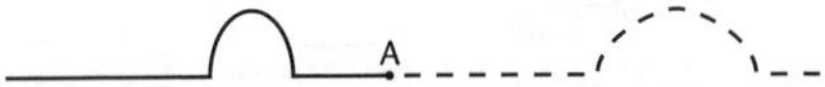 A (**A**)

Unit 12.3 Activity 4B: Multiple choice questions (page 94)

1. B
2. D
3. C
4. D
5. C
6. A
7. B
8. A
9. B
10. B
11. C
12. D

Unit 12.3 Activity 4C: Reflection, refraction and diffraction of waves (page 97)

1. Magnification can be found by putting a 15 cm ruler in the tank and comparing it with the length of its shadow.

The wavelength is found by measuring the stroboscopically frozen waves. (***E***)

2. (***A***)

3. The frequency is unchanged, but the wavelength is reduced, leading to a lower velocity. (***M***)

4. 1.0 cm (***A***)

5. λ_{deep} = 0.067 m, $\lambda_{shallow}$ = 0.033 m

deep
2 m s^{-1}
20°
43°
0.033 m
4 m s^{-1}
shallow
0.067 m

(***A***)

6.

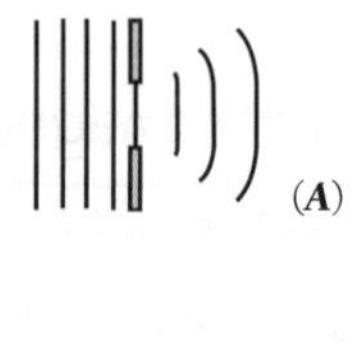

(***A***)

Unit 12.3 Activity 4D: Diffraction, superposition and electromagnetic waves (page 106)

1. C. The sound waves' wavelength is approximately the same size as the window. (***M***)

2. The AM radio waves (wavelength approximately 300 m) diffract around the sides of the hill better than the smaller wavelength (3 m) FM radio waves. (***E***)

3. **a**.

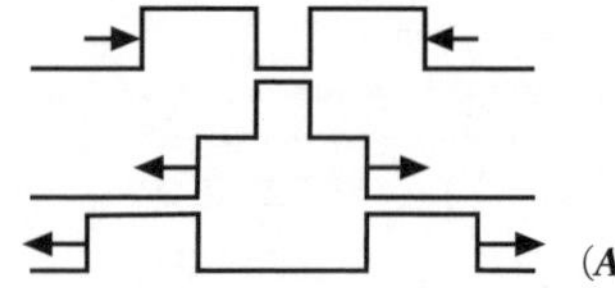

(***A***)

b.

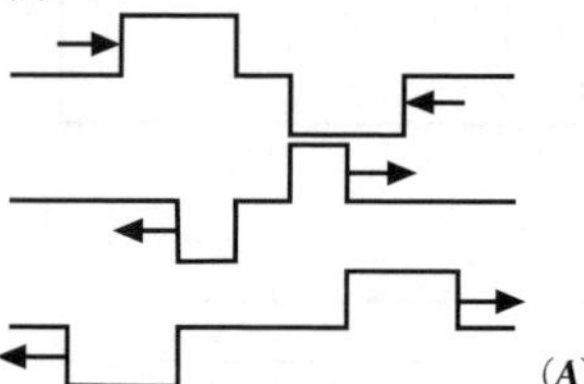

(***A***)

4. Transverse wave. (***A***)

5. Wave speed. (***A***)

6. Visible light. (***A***)

7. 9.03×10^5 times faster.

8. 2.46×10^9 Hz

9. **a**. A = $\frac{1}{2}$ wavelength. (***A***)

B = $\frac{1}{2}$ wavelength. (***A***)

C = 1 wavelength. (***A***)

D = 3 wavelengths. (***A***)

E = 1 wavelength. (***A***)

b. A and B are destructive (***A***); C, D and E are constructive. (***A***)

10. Interference of sound waves. (***A***)

Make the auditorium less box-shaped or place absorbing surfaces around the auditorium to absorb reflected sound. (***M***)

11. Since the wavelength has reduced, the amount of diffraction is less. This results in the bright bands getting closer together. (***M***)

Unit 12.3 Waves

Topic 5: Reflection of light – review and extension

Unit 12.3 Activity 5A: Plane mirrors (page 113)

1. 12° (***A***) **2**. **a**. 0.75 m (***M***) **b**. 0.70 m (***M***)

Unit 12.3 Activity 5B: Reflection of light (page 119)

1.

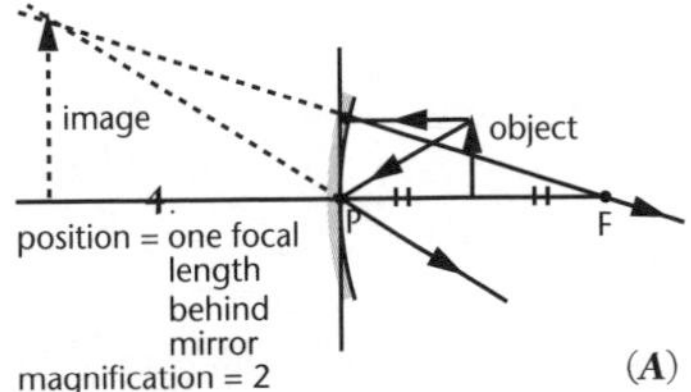

(***A***)

2. **a**. 160 cm in front of mirror. (***A***)

b. Inverted, magnified, real. (***A***)

3. 180 cm (***E***)

a. $\frac{1}{3}$ (***A***)

b. Real, inverted, smaller. (***A***)

c. 3 cm in front of the mirror. (***A***)

d. It will move further away from the mirror and become larger. (***M***)

5. **a**. 9.0 cm behind the mirror (f = 22.5 cm). (***M***)

b. 4.2 cm (***M***)

6. 45 cm (***E***)

7. **a**.

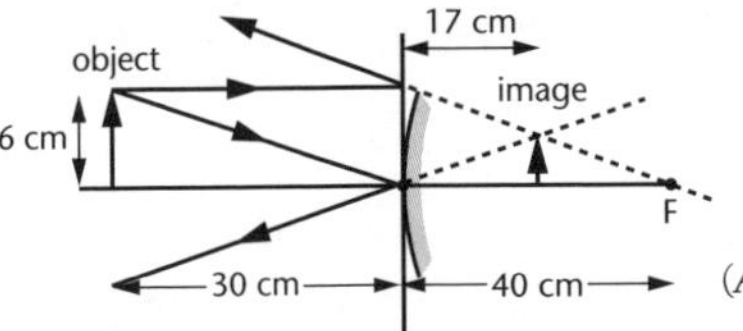

(***A***)

b. Image is 17.1 cm behind the mirror, 3.4 cm high, upright, smaller and virtual. (***A***)

8. 75 cm in front (ie at the focus). (***M***)

9. 3 (virtual). (***M***)

10. Concave, 18 cm. (***M***)

11. **a**. 1.3 m (***A***)

b. 0.5 m (***A***)

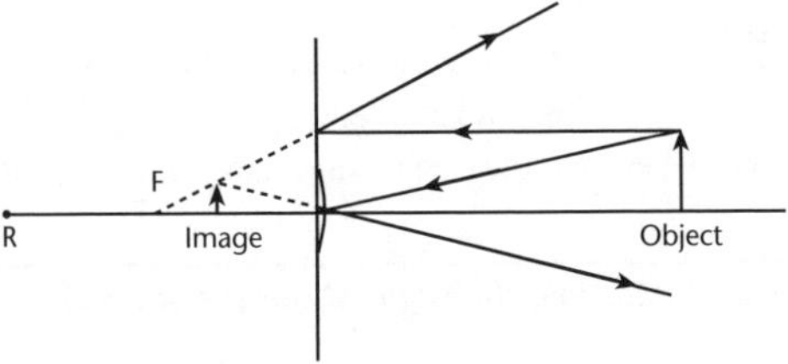

12. Object is 5 cm in front of the mirror. The image is 10 cm behind the mirror. (***E***)

13. **a**. 24 cm (***E***) **b**. 8 cm (***E***)

Unit 12.3 Waves

Topic 6: Refraction – review and extension

Unit 12.3 Activity 6A: Refraction (page 125)

1. Reflected at 60.0° to normal, refracted at 40.6° to normal. (***A***)

2. **a**. 40° (***A***) **b**. 75° (***A***) **3**. 28.9° (***A***)

Unit 12.3 Activity 6B: Critical angle (page 127)

1. 24.4° (***A***) **2**. 1.54 (***A***)

Unit 12.3 Activity 6C: Prisms (page 129)

1.

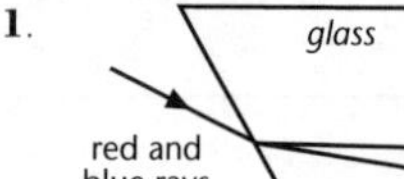
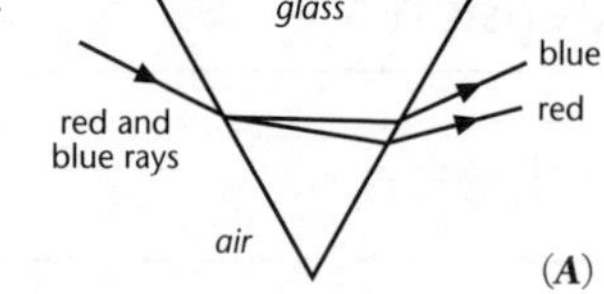

(***A***)

2. Green. (***A***)

3. 38.5° (***E***)

Unit 12.3 Activity 6D: Lenses (page 131)

1. A = concave lens, B = convex lens. (***A***)

2. **a**. Image becomes larger. (***A***) **b**. At A. (***A***)

3. **a**.

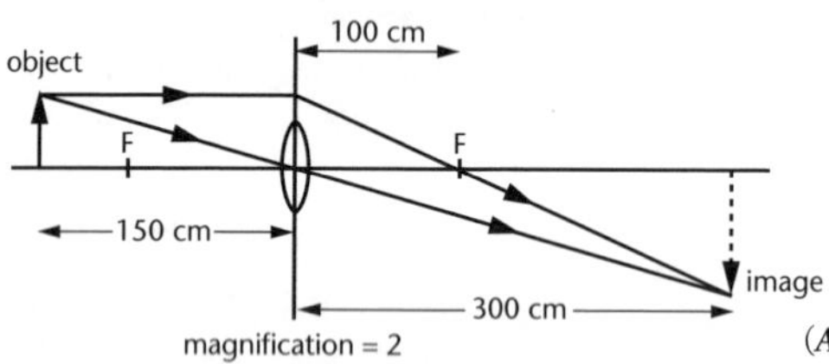

(***A***)

b.

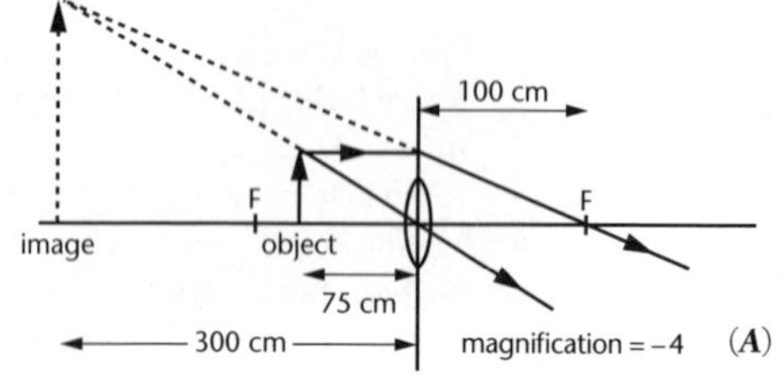

(***A***)

4. 10 cm or 30 cm (***E***)

5. **a**. Convex (converging). (***A***) **b**. 2.25 cm (***E***)

6. **a**. 20 cm (***A***) **b**. 1.8 mm (***E***) **c**. 5.3×10^{-10} (***A***)

7. Real image 10 cm behind lens (20 cm in front of mirror), virtual image 20 cm behind mirror. (***M***)

Unit 12.3 Activity 6E: Multiple choice questions (page 132)

1. D

2. B

3. D

4. C

Unit 12.3 Waves

Enrichment Topic: Vision and the camera

Unit 12.3 Enrichment Topic Activity A: The eye and cameras (page 139)

1. The iris controls the amount of light entering the eyes by making the pupil larger or smaller. (***A***)

2.

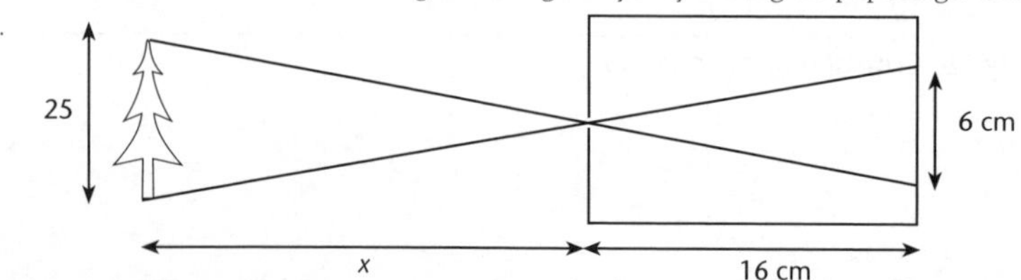

$$\frac{25}{x} = \frac{0.06}{0.16}$$

$$x = \frac{25 \times 0.16}{0.06}$$

$$= 66.7 \text{ m} \quad (\boldsymbol{E})$$

3. C [The size of the aperture should be reduced to decrease the amount of light entering the camera.] (***M***)

4. **a**. Opens and closes the aperture; the shutter speed controls the time the aperture is open. (***A***)
b. Controls the amount of light that enables the camera and determines the sharpness of the image. (***A***)
c. Adjusts the distance between the lens and the film. (***A***)

5. The human eye and a camera have the following corresponding parts:
- Shutter – eyelid.
- Lens – lens.
- Film – retina.
- Aperture – pupil.

The main difference is that the eye can record motion and converts light energy into electrical energy. (***M***)

6. A [The film is a camera and the retina in the eye are both screens on which the image is formed.] (***A***)

7. **a**. **i**. E (***A***) **ii**. C (***A***) **iii**. F (***A***)
b. **i**. The pupil will get larger. (***A***)
ii. The larger pupil will enable more light to enter the eye. (***A***)

8. Longsighted and can be corrected using a convex lens. (***M***)

9. A and D and E. (***M***)

10. **a**.

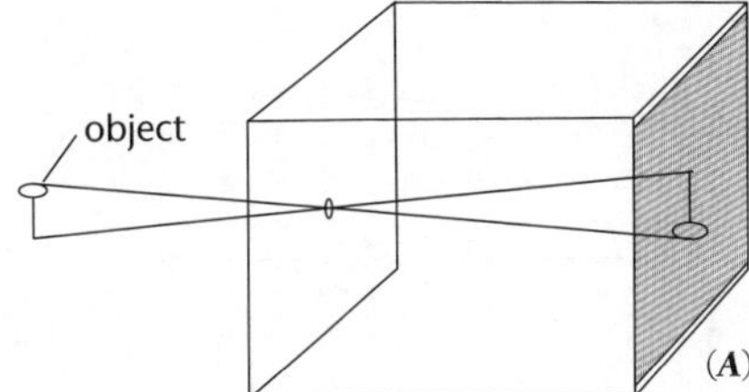

(***A***)

b. **i**. Real. (***A***) **ii**. Inverted. (***A***) **iii**. Magnified. (***A***)
c. **i**. Image gets smaller. (***A***) **ii**. Image gets bigger. (***A***)
iii. Image gets brighter but fuzzy. (***A***) **iv**.

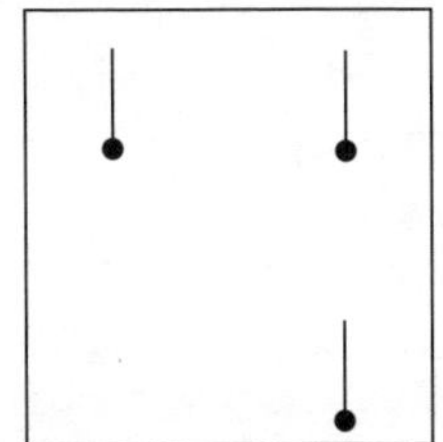

(***A***)

11. **i**. Shutter (***A***) **ii**. Lens (***A***) **iii**. To focus light (***A***) **iv**. Pupil (***A***) **v**. Retina (***A***)
vi. Screen on which image is located (***A***)

12. **a**. A biconcave lens is used to increaase the focal length and focus light onto the retina. (***M***)
    b. A convex lens is used to decrease the focal length and focus light onto the retina. (***M***)

Unit 12.3 Enrichment Topic Activity B: Multiple choice questions (page 142)

1. D
2. B
3. D
4. C
5. B
6. B
7. C

Unit 12.4 Electromagnetism

Topic 1: Magnetic fields

Unit 12.4 Activity 1A: Magnetic fields (page 151)

1. C [Each half will have a N and a S pole. 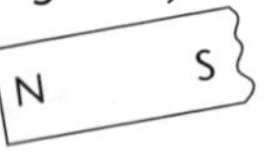

] (***A***)

2. **a**.

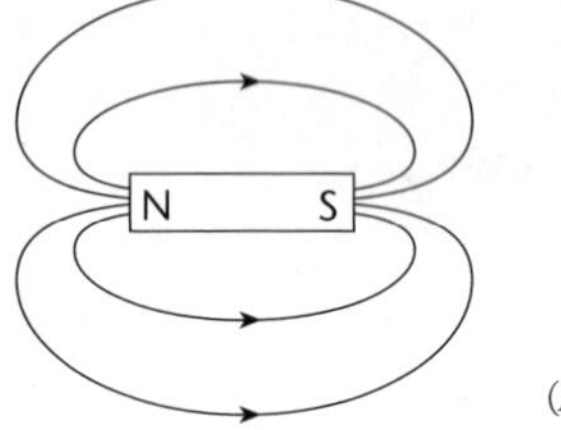

(***A***)

[Magnetic field lines point from N to S. Magnetic field lines do not cross.]

b.

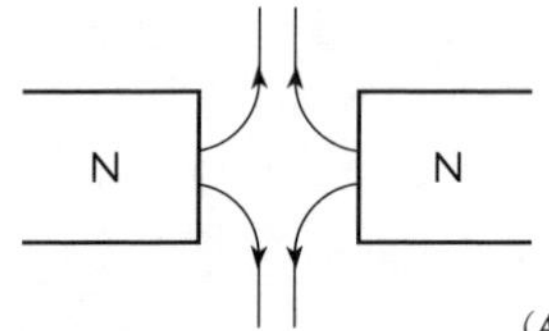

(***A***)

c.

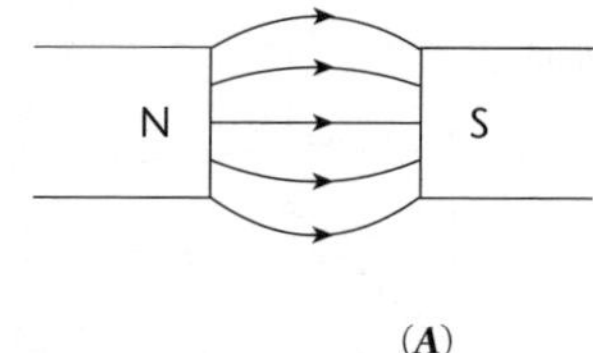

(***A***)

d.

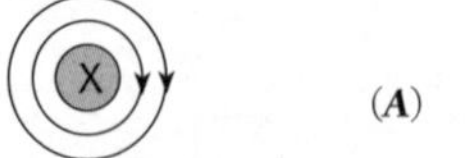

(***A***)

[Use the 'right hand grasp rule' to find the direction of the magnetic field.]

e.

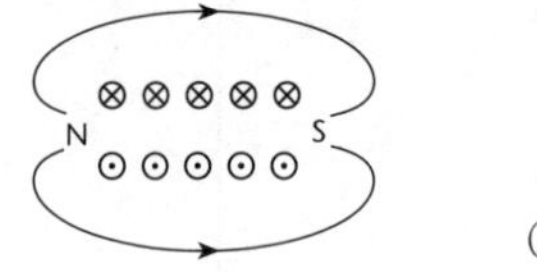

(***A***)

[Use the 'right hand solenoid rule'; your thumb points towards the N pole.]

3. Compass needle will point East, in the opposite direction to the magnetic field lines.

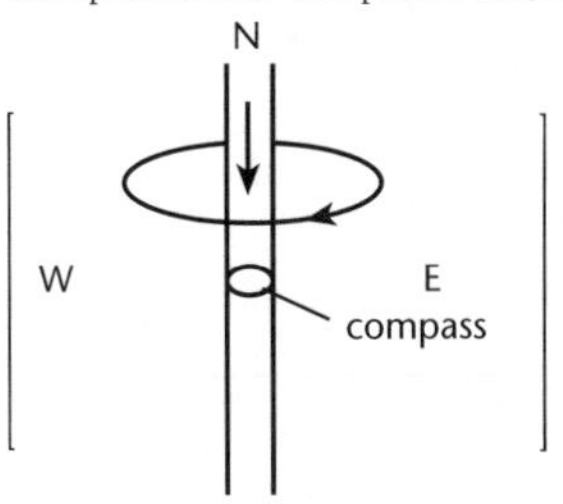

(*A*)

4. Current directed towards South. (*A*)

5. 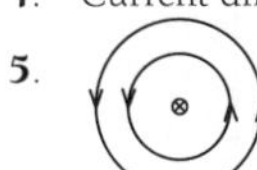

(*A*)

6. When the switch is closed the coil will swing to the right. The left end of the coil is a magnetic North pole when current flows through the coil, so repulsion occurs. (**M**)

7. **a**. Solenoids will be attracted to each other. (**M**)

   b. Solenoids will be repelled from each other. (**M**)

8. The iron becomes magnetised and strengthens the field. (*A*)

9. Iron because it is easily magnetised. (*A*)

10. **a**. When the switch is closed, the soft iron core is magnetised and attracts the arm, which closes the contacts to complete the secondary circuit. (**M**)

    b. When a current flows in the primary circuit, the secondary circuit is switched on. (**M**)

    c.

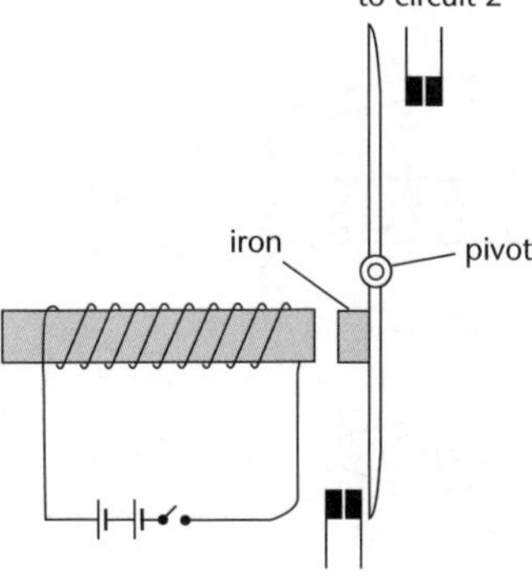

(**M**)

11. **a**. The iron bar moves downwards. (*A*)

    b.

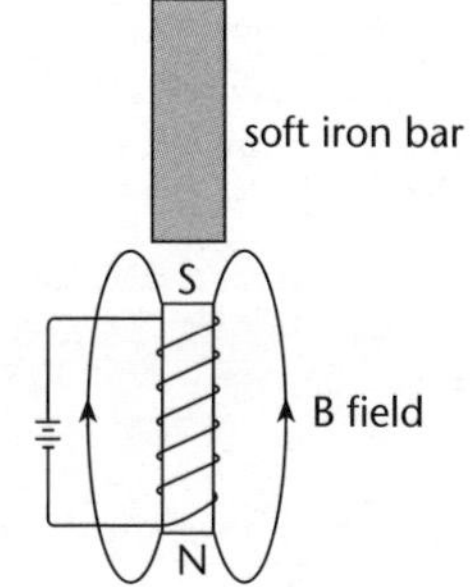

(**M**)

12. a. The cone is attached to a coil. When a current flows in the coil it becomes an electromagnet. This electromagnet is attracted to the permanent magnet when the current is in one direction and repelled when the current is reversed. (***M***)

b. The size of the current through the coil of the speaker. (***A***)

c. 'Squawker' – a speaker that best produces mid-frequency sounds.
'Woofer' – a speaker that best produces low frequency sounds.
'Tweeter' – a speaker that best produces high frequency sounds. (***M***)

13.

a. S N (***A***)	**b.** S N (***A***)
c. (***A***)	**d.** N S (***A***)

14. (***A***)

15. (***A***)

16. N S (***A***)

17. S N S N

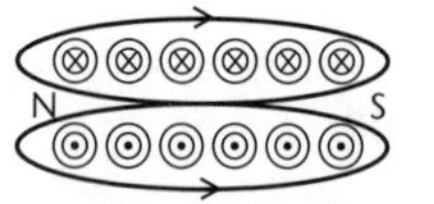

(***A***)

18. i. The magnets will turn. (***A***)

ii. The magnets will push apart. (***A***)

iii. The magnets will turn. (***A***)

iv. The magnets will slide together. (***A***)

19. a. The north and south poles are next to each other, as shown.

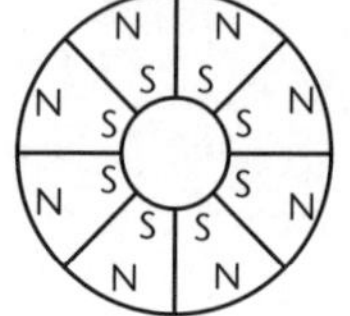

(***A***)

b. Iron (***A***)

Unit 12.4 Activity 1B: Multiple choice questions (page 155)

1. B
2. B
3. D
4. C
5. A
6. B
7. D
8. C
9. B
10. C
11. B and C
12. D
13. A
14. B
15. D

Unit 12.4 Electromagnetism

Topic 2: Magnetism

Unit 12.4 Activity 2A: Magnets (page 164)

1. **a**.

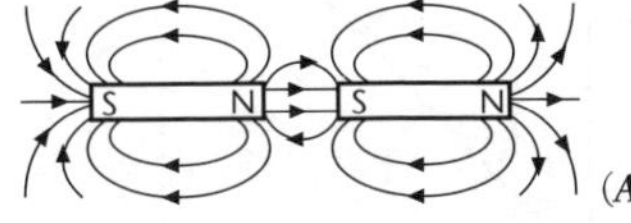

(***A***)

b.

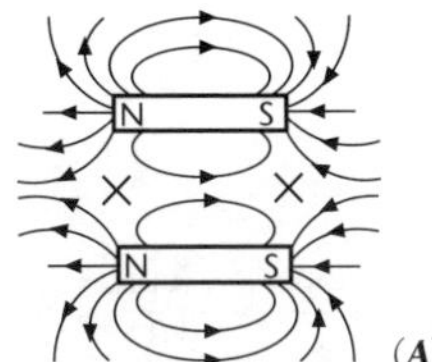

(***A***)

c.

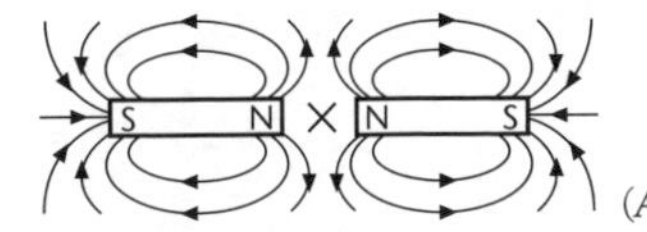

(***A***)

d.

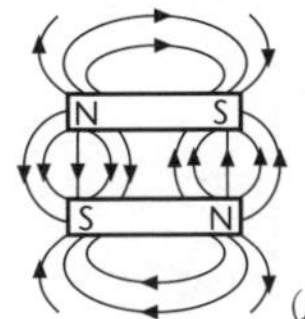

(***A***)

2.

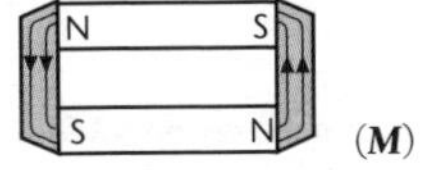

(***M***)

3. **a**.

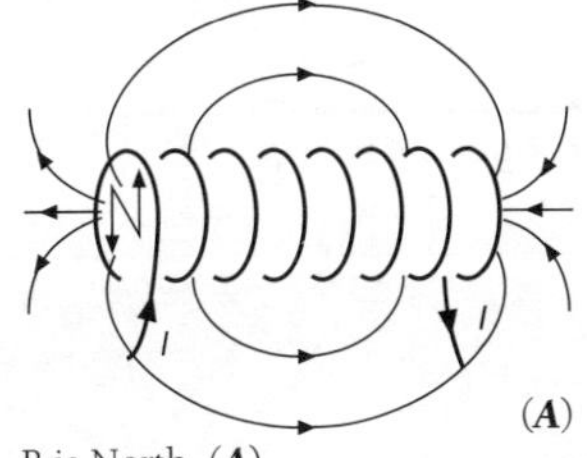

(***A***)

b.

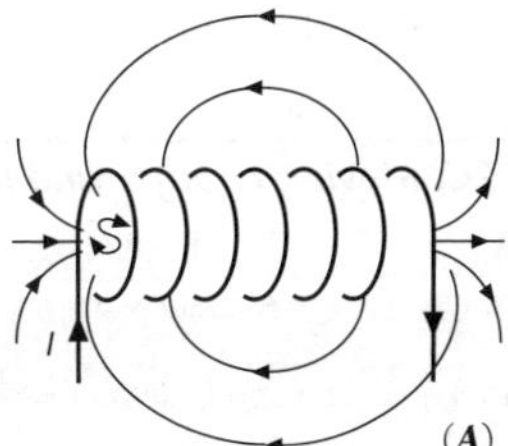

(***A***)

4. **a**. B is North. (***A***)

b. Soft iron becomes a strong temporary magnet and increases the magnetic field through the coils. (***A***)

5. The electromagnet would still work. The poles of the magnet would reverse 50 times a second but the magnetic field in any iron attracted to the electromagnet would also reverse 50 times a second to match. (**M**)

Unit 12.4 Activity 2B: Magnetic forces (page 168)

1. **a.** 0.067 T (**A**) **b.** Up out of the page. (**A**)

2. 0.15 N (**A**)

3.

Magnet position	Switch position	Magnetic field direction	Movement of roller
A	P	*down*	*x*
B	P	*right*	*none*
C	P	*up*	*y*
D	P	*left*	*none*
A	Q	*down*	*y*
B	Q	*right*	*none*
C	Q	*up*	*x*
D	Q	*left*	*none*

(**A**)

4. **a.** Towards the left. (**A**)
   b. The battery connections to the runners can be reversed and the direction of the magnetic field can be reversed. (**A**)
   c. The resistance of the variable resistor could be decreased and the size of the magnetic field could be increased. (**M**)

5. **a.** Downwards. (**A**)
   b. Reverse solenoid battery connections or loop battery connections, but not both. (**A**)
   c. **i.** 0.01, 0.02, 0.03, 0.04, 0.05 N (**A**)
   ii. slope = 0.01 N A^{-1} (**A**)

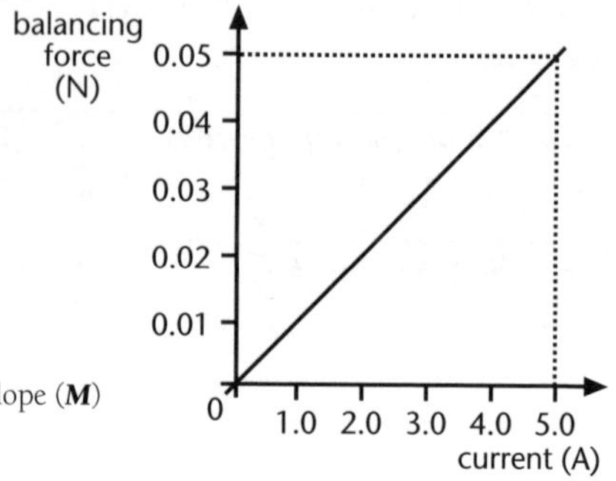

   d. $\frac{F}{I} = Bl = \frac{\text{slope}}{l} = \frac{\text{slope}}{0.02} = 50 \times \text{slope}$ (**M**)
   e. $B = 50 \times 0.01 = 0.5$ T (**A**)
   f.

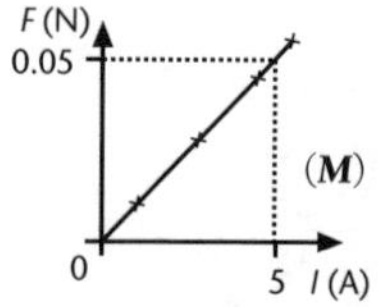

(**M**)

6. **a.** **i.** Clockwise. (**M**) **ii.** 2.4 A (**A**) **iii.** 0.63 T (**E**)
   b. 15 Ω (**E**) **c.** Magnetic field has been doubled in strength and reversed in direction. (**M**)

7. **a.** **i.** B (**M**) **ii.** 0.17 T (**E**) **b.** 3.0 cm (**E**)

Unit 12.4 Activity 2C: Moving charges and magnetism (page 174)

1. **a.** Upwards. (**A**) **b.** To the right. (**A**)

2. **a.** 6.4×10^{-14} N (**A**) **b.** 3.1×10^{-3} T (**A**)

3. **a.** The forces are equal in magnitude, $Bev = \frac{mv^2}{r}$

$$r = \frac{mv^2}{Bev} \quad \text{[rearranging]}$$

$$= \frac{mv}{Be} \quad \text{[cancelling] } (\mathbf{E})$$

Unit 12.4 Activity 2D: Multiple choice questions (page 175)

1. C
2. D
3. B
4. C
5. D
6. C
7. B
8. C
9. B
10. B
11. C
12. C
13. B
14. B
15. D

Unit 12.4 Electromagnetism

Topic 3: The motor effect

Unit 12.4 Activity 3A: Magnetic forces and motors (page 181)

1. $F = BIl = 7 \times 5 \times 1.5 = 52.5$ N (**A**)
2. $B = \frac{F}{Il} = \frac{3}{3 \times 0.2} = 5$ T (20 cm = 0.2 m) (**A**)
3. $l = \frac{F}{BI} = \frac{6}{10 \times 6} = 0.1$ m (**A**)
4. $B = \frac{F}{Il} = \frac{4.8 \times 10^{-3}}{2 \times 0.3} = 8 \times 10^{-3}$ T (**A**)
5. **a.** 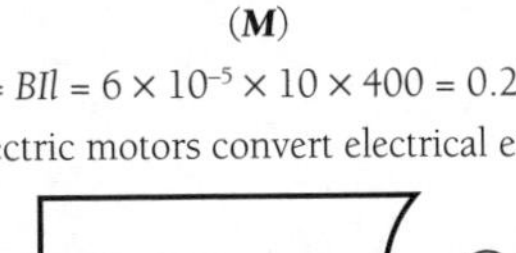 N S F (**M**) **b.** 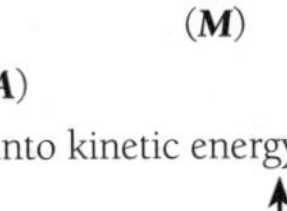 F I (**M**) **c.** 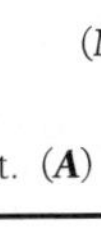 F (**M**) **d.**  N S F (**M**)
6. $F = BIl = 6 \times 10^{-5} \times 10 \times 400 = 0.24$ N (**A**)
7. Electric motors convert electrical energy into kinetic energy and heat. (**A**)
8. **a.** S F F N

   The coil will rotate in an anticlockwise direction. (**M**)

   b. Carbon brushes – connect coil to circuit and lubricate contacts.

   Spring contacts – prevent coil from tangling connection wires when rotating.

   Split ring commutator – reverses direction of current in coil. (**A**)

c. • Stronger magnet.
• Bigger current.
• More windings on coil. (**A**)

9. **a**. $F = BIl = 0.05 \times 10 \times 0.1 = 0.5$ N

$\tau = (F \times d) \times 2$

$= 0.05 \times 0.025 \times 2$ (2.5 cm = 0.025 m)

$= 2.5 \times 10^{-3}$ N m (**M**)

b. $\tau = 200 \times 2.5 \times 10^{-3}$

$= 0.5$ N m (**A**)

10. Carbon is easily shaped, is self-lubricating and conducts electricity.

11.

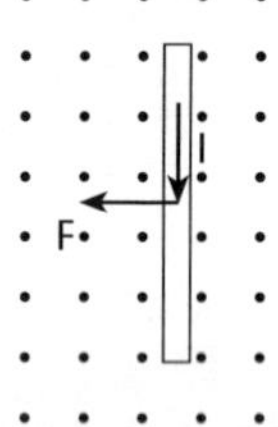

Current is down (from + to –); use 'right hand slap rule'; the conductor moves towards the left.

(**M**)

12. Charges (i) and (iv) (**M**)

13. **a**. **i**. Maintain contact between power supply and coil. (**A**)

ii. Reverses the direction of the current every half turn of the coil. (**M**)

b. **i**.

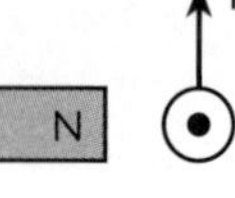

S

[Use the 'right hand slap rule'.] (**M**)

ii. Clockwise. (**A**)

c. **i**. $F = BIl$

$= 1.5 \times 8 \times 0.15$ (15 cm = 0.15 m)

$= 1.8$ N (**A**)

ii. $\tau = 2Fd$

$= 2 \times 1.8 \times 3$

$= 7.8$ N m (**A**)

d. Increasing the number of turns; increasing the current; increasing the strength of the magnetic field. (**A**)

e. **i**. $P = VI; I = \frac{P}{V} = \frac{12\,000}{600} = 20$ A (**A**)

ii. $V = IR; R = \frac{V}{I} = \frac{600}{20} = 30\ \Omega$ (**A**)

iii. $P = \frac{E}{t} = \frac{mgh}{t}$

$h = \frac{Pt}{mg} = \frac{12\,000 \times 10}{150 \times 10} = 80$ m (**E**)

14. **a**. **i**. = North; (**A**) **ii**. = South. (**A**)

b. It reverses the direction of the current every half turn of the coil. (**M**)

c. It is a conductor and it is self-lubricating. (***A***)

d. Increase the number of turns; use a stronger magnet. (***A***)

15. **a**. $P = VI;\ I = \frac{F}{Il} = \frac{3}{12} = 0.25\ \text{A}$ (***A***)

b. $\Delta Ep = mg\Delta h = 0.2 \times 10 \times 1.2 = 2.4\ \text{J}$ (***A***)

c. $P = \frac{E}{t};\ E = P \times t = 3 \times 5 = 15\ \text{J}$ (***A***)

d. The electical energy produced by the motor is changed to kinetic and gravitational potential energy as the lift rises. (***A***)

16.

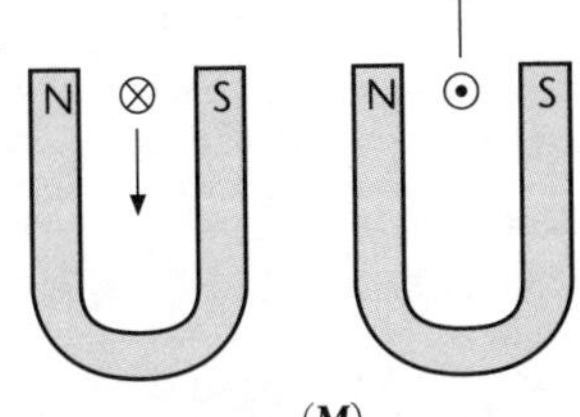

(***M***) (***M***)

17.

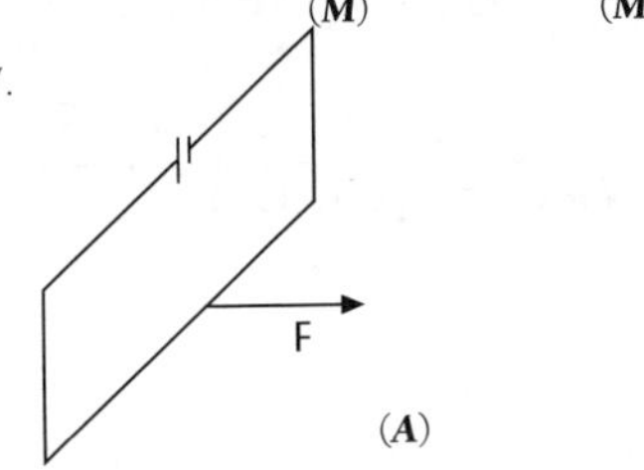

(***A***)

18.

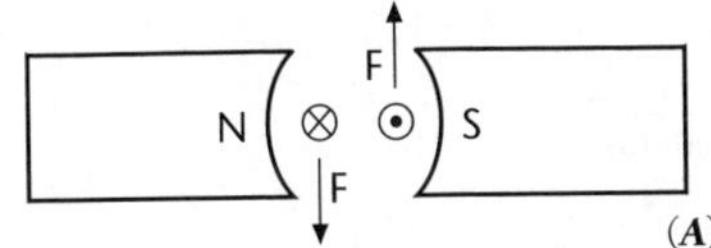

(***A***)

19. A = permanent magnet; B = coil; C = cone. (***A***)

b. **i**. AC; (***A***) **ii**. small current; (***A***) **iii**. high frequency. (***A***)

20. **a**.

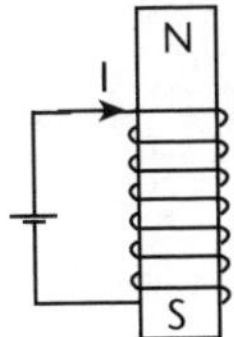

b. The right hand end moves down (since like poles repel). (***A***)

21. **a**.

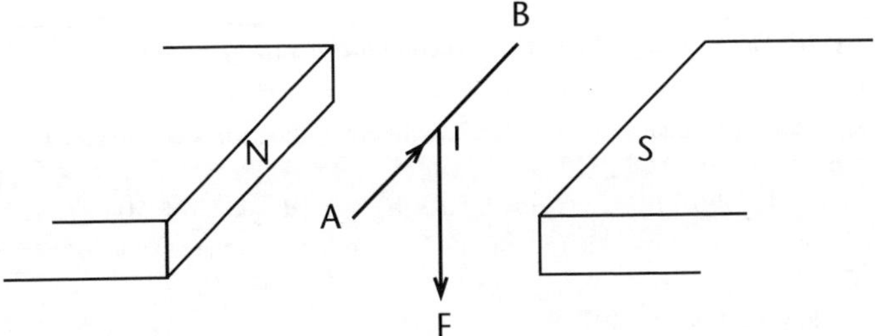

Force is downwards. (***M***)

b. $F = BIL$

$$B = \frac{F}{IL} = \frac{0.050}{2 \times 0.1} = 0.25 \text{ N A}^{-1} \text{ m}^{-1}$$ (**A**)

c. Using a stronger permanent magnet will increase the torque acting on the coil.

OR

Increasing the number of windings on the coil will increase the current and torque acting on the coil. (**A**)

Unit 12.4 Activity 3B: Multiple choice questions (page 186)

1. A
2. A
3. B
4. D
5. C

Unit 12.4 Electromagnetism

Topic 4: Electromagnetic induction

Unit 12.4 Activity 4A: An introduction to electromagnetic induction (page 191)

1. a. C (**A**)
   b. The magnet is travelling faster and takes less time to leave the solenoid. (**E**)
   c. Induced voltage is proportional to the magnet's velocity ($V = Bvl$). (**E**)
2. a. Anticlockwise when viewed from the magnet. (**M**)
   b. Ring moves away from the approaching magnet. (**M**)
   c. i. Clockwise when viewed from the magnet. (**M**)
      ii. Ring swings to the left as the magnet is withdrawn. (**M**)
   d. i. Clockwise when viewed from left. (**M**)
      ii. Ring swings to the right as the magnet is withdrawn. (**M**)
3. a. i. P sets up an N pole and the yellow LED lights. (**M**)
      ii. Nothing happens. (**A**)
      iii. P sets up an S pole and the orange LED lights. (**M**)
   b. i. P sets up an S pole and the yellow LED lights. (**M**)
      ii. Nothing happens. (**A**)
      iii. P sets up an N pole and the orange LED lights. (**M**)
4. a. NONE, ORANGE, RED, NONE, GREEN, YELLOW, NONE. (**M**)
   b. NONE, RED, ORANGE, NONE, YELLOW, GREEN, NONE. (**M**)
   c. For both LEDs to be on, current must flow in two directions simultaneously, which is impossible. (**M**)

Unit 12.4 Activity 4B: Electromagnetic induction calculations (page 195)

1. a. 2.4 V (**M**)
   b. The left wing when looking from above. (**M**)
   c. Neither, since an equal voltage will be induced in the connecting wires and no current will flow. (**E**)
2. a. D2 (**M**)
   b. 3.0 V (**A**)
   c. 23 mA (**A**)
   d. 0.023 N (**M**)
   e. Converted to electrical energy which is converted to heat energy in the resistor and heat and light in the LED. (**M**)
   f. i. D2 will glow more brightly (0.069 A flows). (**A**)
      ii. D2 will not light, D1 will light instead. (**M**)

3. a. Towards S (right-hand slap rule is for conventional current). (**M**) b. 0.8 V (**A**)
4. a. 0.60 A (**M**) b. 0.60 N (**M**)
   c. Converted to electrical energy which is converted to heat energy in the resistor. (**M**)
5. a. area = $vl\Delta t$ (**E**) b. magnetic flux = $Bvl\Delta t$ (**E**)
   c. $V = \dfrac{\Delta(BA)}{\Delta t}$

   $= \dfrac{\Delta Bvl\Delta t}{\Delta t}$ $\quad [\Delta(BA) = Bvl\Delta t]$

   $V = Bvl$ (**E**)
6. a. 5.25×10^{-4} V (**A**)
   b. Since the piece of wire is also in the magnetic field, the induced voltage across the mast will equal the induced voltage across the wire. This will effectively lead to no net voltage existing – therefore, no current delivered. (**E**)
   c. The boat now travels along the magnetic field lines. For a voltage to be induced, the field lines need to be cut by the conductor. In this case this is not occurring, therefore no induced voltage. (**E**)
7. a. 83 m s^{-1} (**A**) b. A (**A**)
   c. i. Low resistance. Power is inversely proportional to resistance. (**M**)
      ii. 8.7×10^{-6} W (**M**)

Unit 12.4 Activity 4C: Multiple choice questions (page 197)

1. D
2. C
3. B
4. A
5. D
6. C
7. B
8. A

Unit 12.4 Electromagnetism

Topic 5: More about induction

Unit 12.4 Activity 5A (page 204)

1. a. Y b. 12 V c. 6.0 A
2. a. There is a changing magnetic field in the coil.
   b. When the magnet leaves the coil, the change in magnetic field is opposite to when the magnet enters.
   c. The magnet is accelerating, so the flux changes more quickly.
   d. While the magnet is inside the coil there is no change of flux.
   e. From right to left.
3. a. i. loop is entering the field. ii. loop is leaving the field.
      iii. loop is completely in the field.
   b. 1 mV
   c. i. 0.2 m s^{-1} ii. 0.3 m
   d. 0.05 T

4. **a.** Induced current in the pipe creates an opposing magnetic field which tends to repel the magnet.
   b. Induced current flows in the moving coil when the terminals are connected. This induced current causes an opposing force which "damps" the motion of the coil.
   c. "Eddy" currents are induced in the wheel producing an opposing magnetic field.

Unit 12.4 Activity 5B (page 208)

1. **a.** 1.3×10^{-3} Wb **b.** 0.63 V **c.** clockwise
2. A **3.** C **4.** B **5.** D

Unit 12.4 Activity 5C (page 213)

1. **a.** 4×10^{-4} Wb **b.** 0.02 s **c.** 310 rad s^{-1} **d.** 13 V
e.

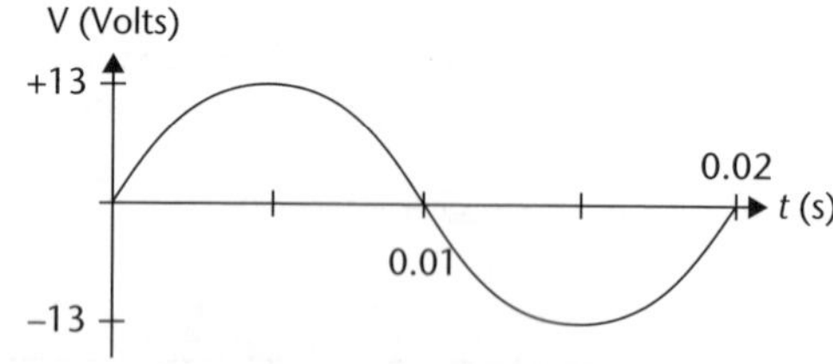

2. C **3.** D

Unit 12.4 Electromagnetism

Topic 6: Transformers and inductance

Unit 12.4 Activity 6A (page 218)

1. 15 turns **2.** 1 200 V
3. **a.** **i** 275 turns **ii** 600 turns
b. To protect against electric shock. **c.** **i** 0.45 A **ii** 0.21 A
4. **a.** 2 400 W **b.** 240 V **c.** Power = 2 300 W, Current = 9.7 A

Unit 12.4 Activity 6B (page 221)

1. **a.** Current = 20.8 A, Power loss = 4 330 W
b. Current = 0.45 A, Power loss = 2.1 W

Unit 12.4 Activity 6C (page 222)

1. **a.** So that the magnetic field is changing which will induce a voltage in the search coil.
b. AC since the magnetic field is alternating direction.
2. 12 H
3. **a.** 2 V **b.** The induced voltage would be greater.

Unit 12.4 Activity 6D (page 226)

1. The change in current can causes an induced voltage.
2. **a.** 40 V **b.** 100 J
3. **a.** 0.25 H **b.** 2 J

Unit 12.4 Activity 6E (page 228)

1. E
2. a. As the current builds up a changing magnetic field is produced in the inductor. This causes an opposing EMF which slows the build up of the current.
   b. 150 Ω
3. a. 0.40 s b. 1.2 A c. 0.76 A
4. a.

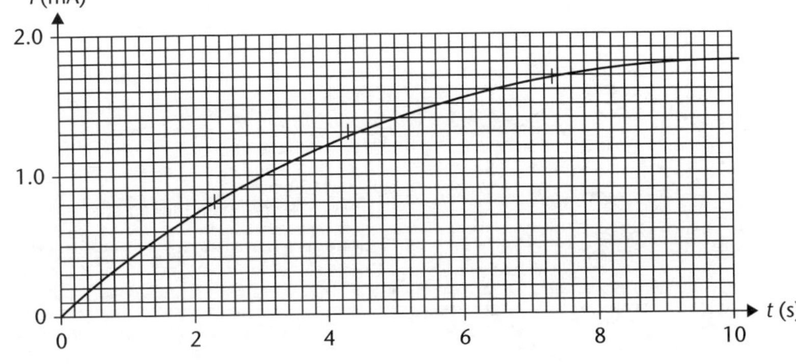

   b. 1.8 mA c. 6 700 Ω d. 3.5 s e. 23 000 H
5. i. D ii. A iii. E

Unit 12.4 Activity 6F: Multiple choice questions (page 230)

1. B
2. B
3. A
4. B
5. D
6. C
7. D

Unit 12.5 Radioactivity and nuclear energy

Topic 1: The atom and nuclear reactions

Unit 12.5 Activity 1A: The atom (page 234)

1. a.

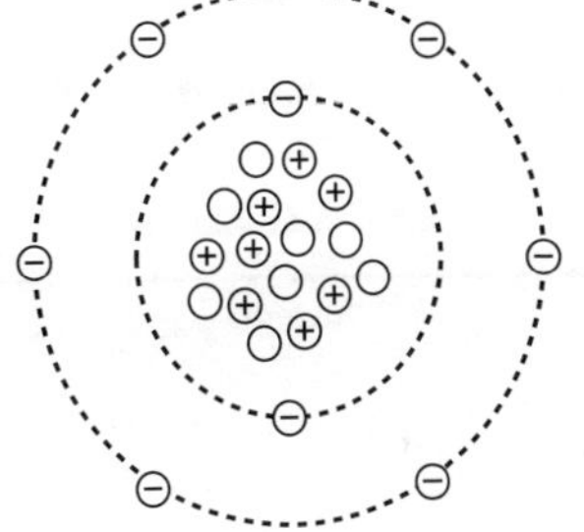

$^{16}_{8}O$

nucleus contains

8 protons ⊕ and 8 neutrons ○

8 electrons ⊖ surrounding the nucleus (**M**)

b.

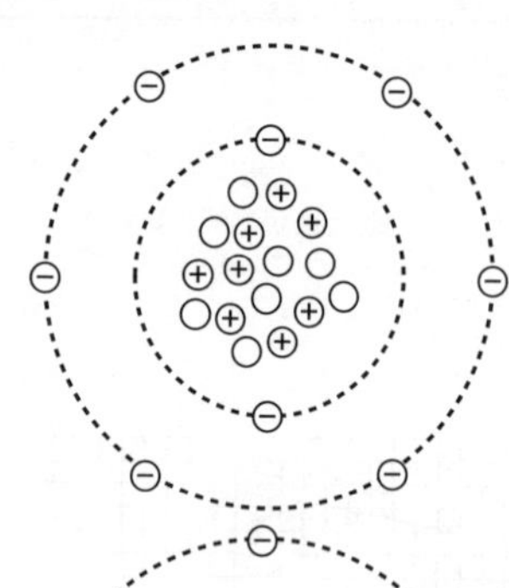

$^{14}_{7}N$
nucleus contains
7 protons ⊕ and 7 neutrons ○
7 electrons ⊖ surrounding the nucleus (**M**)

c.

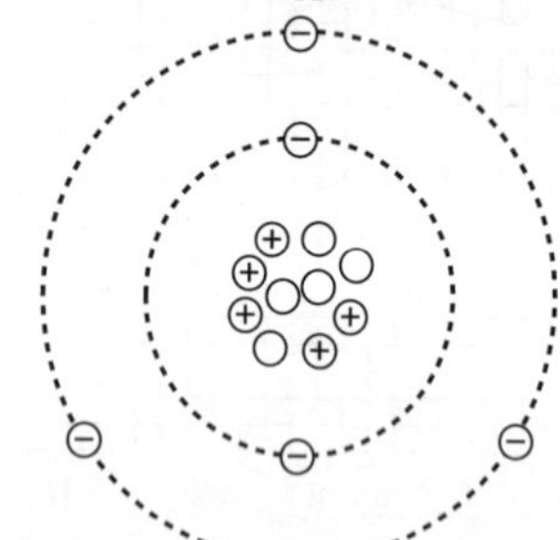

$^{10}_{5}U$
nucleus contains
5 protons ⊕ and 5 neutrons ○
5 electrons ⊖ surrounding the nucleus (**M**)

2. **a**. 84 (**A**) **b**. 84 (**A**) **c**. 125 (**A**) **d**. 99 (**A**) **e**. 155 (**A**) **f**. 99 (**A**)
   g. 210 (**A**) **h**. 85 (**A**) **i**. 85 (**A**) **j**. 96 (**A**) **k**. 96 (**A**) **l**. 96 (**A**)
   m. 258 (**A**) **n**. 101 (**A**) **o**. 101 (**A**) **p**. 92 (**A**) **q**. 143 (**A**) **r**. 92 (**A**)
3. A proton (**A**)
4. **a**. $^{238}_{92}U$, 90 protons, 92 electrons, 146 neutrons. (**A**)
   b. $^{229}_{90}Th$, 90 protons, 90 electrons, 139 neutrons. (**A**)
   c. $^{4}_{2}He$, 2 protons, 2 electrons, 2 neutrons. (**A**)
   d. $^{225}_{89}Ac$, 89 protons, 89 electrons, 136 neutrons. (**A**)
5. Two smaller nuclei and 3 neutrons. (**A**)
6. $^{235}_{92}U$, 92 protons, 92 electrons, 143 neutrons;
   $^{235}_{92}U$, 92 protons, 92 electrons, 146 neutrons. (**A**)
7. B [$^{156}_{66}X$ and $^{156}_{66}X$ are isotopes because they have the same atomic number.] (**M**)
8. 85 – 36 = 49 (**A**)

Unit 12.5 Activity 1B: Nuclear reactions (page 236)

1. **a**. a = 90 [92 – 2]; b = 4 [238 – 234] (**A**)
   b. c = 14 [14 + 0]; d = C [Carbon has atomic no. 6] (**M**)
   c. e = 221 [225 – 4] (**A**)

2. a. $^{1}_{1}P$ (**M**) b. $^{0}_{-1}e$ (**M**) c. $^{1}_{0}n$ (**M**) d. $^{4}_{2}He$ (**M**)
3. In nuclear reactors. (**A**)
4. a. Isotopes are atoms with the same atomic number but a different mass number. (**M**)
   b. Fission occurs when a large nucleus splits into two smaller nuclei and energy is released. (**M**)
   c. X = 92 Y = 56 Z = 92 (**A**)
   d. nuclear power generation, scientific research. (**A**)
5. a. Isotopes have the same atomic number but a different mass number, ie they have the same number of protons but a different number of neutrons. (**A**)
   b. Number of protons = 92
      Number of neutrons = 146 (**A**)
   c. Value of $x = 56 + 38 = 94$ (**A**)
      Law: Conservation of atomic number. (**A**)
   d. $239 + 1 = 147 + 90 + y$
      $y = 240 - 237 = 3$ (**A**)
      Law: Conservation of mass number. (**A**)
   e. $^{15}_{7}N + ^{[1]}_{[1]}[H] \rightarrow ^{[12]}_{6}C + ^{4}_{[2]}[He] + \text{energy}$ (**M**)
   f. Because of the high temperatures required and finding a suitable vessel for fusion to occur in. (**M**)
6. a. The nucleus of the gold atom is small compared to the size of the atom. (**M**)
   b. Protons and neutrons. (**A**)
   c. 79 (**A**)
   d. 118 (**A**)
   e. Electrons. (**A**)
   f. 79 (**A**)
   g. Isotopes have the same atomic number but different mass numbers. Isotopes belong to the same element but have a different number of neutrons in their nuclei. (**A**)
   h. Fission occurs when a large atom splits into smaller atoms. It is accompanied by a release of energy. (**A**)
   i. X = 92
      92 = Y + 36
      Y = 56
      $235 + 1 = 141 + Z + 3, \therefore Z = 92$ (**M**)
   j. Uranium 235 is used in nuclear reactors for the generation of electricity. It is also used to power nuclear ships and in nuclear medicine. (**M**)

Unit 12.5 Radioactivity and nuclear energy

Topic 2: The nuclear reactor and safety

Unit 12.5 Activity 2A: The nuclear reactor (page 241)

1. a. Uranium-235 (**A**) b. Nuclear fission (**A**) c. Boron or cadmium (**A**)
   d. Absorbs neutron (**A**) e. Heavy water or graphite (**A**) f. Slows neutron (**A**)
   g. Water (**A**) h. Absorbs heat energy (**A**) i. Concrete/lead (**A**)
   j. Stops radiation escaping (**A**)
2. a. To prevent radiation from escaping. (**A**)
   b. Water absorbs heat energy, turns into steam and drives the turbine(s) (**A**)
   c. Uranium-235. (**A**)
   d. No, eventually they need to be replaced. (**A**)

e. Boron. (**A**)

f. To slow them down to the right speed for fission to occur when a neutron hits a uranium-235 nucleus. (**M**)

g. If the temperature increases too much, the reaction is going to turn fast and too much energy is released and the reactor could 'melt' down. (**M**)

Unit 12.5 Activity 2B: Multiple choice questions (page 243)

1. D

2. D

3. A

4. C

5. C

6. B

7. D

8. C

9. C

10. D

11. D

12. D

Unit 12.5 Radioactivity and nuclear energy

Topic 3: Nuclear reactions

Unit 12.5 Activity 3A (page 252)

1.

	p	n	e
a	92	146	92
b	26	30	26
c	56	85	56
d	6	8	6

2. Alpha and beta radiations are bent in opposite directions when passing through a magnetic field while gamma rays are unaffected. Alpha particles have little penetrating power, beta have more than alpha, while gamma rays are very penetrating.

3. **a**. $^{241}_{95}\text{Am} \rightarrow {}^{237}_{93}\text{Np} + {}^{4}_{2}\text{He}$ **b**. A photon of gamma radiation.

4. **a**. $^{239}_{92}\text{U} \rightarrow {}^{239}_{93}\text{Np} + {}^{0}_{-1}\text{e}$ **b**. $^{239}_{93}\text{Np} \rightarrow {}^{239}_{94}\text{Pu} + {}^{0}_{-1}\text{e}$ **c**. Plutonium

5. B **6**. C **7**. C

Unit 12.5 Activity 3B (page 255)

1. **a**. **i**. 0 **ii**. 0 **b**. 52 v

c. **i**. zero **ii**. 5 512 energy units.

d. The mass of the products must be less than the mass of the reactants. The loss in mass corresponds to a gain in energy.

2. **a**. 3 protons and 4 neutrons.

b. $E_K = \frac{1}{2} \times 6.646 \times 10^{-27} \times (2.111 \times 10^7)^2$

c. 3.09×10^{-29} kg

d. 1.46×10^7 m s^{-1}

e. momentum, mass number, atomic number

3. **a**. $r = 30$, $s = 15$ **b**. 49.7744×10^{-27} kg

c. The mass is greater than expected so the products must have less energy than the reactants. The decrease in energy is 4.23×10^{-13} J

4. **a**. **i**. photon of high energy electromagnetic radiation.

ii. a hydrogen isotope with one proton and one neutron in the nucleus.

iii. an uncharged particle of mass number 1.

iv. a positively charged particle of mass number 1.

b. 1.6750×10^{-27} kg

5. C

Unit 12.5 Activity 3C (page 264)

1. **a**. Iron has the highest binding energy per nucleon.

b. **i**. Heavy nuclei splitting in smaller more stable nuclei.

ii. Small nuclei join together to form heavier nuclei.

c. **i**. 2 **ii**. 2

d. **i**. 0.0483×10^{-27} kg **ii**. 4.35×10^{-12} J **iii**. 1.09×10^{-12} J

2. False. Energy can change to mass and mass to energy in a nuclear reaction.

3. Neutrons are uncharged, so are not repelled by the nucleus.

4. 1.8×10^{13} J

5. **a**. ${}^{239}_{94}\text{Pu} + {}^{1}_{0}\text{n} \rightarrow {}^{93}_{38}\text{Sr} + {}^{142}_{56}\text{Ba} + 5\,{}^{1}_{0}\text{n}$

b. **i**. 0.3095×10^{-27} kg **ii**. 2.7855×10^{-11} J

6. B

Unit 12.5 Activity 3D: Fundamental particles, isotopes (page 268)

1. 92 protons, 92 electrons, 143 neutrons. (**A**) **2.** 146 neutrons. (**A**)

3. Most of the particles pass straight through the metal, but a small proportion are scattered. (**A**)

4. (**A**)

5. Two nuclei with the same number of protons but different number of neutrons, eg ^{13}C and ^{14}C. (**A**)

6. **a**. The material of the pocket would stop alpha-particles. (**A**)

b. Section X would be dark when developed. (**A**)

c. Section Z would be white when developed. (**A**)

d. Light will also make the film go dark when developed. (**A**)

7. **a**. Stage 1, α; stage 2, β; stage 3, β; stage 4, α; stage 5, α. (**A**) **b**. γ-rays and energy. (**A**)

8. ${}^{222}_{86}\text{Rn}$ (**A**) **9.** α-particle (**A**)

10. **a**. ${}^{1}_{1}\text{p}$ or ${}^{1}_{1}\text{H}$ (**A**) **b**. ${}^{1}_{0}\text{n}$ (**A**) **c**. ${}^{0}_{-1}\beta$ (**A**) **d**. ${}^{4}_{2}\alpha$ (**A**)

11. **a**. Nothing. (**A**) **b**. One-eighth. (**A**)

12. **a**. 40 (**A**) **b**. 800 (**A**)

13. 320 hours. (**A**)

14. a.

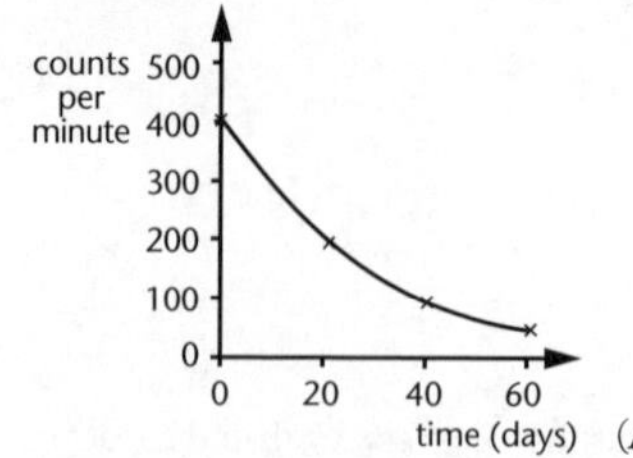

(*A*)

b. 5 days: 340 counts min^{-1}
10 days: 280 counts min^{-1}
25 days: 170 counts min^{-1}
50 days: 70 counts min^{-1} (***A***)

c. 60 days. (***A***)

15. a. Alpha particles cannot penetrate, gamma rays are too penetrating. (***A***)
b. Do not get within 30 cm of the source or use metal-lined clothing. (***A***)
c. It is not likely since beta particles are simply high-speed electrons. (***A***)
d.

count in 10 seconds	50	50	50	45	40	40	40

(***A***)

e. The paper has thickened. Because more beta particles have been absorbed. (***A***)
f. i. 30 s approximately. (***A***) ii. 90 m approximately. (***A***)

16. Gamma rays from source A, alpha particles and gamma rays from source B, beta particles from source C. (***E***)

Glossary/Index

absolute refractive index (71, 123): ratio of the speed of light in a vacuum to that in a medium.

accelerate (87): when something changes its velocity.

acceleration due to gravity (14): rate at which objects change their velocity when they move under the influence of gravity.

accurate (20): when a measurement is close to the actual value.

alloy (159): mixture of metals.

alpha particle (237): helium nucleus emitted at high speed from a much larger unstable nucleus.

alternating current (209): electric current that changes its direction periodically.

ammeter (229): device used to measure electrical current in a circuit.

ampere (2): symbol **A** – the SI unit of current ($1 A = 1 C s^{-1}$).

amplitude (46): the maximum distance a particle in a medium moves from its at-rest position; the height of a wave.

angle of reflection (65): The angle between the reflected ray and the normal.

antinodal line (101): line along which waves reinforce each other.

antinode (100): place of maximum displacement in a standing wave.

aperture (138): hole that allows light to enter a camera.

aqueous humour (135): watery liquid between the cornea and the lens of the eye.

atom (87, 233): smallest particle of matter that cannot be broken down chemically.

atomic bomb (235): a bomb that releases a large amount of energy produced by an uncontrolled nuclear reaction.

atomic number (233): number of protons in a particular atom.

average speed (52): symbol v – total distanced travelled divided by total time taken.

battery (193): power source that converts chemical potential energy to electrical energy; two or more cells in series.

beta particle (249): high-speed electron emitted from an unstable nucleus.

binocular vision (113): using two eyes to focus on something.

blind spot (135): part of the retina where the optic nerve leaves the eye; not sensitive to light.

breeder reactor (240): nuclear reactor that produces fuel for further nuclear fission reactions.

carbon brushes (181): electrical contacts that push against the rotating commutator of a DC electric motor.

centre of curvature (113): centre of an imaginary circle of which the mirror is an arc.

centripetal force (174): inwards force necessary to keep an object moving in a circle.

chain reaction (235): nuclear fission reaction that produces neutrons that trigger other nuclear fission reactions.

ciliary muscles (135): small muscles in the eye that are used to alter the shape of the lens to enable it to focus light on the retina.

circuit (150): a conducting path for conventional current to flow from the positive side of a cell around to the negative side of a cell.

collisions (243): when two or more objects interact.

commutator (181): parts of a DC electric motor that connect the rotating coil to the outside circuit.

compass (146): device that can be used to show the direction of a magnetic field; consists of a small pivoted magnet that can rotate freely.

components (165): vectors – two vectors at right angles which when added together are equivalent to a single vector; electricity – the different parts of an electrical circuit.

compressions (98): where particles in a medium are bunched together due to longitudinal wave motion.

concave (16, 113): curving inwards; used in describing mirrors and lenses.

conduction (33): transfer of heat energy by particle vibration.

conductor (33): a substance that allows an electric current (*electrical energy*) or heat (*thermal energy*) to travel through it.

cones (135): cells on the retina of the eye that register light and colour.

conserved (251): when something stays constant.

constant proportion (11): when two variable quantities have a constant ratio and form a linear graph.

constant speed (2): time during which an object does not change it speed throughout its journey.

constructive interference (83): when two or more waves add together to give a resultant wave with maximum amplitude.

constructive superposition (85): when two or more waves add together to give a larger resultant wave.

control rods (239): rods of neutron-absorbing material (eg boron or cadmium) that can be raised or lowered into the core of a nuclear reactor to control the rate of the chain reaction.

convection (33): transfer of heat energy by the transfer of material.

conventional current (161): current which travels from the positive (red) to the negative terminal (black) of a DC power supply.

converging (92, 113): curved mirror or lens that brings parallel rays of light together to a point.

convex (113): curved outwards; used in describing mirrors and lenses.

coolant (239): a fluid that removes heat energy from a source, such as the core of a nuclear reactor.

core (239): the part of a nuclear reactor in which the chain reaction occurs.

cornea (135): the transparent outer cover of the front of the eye.

couple (180): two equal and opposite forces that act at a perpendicular distance apart to cause rotation.

crest (88): highest point on a transverse wave.

critical angle (126): angle of incidence where light entering an optically less dense medium refracts at right angles.

critical mass (235): the minimum mass of a nuclear isotope that can sustain a nuclear chain reaction.

current (29): rate of movement of charge in a closed circuit; amount of charge passing a point in a circuit per second.

current balance (164): device used to measure the force on a current carrying wire in a magnetic field.

density (13): symbol ρ – mass per unit volume.

Descartes' formula (117): mathematical formula to use with mirror and lens problems.

destructive interference (101): when two or more waves add to give a zero resultant wave.

destructive superposition (85): when two or more waves add to give a zero resultant wave.

diaphragm (138): part of a microphone that vibrates when sound waves hit it.

diffraction (97, 102): bending of waves around a barrier or out through a gap.

diffuse reflection (64): occurs when parallel incident rays are reflected and scattered in all directions.

diode (190): electric component that allows current to flow in one direction only.

direct current (183): electric current that always moves in the same direction around a circuit (away from the positive terminal of the supply towards the negative terminal) – this is in the opposite direction to electron flow.

dispersion (76): splitting up of white light into the colours of the spectrum.

displacement (46): distance and direction from a known origin.

distance (2): length between two positions.

diverging (92): curved mirror or lens that spreads parallel light out as if from a point.

earthquake (47): vibration of the Earth's surface as a result of movement along a fault line.

elastic (32): kinetic energy is conserved.

electric circuit (171): unbroken pathway through which electricity can flow.

electric field (175): region in which a charged object experiences a force.

electromagnet (149, 163): coil of wire that becomes a magnet when a current passes through it.

electromagnetic induction (189): voltages and currents produced by changes in magnetic fields.

electromagnetic radiation (51, 63, 109): whole spectrum of waves, from cosmic and gamma rays to visible light to long-wave radio waves.

electromagnetic spectrum (35, 47, 104): family of waves that travel at the speed of light (3×10^8 m s^{-1}).

electromagnetic waves (47, 87): waves caused by the vibration of charged particles and which are able to pass through a vacuum at the speed of light.

electron (87): small negatively-charged particle that orbits the nucleus of an atom.

element (233): substance that consists of one kind of atom only; cannot be separated into simpler substances by chemical or physical means.

emf (194): electromotive force; a source of voltage.

endoscope (126): medical instrument used to examine interior organs.

energy (6): symbol **E** – capacity to do work.

epicentre (49): point on the surface of the Earth directly above the focus of an earthquake.

equilibrium (31): situation when an object is at rest or moving uniformly as described in Newton's First Law.

error (113): word used to describe how a measurement could differ from the true value.

explosion (259): when an object breaks up into two or more parts violently.

f number (138): the ratio of the focal length to the aperture diameter of a camera.

fault line (49): break in layers of rock of the Earth's crust along which movement can occur.

filter (64): device that passes only some frequencies of radiation and blocks all other frequencies, eg a red filter transmits red light but blocks all the other colours.

fission (234): nuclear reaction in which a large nucleus splits into two or more fragments.

focal length (114, 130): distance from the focus to the pole or optical centre.

focus (49): point below the surface of the Earth where an earthquake originates.

force (6): a push or a pull in a particular direction.

frequency (46, 88): number of revolutions, vibrations or events that occur in one second; number of waves that pass a point per second.

friction (19): force produced when two surfaces come in contact or slide past each other.

fringes (102): alternating dark and bright bands of light.

fuel rods (239): rods of fuel (eg uranium) in the core of a nuclear reactor.

fundamental units (2): the seven units in the SI System from which all others can be derived.

fusion (234): nuclear reaction in which two small nuclei combine to form a larger nucleus.

galvanometer (189): very sensitive ammeter for measuring small currents.

gamma (249): high energy electromagnetic radiation.

Geiger counter (240): device for measuring radioactive radiation.

generator effect (189): electromagnetic induction.

geothermal power (239): electrical power generated by turbines that are driven by naturally occurring steam that escapes through the surface of the Earth.

graph (7): visual way of displaying data using axes.

gravitational potential energy (219): energy stored in a gravitational field when an object is raised against the field.

half-life (249): time taken for half the radioactive atoms in a sample to decay.

heat (25): energy transferred when an object changes temperature or state.

heat exchanger (239): device for transferring heat energy from one place to another, eg refrigerator and car radiator.

heat shimmer (75, 122): refraction effect caused by air currents.

hertz (88): SI unit of frequency.

hydroelectric power (219): electrical power generated by turbines which are driven by falling water.

illumination (110): describing how intense a light source is at a distance away from it.

image (64, 111): what is seen in a mirror or produced by a curved mirror or lens.

implosion (235): bursting inwards.

induction (160): when a magnet causes a material in its own magnetic field to behave as though it was a magnet.

insulator (32): substance that does not permit electrons (electrical energy) or heat (thermal energy) to flow through.

intensity (98): rate of flow of energy associated with wave motion.

interference pattern (101): produced by the superposition of waves.

inverse (11): opposite mathematical operation, eg division by 3.6 instead of multiplication by 3.6.

inverse square law of illumination (110): law that describes the variation of illumination mathematically.

ions (174): atoms or groups of atoms that have gained or lost electrons.

iris (135): part of the eye that controls the amount of light that enters the eye.

isotopes (233): atoms of the same element having a different number of neutrons.

joule (29): symbol **J** – SI unit of energy and work.

Kelvin scale (25): symbol **K** – temperature scale beginning at absolute zero (–273 °C).

kilogram (2): symbol **kg** – SI unit of mass.

kilowatt (4): 1 000 watts.

kinetic energy (19): symbol **Ek** – energy possessed by a moving object.

L-wave (50): slow earthquake waves that travel along the surface of the Earth.

latent heat (30): symbol **L** – heat energy absorbed or released when a substance melts or boils.

latent heat of fusion (30): heat energy required to change the state of 1 kg of a substance from solid to liquid at a constant temperature.

latent heat of vaporisation (30): heat energy required to change the state of 1 kg of a substance from liquid to gas at a constant temperature.

lateral inversion (66): effect produced by a mirror in reversing images from left to right.

laws of reflection (92, 111): laws that describe how reflection takes place.

laws of refraction (123): laws that describe how refraction takes place.

lens (129): piece of glass with curved surfaces.

Lenz's law (190): rule for remembering the direction of the induced current.

light (63): narrow band of frequencies of electromagnetic radiation visible to our eyes.

light-emitting diodes (190): LED; special diodes that give off light when current passes through them.

linear (28): forming a straight line.

lines of force (153): direction followed by a small positive charge or north magnetic pole when subjected to an electric or magnetic force field.

longitudinal wave (45): wave in which the particles of the medium vibrate in a direction parallel to the direction of energy transfer.

magnet (146, 159): an object that can attract iron or steel.

magnetic deviation (147): angle between geographic North and magnetic South.

magnetic field (105, 145, 159): region around a magnet where iron or other magnets experience a force.

magnetic field lines (145): lines drawn to graphically represent the strength and direction of a magnetic field.

magnetic force (179): a field force – for a magnetic pole has on it when it enters magnetic field.

mass (2): symbol **m** – amount of matter in an object.

mass number (233): sum of the number of protons and neutrons in an atom.

mechanical energy (20): kinetic or potential energy.

mechanical waves (87): waves that need a medium in which to move, eg sound.

medium (33, 45): material through which waves or light can pass, eg, water, air.

metric (3): related in multiples of ten.

moderator (239): substance (such as graphite or heavy water) that can slow down neutrons to the correct speed for fission to occur in a nuclear reactor.

moment (183): alternative name for torque.

momentum (254): physical quantity defined as the mass times the velocity of an object.

motion (20): movement.

motor effect (179): a current-carrying wire placed at right angles to a magnetic field has a force exerted on it that moves the wire.

nature (115): description of the characteristics of an optical image.

neutral (233): situation where the number of positive and negative charges are equal.

neutron (233): uncharged particle in the nucleus of an atom.

newton (173): symbol **N** – SI unit of force.

Newton's formula (119): mathematical formula to use with mirror and lens situations.

nodal line (101): line along which waves cancel each other out.

node (100): place of zero displacement in a standing wave.

normal (65): line perpendicular to the surface.

north pole (145, 159): the end of a magnet that would point towards the Earth's north pole.

nuclear reaction (234, 247, 251): a reaction in which particles and/or energy are absorbed or released by an atomic nucleus.

nuclear reactor (234, 239): device to obtain energy from controlled nuclear chain reactions.

nucleon (247): particle in the nucleus – neutron or proton.

nucleus (233, 247): collection of protons and neutrons at the centre of an atom; tiny, dense and positively charged.

nuclide (247): particular isotope of an element.

ohm (196): symbol Ω – SI unit of electrical resistance.

Ohm's Law (191): relationship between voltage, resistance, and current; $V = IR$.

opaque (64): not able to be seen through.

optic nerve (135): nerve that conducts the electrical impulses from the retina to the brain.

optical centre (130): common term for pole.

optical density (71): a measure of how easily light passes through a substance.

optical fibre (126): thin strands of plastic or glass used to transmit light.

optical illusion (135): occurs when the brain perceives a different image than that produced on the retina.

oscilloscope (49, 99): electronic device that displays vibrations in graphical form on a screen.

P-wave (50): a longitudinal, fast earthquake wave.

parabolic (114): U-shaped, like a parabola.

parallax (113): apparent movement of two objects due to the movement of the observer.

parallax error (113): error produced by reading a scale from an angle.

parallel (45): connected 'side by side' in a circuit.

particle model (91): idea in which light consists of fast moving particles called photons.

pascal (14): symbol **Pa** – SI unit of pressure; 1 Pa = 1 N m^{-2}.

period (46): time taken for one vibration or for one wave to pass a point; the time taken for one revolution or event.

periodic (87): wave motion that repeats regularly.

phase (30): whether a substance is a solid, liquid or gas.

phase (89): measure of how much one wave is out of step with another.

pinhole camera (64): simple camera with a pin hole instead of a lens.

pitch (48, 98): frequency of a sound wave.

plane (89, 111): flat.

point source (89): source of waves from a point.

pole (113, 130): centre of symmetry of a mirror; optical centre.

potential difference (209): difference in potential energy per coulomb between parts of a circuit.

power (42): symbol **P** – the rate of doing work.

prefix multipliers (3): letters in front of a unit that stand for various powers of ten.

pressure (13): symbol **P** – the force per unit area.

principal axis (113, 129): line of symmetry in a lens or mirror from the pole to the centre of curvature.

principal focus (114, 129): point which rays parallel to the principal axis are made to converge to or diverge from, by curved mirrors and lenses.

principle of superposition (99): method by which waves add together to give a resultant wave.

prism (77, 127): a triangular block of glass or other transparent material.

propagation (88): direction in which a wave travels.

proportional (11): when two quantities are related by a constant ratio.

proportionality constant (11): value of the ratio between two quantities.

proton (172, 233): a positively-charged particle found in the nucleus of an atom.

pulse (45, 87): a single disturbance that travels through a medium; a wave consists of a series of pulses.

radiation (33, 35): transfer of heat energy by electromagnetic radiation; radiation requires no medium to travel through.

radioactive (240, 247): atom whose nucleus can spontaneously emit a particle or radiation.

radioactive decay (248): nucleus that breaks apart and releases an α-particle, β-particle or γ-radiation.

radius of curvature (113): distance from the centre of curvature to the pole of the mirror.

rarefactions (98): where particles in a medium are separated further apart due to wave motion.

ratio (11): the quotient of two quantities.

ray (63): narrow beam of light represented by an arrow drawn in the direction that the light travels.

ray diagrams (65, 115): scale drawings to find the details of the images formed by mirrors and lenses.

reaction (138): name given to the force that is equal in size and opposite in direction to an action force.

real (115): images that can be focused on a screen.

reflected (35, 111): when light or a wave 'bounces' off a surface or boundary.

reflection (64): wave motion that 'bounces back' at a boundary.

refraction (72, 121): change in speed and direction of light when it travels from one medium into another; the bending of light when it passes at an angle from one transparent medium into another.

refractive index (71, 123): ratio of the speed of light in one medium to another.

relationships (8, 11): how different quantities or ideas are connected.

relative refractive index (123): proper term for refractive index.

relay (150): a magnetic switch for switching a circuit on or off.

resistance (191): the ratio of voltage across a component to the current through the component; a measure of how difficult it is for charges to flow in a conductor.

resistor (194): electrical component that converts electrical energy to heat.

resultant (16): vector obtained by the addition or subtraction of other vectors.

retina (135): part of the eye that acts as a screen onto which light rays are focused and images are formed; contains the rods and cones.

revolution (213): motion once around a circle.

rheostat (182): a variable resistor.

right hand grasp rule (148): rule for determining the direction of the magnetic field around a current-carrying wire.

right hand solenoid rule (148): rule for determining the direction of the magnetic field around a solenoid.

right-hand grip rule (161): easy way of remembering how the magnetic field is directed around a wire carrying a current.

right-hand slap rule (165): easy way of remembering how the direction of magnetic fields, currents and resultant forces are related.

ripple tank (92): tank of water used to demonstrate the properties of waves.

rods (135): cells on the retina of the eye that are used for nighttime and peripheral vision.

rotation (180): motion once around a circle.

S-wave (50): secondary, transverse earthquake wave.

secondary circuit (150): circuit that is not physically connected to the primary circuit that contains the power supply.

seismograph (49): device for measuring the strength of an earthquake.

sense (163): anticlockwise or clockwise direction.

shutter speed (138): time that the aperture of a camera is open.

SI (2): the international system of units.

sleepers (147): soft iron bars placed across the poles of a permanent magnet to prevent the magnet losing strength during storage.

slope (170): quantity used to describe how steep a straight line is – vertical change divided by the horizontal change.

Snell's law (73, 124): formula that links refractive indices and refraction angles for different mediums.

solenoid (148, 162): length of wire wound into a circular coil.

south pole (147, 159): the end of a magnet that would point towards the Earth's south pole.

specific heat capacity (29): symbol **c** – amount of energy required to raise the temperature of 1 kg of a substance by 1 °C.

spectrum (76, 127): range of colours that white light is made up of.

specular reflection (64): reflection in which parallel incident rays are reflected parallel.

speed (2): symbol **s** – rate of change of distance.

speed of light (35, 109): fundamental constant in physics; the speed of all electromagnetic radiation in a vacuum.

standing wave (100): produced by the superposition of two equal but opposite waves.

subatomic (245): smaller than an atom.

sublimation (31): change of state directly from solid to gas without going through the liquid stage.

superimpose (99): when two or more waves in the same region add together to give a resultant wave.

surface waves (92): transverse waves that move along the surface of a medium, eg water ripples.

temperature (25): symbol **T** – measure of the average kinetic energy of the particles in a substance.

tension (16): force in connecting strings and ropes that tries to stretch them.

tesla (162): SI unit used to record the size of a magnetic field.

thermometer (25): device for measuring temperature.

torque (180): turning or twisting effect about a pivot; defined as the product of force and perpendicular distance to the pivot.

total internal reflection (96, 125): where light is totally reflected instead of being refracted.

transmission (33): wave motion that continues through a boundary.

transparent (64): a substance that can be clearly seen through.

transverse (87): where particles of the medium vibrate at right angles to the direction of motion of the wave or pulse.

transverse wave (45): a wave in which the particles of the medium vibrate at right angles to the direction of energy transfer.

trough (88): lowest point on a wave.

ultrasound (51): sound of frequencies higher than the upper limit of human hearing (20 kHz).

uniform field (196): where the field lines are parallel and equally spaced; a constant field.

unit (2): reference standard for measuring a quantity.

unstable (248): atom whose nucleus can spontaneously emit a particle or radiation.

variable (10): physical quantity that can have a range of values and can be drawn on a graph.

vector (11): quantity that possesses both size (magnitude) and direction.

vector diagram (11): scale diagram showing vector magnitudes and directions.

velocity (19): symbol **v** – speed in a stated direction.

virtual (65, 112): image that cannot be projected on a screen, it appears 'behind' the mirror or on the object side of a lens.

volt (143): symbol **V** – SI unit of potential difference.

voltage (99): common term for potential difference.

voltmeter (194): instrument for measuring voltage.

watt (3): symbol **W** – SI unit of power; 1 W = 1 J s^{-1}.

wave (45): vibration that transmits energy.

wave equation (46, 88): the formula, $v = f\lambda$ which relates wave velocity to frequency and wavelength.

wave model (92): an idea in which light consists of electromagnetic waves.

wave motion (45, 87): disturbance in a material which transfers energy from one point to another without an overall displacement of the material.

wave velocity (88): speed at which a wave or pulse moves through a medium.

wavefront (89): line drawn along the crest of a wave; is at right angles to the wave's direction of travel.

wavelength (46): distance between adjacent corresponding points in a wave that are vibrating in the same manner.

weight (13): symbol **W** – force of gravity acting on an object.

work (19): process of transforming energy from one form to another, defined as force times distance moved in the direction of the force.